Nanoparticle Drug Delivery Systems for Cancer Treatment

Nanoparticle Drug Delivery Systems for Cancer Treatment

edited by
Hala Gali-Muhtasib
Racha Chouaib

JENNY STANFORD
PUBLISHING

Published by

Jenny Stanford Publishing Pte. Ltd.
Level 34, Centennial Tower
3 Temasek Avenue
Singapore 039190

Email: editorial@jennystanford.com
Web: www.jennystanford.com

British Library Cataloguing-in-Publication Data
A catalogue record for this book is available from the British Library.

Nanoparticle Drug Delivery Systems for Cancer Treatment

ISBN 978-981-4800-54-9 (Hardcover)
ISBN 978-0-429-34125-0 (eBook)

Contents

Preface

Nanoparticles are carriers for a wide range of drugs used for various therapeutic applications including cancer treatment. The use of nanoparticle formulations in the cancer clinical setting is rapidly progressing as they are being implemented to solve several limitations of conventional drug delivery systems, namely nonspecific targeting and biodistribution, reduced efficacy, poor oral bioavailability, and low therapeutic indices. This book presents recent advances in the field of nanomedicine and nano-based drug delivery systems and brings together the collective works of many researchers in the area of nanoparticle drug delivery for cancer treatment. Various nanostructures, including liposomes, polymers, dendrimers, silicon or carbon materials, magnetic nanoparticles, gold, metal, copper nanoparticles, and plant-derived edible nanoparticles, have been covered in this book. The topics include the methods of preparation of these nanoparticles, their physicochemical characteristics, and the implications of using them against cancer treatment by adopting different strategies in a multitude of cancer systems. Chapter 1 includes introductory information about the various types of nanoparticles and their applications in targeted therapy and personalized medicine. While many of key areas in nanoparticle drug discovery and delivery are covered in depth, the book does not represent a comprehensive coverage of all developments in the field. Yet, it is hoped that the diversity of the topics covered will provide insights to the researchers in the area of nanomedicine. The extensive references at the end of each chapter will encourage further research and provide a rich library of resources that will deepen the understanding of researchers working in this growing field.

This book would not have been made possible without the valuable contributions of the authors, who are experts in their respective fields. We are also indebted to Jenny Stanford Publishing for inviting us to write this book: special thanks to Arvind Kanswal, Jenny Rompas, and Stanford Chong for closely working with us to ensure the timely publication of this book.

Hala Gali-Muhtasib
Racha Chouaib

Chapter 1

Nanoparticles as Drug Delivery Systems for Cancer Treatment: Applications in Targeted Therapy and Personalized Medicine

Racha Chouaib,[a,b] **Rana Sarieddine,**[b] **and Hala Gali-Muhtasib**[b]

[a]*Faculty of Sciences, Section 1, Lebanese University, Lebanon*
[b]*Department of Biology and Center for Drug Discovery, American University of Beirut, Riad El Solh 1107 2020, Beirut, Lebanon*

racha-chouaib@hotmail.com, amro@aub.edu.lb

Nanoparticles are bio-nanomaterials with specific physicochemical properties. Therapeutic drugs are formulated in nanoparticles to enhance their pharmacokinetic and pharmacodynamic properties and to improve targeted delivery to the relevant tissue. There are several types of nanoparticles, including liposomes, polymeric nanoparticles, dendrimers, ceramic nanoparticles, carbon nanomaterials, and magnetic nanoparticles. After synthesis, nanoparticles are evaluated for their surface charge, particle size and shape, efficiency, loading capacity, and stability. Nanoparticle

Nanoparticle Drug Delivery Systems for Cancer Treatment
Edited by Hala Gali-Muhtasib and Racha Chouaib

ISBN 978-981-4800-54-9 (Hardcover), 978-0-429-34125-0 (eBook)
www.jennystanford.com

formulated drugs have provided massive advantages in personalized medicine for the treatment of diseases including the molecular targeting of cancer. In this chapter, we review some kinds of nanoparticle formulations, their characterization, advantages, limitations, and potential use in nanomedicine.

1.1 Introduction

Nanotechnology or nanoscience covers a broad multidisciplinary field where research progressed rapidly worldwide during the last decade. Nano drug delivery systems have marked a new era of industrial revolution. Physicists, chemists, biologists, engineers, material scientists and medical specialists have worked together on nanometric dimensions.

The word nanoparticle comes from the Greek word "nanos," which means dwarf particles. In addition, the prefix nano means one-billionth [1]. Using nanotechnology, scientists manufacture particles that have at least one nanometric dimension and sizes ranging 10–1000 nm. These materials are produced from the structured organization of groups of atoms and molecules or by nanoscale reduction of macroscopic materials [2, 3]. Today's nanotechnology has allowed real progress to achieve temporal and spatial site-specific drug delivery, especially in the pharmaceutical industry. Nanocarriers, with optimized physicochemical and biological properties, are taken up by cells more efficiently than larger molecules. Thus, they could be used as delivery tools and effective transporters for currently available bioactive compounds [4].

1.2 Goals of Nanoparticle Systems

The predominant method to deliver drugs is either orally or by injection, using different conventional preparations that include solutions, suspensions, or emulsions. Conventional preparations suffer from certain limitations, namely their reduced efficacy due to a limited potential to reach the site of therapeutic action. Moreover, the drug circulates in the whole body affecting the diseased and the healthy cells at the same time. This may cause serious side effects. Therefore, developing the nanoparticle system that optimizes drug delivery and the action of the drug while reducing side effects in vivo is an exciting approach [5, 6].

Nanoparticles are ideal drug delivery and carrier systems that have a great potential of tackling different limitations like high dose, low bioavailability, intolerance and instability. This is because nanoparticles allow the control of particle size, surface properties, and release of pharmacologically active agents in order to achieve site-specific action [5].

For therapeutic applications, the drug is dissolved, entrapped, encapsulated, or attached to a nanoparticle matrix. A drug targeting system should be able to control the fate of a drug entering the biological environment [4, 7]. This enables the safe delivery of toxic therapeutic drugs and protection of non-target tissues and cells, thus lowering toxicity and severe side effects [8].

Different examples of nanocarriers are used as drug delivery systems that include organic (polymers, liposomes and dendrimers and inorganic nanosystems (magnetic, silicon and metallic) (Fig. 1.1) [7, 9].

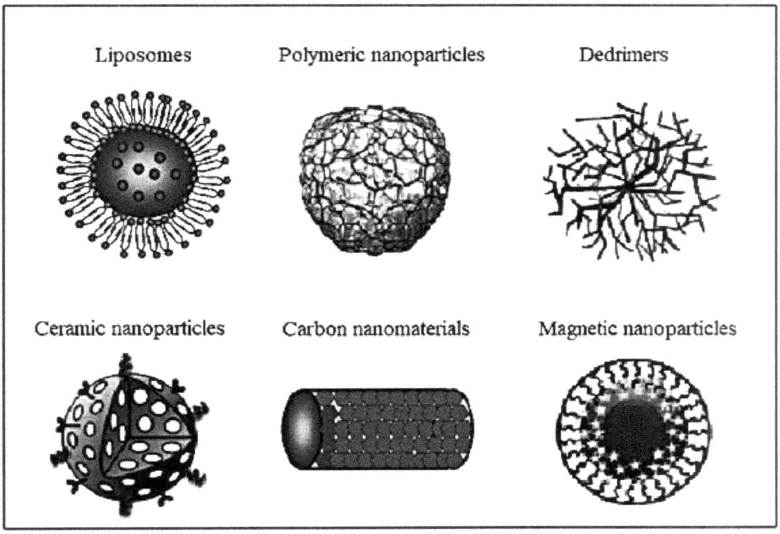

Figure 1.1 Schematic representation of certain types of nanoparticles.

1.3 Types of Nanoparticles

The field of nanotechnology is under rapid growth and several products have been introduced into the pharmaceutical market. Several kinds of nanoparticles have already been manufactured and are widely investigated as drug delivery systems. The selection of materials used is based on different factors that include (a) size of nanoparticles required and surface characteristics (charge and permeability); (b) drug properties (solubility and stability); (c) degree of biodegradability, biocompatibility and toxicity; and (d) the desired drug release profile [5, 6, 10].

The variety of nanoparticles include liposomes, polymeric nanoparticles, dendrimers, ceramic nanoparticles, carbon nanomaterials, and magnetic nanoparticles [6, 9, 11, 12]. In the next sections, we will discuss some of these types of nanoparticles (Table 1.1).

Table 1.1 Properties of some nanoparticle types

Nanoparticle type	Characteristics	Advantages
Liposomes [13]	80–300 nm, lipid vesicles composed of natural or synthetic phospholipids and steroids having at least one lipid bilayer	Increase drug stability and solubility, protect drug from degradation, reduce systemic toxicity, non-toxic, biodegradable, and biocompatible
Polymeric nanoparticles [14, 15]	10–100 nm, composed of either synthetic or natural polymers	Increase the stability of volatile agents, control drug release, stable, biodegradable, biocompatible, non-toxic, easy fabrication
Dendrimers [16]	<200 nm, highly branched globular macromolecules made up from a core and layers of branched repeated units and functional end groups	Enhance the oral bioavailability of the drugs, act as image contrast agents, have large surface area to size ratio, targeted drug delivery, high solubility, reduced side effects, accumulation at the targeted site

Ceramic nanoparticles [17]	Inorganic, made up of metallic and non-metallic compounds	Highly stable, known and cheap synthesis routes, biocompatible, highly porous
Carbon nanoparticles/ nanotubes [15]	Hollow sphere/ cylindrical nanostructures of pure carbon	Enhanced biocompatibility and thermal and electronic conductivity depending on the chemical modifications, protect the drug from degradation, controlled drug release profile
Magnetic nanoparticles [18]	Superparamagnetic, high magnetic susceptibility, low curie temperature	Easy to handle and visualize, enhanced, biosensors, act as image contrast agents for magnetic resonance imaging, non-toxic and biocompatible

1.3.1 Liposomes

Liposomes are lipid-based vesicles mainly composed of natural or synthetic phospholipids. They are widely used as antimicrobial drug delivery systems [19].

These vesicles are the first nanoparticles that have been used, usually with 80–300 nm size ranges [20]. They are spherical lipid vesicles composed of phospholipids and steroids, with a bilayer membrane structure consisting of amphiphilic lipid molecules with characterized size, surface charge, and number of bilayers. These potential carriers are important for protecting the drug from degradation, increasing their solubility, improving their pharmacokinetic properties and reducing side effects due to the targeting of the site of therapeutic action [7, 21].

In liposomes, the drug is incorporated by the encapsulation process. Anticancer drugs, neurotransmitters, antibiotics, and anti-inflammatory molecules are examples of drugs encapsulated in liposomal formulations [9].

1.3.2 Polymeric Nanoparticles

Polymeric nanoparticles (PNPs) are structurally stable, with a diameter ranging from 10–100 nm. The PNPs are obtained from synthetic polymers, such as polyacrylamide, or natural polymers, like albumin and gelatin [22–24]. They consist of biocompatible and biodegradable polymers and are used for controlled drug release. These nanosystems are stable, non-toxic, non-immunogenic, and non-inflammatory [6, 25].

The drugs can be either adsorbed to the nanocapsules after a polymerization reaction or covalently encapsulated to their surface after formation [26, 27]. These types of nanoparticles have multiple advantages, including (1) easier fabrication in large quantities by different methods, (2) increased stability of any volatile pharmaceutical agent, and (3) delivery of higher quantities of pharmaceutical agents to a desired location. The functional groups presented on their surface can be modified with targeting ligands [12, 28].

1.3.3 Dendrimers

Dendrimers are highly branched globular macromolecules fabricated from monomers. Their sizes and the shapes are well ordered and controlled. Dendritic architecture is one of the most popular structures observed in the biological systems. Glycogen and amylopectin are examples of nanometric molecules possessing dendritic structures [29]. They are made up of a core and layers of branched repeated units (dendrons or dendrimer arms) and functional end groups.

Dendrimers have a large surface area to size ratio due to their highly branched nature. This ratio, in addition to the surface functionalization capability and stability make them important drug delivery systems, as well as gene delivery systems [6, 30, 31]. In general, the drug can be encapsulated in the dendrimer's internal structure or attached on dendrimer surfaces [32, 33].

1.3.4 Ceramic Nanoparticles

Ceramic nanoparticles are inorganic systems for which there is growing interest in using them as carrier systems for drug delivery. These nanoparticles can be engineered with the desired size and porosity. Thus, they have different features, including biocompatibility and a highly porous framework [34]. Most biological research has explored silica materials as typical biocompatible ceramic nanoparticle, due to its amorphous structure, high porosity, and enormous surface area [35].

Because of their physiological stability, silica nanoparticles have a great potential for a variety of therapeutic applications in medicine. However, recent studies have revealed in vitro and in vivo toxicity of silica materials, but this effect depends on the cell type and nanoparticle size [9, 34].

1.3.5 Carbon Nanomaterials

Carbon nanomaterials (nanotubes) are characterized by a unique architecture formed by rolling of single or multilayers of graphite. The surface area is enormous containing chemical modifications that improve the biocompatibility and the electronic and thermal conductivity of the nanoparticle [36, 37].

Drugs can be immobilized in carbon nanocarriers by encapsulation. This can protect the drug from degradation during its transport to the cells. The drug is then released only in electrically or chemically controlled conditions. In some cases, drugs are adsorbed chemically or immobilized by attachment of active agents to functionalized carbon nanotubes [38–40].

Nanotubes have a toxic potential, which depends on their unique well-defined geometric structures. The toxicity can result from the high length to diameter ratio and the toxicity of graphite [9].

1.3.6 Magnetic Nanoparticles

Magnetic nanoparticles are highly promising carriers for drug delivery, due to their specific features. These nanocarriers are easy to handle (using an external magnetic field), can use passive and active drug delivery strategies, can be visualized, and have an enhanced uptake by the target tissue. Drugs can be connected with the magnetic nanoparticles by covalent binding, electrostatic interactions, adsorption, or by encapsulation [9, 41].

1.4 Evaluation Parameters of Nanoparticles

There are multiple features needed to characterize and evaluate nanoparticles which are summarized in Table 1.2. These include the particle size, surface morphology, zeta potential, entrapment efficiency, loading capacity and stability of nanoparticles.

Table 1.2 Techniques used to study the parameters of the nanoparticles

Parameter	Techniques
Particle size	Scanning electron microscope Dynamic light scattering Transmission electron microscope
Molecular weight	Matrix-assisted laser desorption/ionization time-of-flight mass spectrometry (MALDI-TOF MS) Size exclusion chromatography
Shape/Morphology	Scanning electron microscope Transmission electron microscope
Surface charge	Zeta potential measurement by DLS and electrophoresis Electro kinetic sonic amplitude technique Laser Doppler Anemometer
Entrapment efficiency	HPLC NaOH-based extraction method Radioactive method Fluorescent method
Stability	Analyzing drug release profile by HPLC Monitoring changes in the physical appearance
Surface hydrophobicity	Hydrophobic interaction chromatography Contact angle measurement
Surface element analysis	Nuclear magnetic resonance X-ray photoelectron spectroscopy Secondary ion mass spectroscopy

Particle size is one of the most important measurements for nanoparticle characterization. It can influence the drug release, drug loading, and stability of nanoparticles. The particle size and distribution are most commonly measured using electron microscopy. Scanning electron microscopy (SEM) images characterize the surface morphology. The measurement of the hydrodynamic diameter can be done by dynamic light scattering (DLS). These features determine the in vivo biological fate, distribution, toxicity and targeting ability of the nanoparticle system [42].

The surface morphology of nanoparticles varies and plays a key role in exploiting its properties. The shapes of nanoparticles can be spherical, flat, cylindrical, tubular, conical, or irregular. The surface is generally determined by electron microscopy imaging techniques [43].

The measurement of the effective electric charge on the nanoparticle system in solution (known as Zeta potential) is extremely important to characterize the electrical potential and to predict the long-term stability of nanoparticles. The magnitude of the zeta potential provides information about particle stability and the electrostatic repulsion between particles. If the nanoparticles have a zeta potential above (±) 30 mV, they are considered stable in suspension and are unlikely to aggregate [44, 45].

The entrapment efficiency (%EE) is a very important physicochemical characteristic of nanoparticles. It provides an idea about the percentage of drug that is successfully entrapped/adsorbed in the nanoparticles. It is determined as the difference between the total amount of drug used to prepare the nanoparticles and the amount of drug present in the aqueous medium. %EE = [(amount of drug initially taken to prepare the nanoparticles – Free "unentrapped drug")/amount of drug initially taken to prepare the nanoparticles] X 100. If the %EE is 20%, it means that 20% of the drug is adsorbed into the nanoparticles [7].

The loading capacity indicates the percentage of mass of the nanoparticle that is due to the encapsulated drug. It is the amount of drug loaded per unit weight of the nanoparticle [46]. The stability of nanoparticles depends on their ability to conserve their chemical composition and their physical appearance, and on the level of aggregation of nanoparticles in physiological conditions [47].

After the measurement of parameters mentioned earlier—particle size, surface morphology, zeta potential, entrapment efficiency and loading capacity, the assessment of the cellular uptake and the release profile of the drug from the nanoparticle is necessary to test the efficacy of the drug-loaded nanoparticle. The physiochemical properties of the nanoparticle impact its interaction with the cells, its cellular uptake, intracellular fate and response. There are several methods that enable us to quantify the nanoparticle uptake. The choice of the method depends on several factors, including the type and properties of the nanoparticle, the research question, and the quality of the parameters of the analytical device, including its accuracy, precision, and resolution. Most analytical techniques for cellular uptake studies include stereoscopic and imaging methods [48]. In addition, studying the drug release profile is important in determining the release mechanism/kinetics, behavior, efficacy, and safety of the formula. Then, the pharmacokinetics and pharmacodynamics should be investigated in vivo to assess and quantify nanoparticle absorption, distribution, metabolism, and excretion (ADME). The latter is essential for studying the mechanism of action of the drug-loaded nanoparticle and its effect on the organs. Finally, in vitro and in vivo toxicological studies should be performed to document No Observed Adverse Effect Level (NOAEL), which represents the allowed exposure level of a certain compound at which no toxic effects on various biological systems are seen [49]. The aim is to demonstrate sustained drug release levels and extended efficacy. Final therapeutic use in humans is dependent upon toxicity measures, pharmacodynamic and pharmacokinetic responses (the effect of the drug on the organism and vice versa), hypersensitivity, and immunogenicity (Table 1.2).

1.5 Advantages of Nanoparticles as Drug Delivery Systems

Nanocarriers as drug delivery systems are designed to improve the pharmacological and therapeutic properties of conventional drugs, providing many significant advantages. The pharmacokinetics and therapeutic index of the drugs can be significantly improved in contrast to the free drug counterparts [7, 12, 47].

The first advantage is that nanoparticles are fairly easy to prepare which justifies why they are used as drug delivery systems. Also, their small particle size increases the surface area, the rate of dissolution, the solubility, and the oral bioavailability of the drug. They can be easily manipulated to achieve both passive and active drug targeting, with more rapid onset of therapeutic action and less amount of dose required. They easily penetrate to cell walls, stomach epithelium, blood vessels and blood–brain barrier. In addition, nanoparticles have the ability to control the release of the drug during transportation, which leads to an increase in drug therapeutic efficacy and reduction in side effects and toxicity. Moreover, they have a relatively higher intracellular uptake than microparticles, and they are characterized by site-specific targeting because the targeting ligands can be attached on their surface or a magnetic guidance can be used.

Other advantages include the relatively high drug loading of nanoparticles. There is no need for any chemical reaction to incorporate the drugs into the systems. Nanoparticles can be used for different routes of administration including oral, nasal, parenteral, and intra-ocular administration, among others. An added advantage is that it is possible to change the drug delivery method in order to improve compliance or reduce manufacturing costs.

1.6 Limitations of Nanoparticles

Nanoparticles have some practical limitations, which have to be overcome before they can be used clinically or made commercially available. Due to their small size, nanoparticles can manifest particle-particle aggregation, making their physical handling difficult in liquid and dry forms. Second, producing nanoparticles as drug delivery systems requires the extensive use of polyvinyl alcohol as a detergent that creates an issue of toxicity. Third, nanoparticles show particle growth, unpredictable gelation tendency, and sometimes limited drug loading and burst release [10].

1.7 Applications of Nanoparticles

Nanoparticles can be used in a variety of applications. They have been increasingly used in medicine for their ability to deliver drugs, often resulting in increased therapeutic efficacy of the drugs, weakened side effects, improved patient compliance, and controlled release of pharmacologically active drugs to the precise action site. The bioavailability of bioactive molecules is limited by the epithelial barriers of the gastrointestinal tract. Polymeric nanoparticles allow encapsulation of these molecules and protect them against hydrolytic degradation [50].

These characteristics reduce the cost and side effects [51]. Nanotechnology can be applied in tissue engineering research and in repair of damaged tissue. The traditional treatments such as organ transplants can be replaced by tissue engineering [52].

In addition, nanoparticles display some characteristics inducing unique mechanical, optical, and imaging properties that are used in certain applications within the medical and commercial sectors. The potential benefits of nanotechnology have been documented by many marketable products such as microelectronics, aerospace, and pharmaceutical industries [5].

Nanotechnology also plays a role in the improvement of the production, processing, protection, and packaging of food. In the food packaging process, the anti-microbial substances can be introduced via nanocomposite coating [53].

Furthermore, environmental remediation using nanoparticles is being used to decontaminate (by redox reactions) the air, water and soil. The removal of heavy metals, pathogens and organic contaminants from natural water using nanoparticles—by disinfection, purification, and desalination—has attracted considerable attention because of their adverse effects on environmental health [43, 54].

1.8 Nanoparticle Carriers for Cancer-Targeted Therapy

Cancer is the most challenging disease. It is characterized by the development of abnormal cells that divide uncontrollably. Most of the current chemotherapeutic drugs should be applied in a high volume of distribution; as a result, they become cytotoxic. They have poor solubility and are formulated with toxic solvents. Drug solubility is one of the challenges in pharmaceutical formulations, and hydrophobic drugs show low solubility and thus the necessity to use higher doses to reach the required concentrations that exhibit a pharmacological response [55]. Low solubility leads to reduced bioavailability of the drug and increased systemic toxicity. Generally, these drugs lack specificity and cause significant damage to noncancerous tissues, leading to different side effects [56, 57].

The combination of targeted drug delivery and controlled release technology may provide solutions to overcome the limitations of conventional chemotherapy. Thus, the development of nanocarriers is important, allowing the preparation of low water-soluble cancer medications in solid or liquid formulations and controlling their release directly into cancer cells [58, 59]. Nanoparticles serve as the cargo for the drug, thus enhancing its biodistribution, bioavailability, and retention [60].

Each nanocarrier transports more than one drug molecule, thus increasing the concentration of the drug delivered to the tumor site without affecting normal tissues. Moreover, it is possible to encapsulate several drugs into a nanocarrier which leads to synergistic anticancer effects and thus the possibility of decreasing the concentration of each drug and avoiding toxicity and drug-resistance of the targeted tumor [61–63].

Drug-loaded nanoparticles can target the tumor either passively or actively. Since highly aggressive tumors form fenestrated vasculature having deregulated and leaky nature with poor lymphatic drainage near the tumor sites, the nanoparticles are retained and accumulate near the tumor sites, thus decreasing their exposure to normal tissues and reducing side effects. This enhanced permeability and retention effect (EPR) results in passive drug targeting to the tumors which ensures an advantage for nanoparticle encapsulated drugs over free drugs [64, 65]. A more efficient way is the active targeting where a ligand is conjugated on the surface of the nanoparticles; this ligand binds specifically to the receptors or antigens on the cancer cell surface [66]. This approach enhances the specificity and thus the uptake and retention of the drug at the cancer site while decreasing systemic toxicity. Currently, the Food and Drug administration (FDA) has approved fifty types of nanoparticles for clinical use including liposomal nanoparticles, polymers, and nanocrystals, among others. About 60 nanoparticle types are classified as investigational drugs [67].

Biodegradable polymer nanoparticles have been extensively investigated for cancer therapy. They can be formulated to encapsulate hydrophilic or hydrophobic small drug molecules, and macromolecules such as proteins and nucleic acids. These nanoparticles are used to release the encapsulated drugs at a controlled rate in a time- or condition-dependent manner, thus increasing the efficacy of the drug [68].

1.9 Nanotheranostics in Personalized and Precision Medicine

It is well known that cancer is not a single disease; rather it is a combination of multiple diseases resulting from different genetic and epigenetic factors, and a wide variation of mutations and deregulations. As a result, no single therapeutic agent could be given to cancer patients, even to those with the same diagnosis. Therefore, to avoid the trial and error of the empirical medicine, personalized medicine has been considered the breakthrough of the 21st century.

Personalized and precision medicine refers to the individualized therapeutic intervention based on the individual's molecular, genetic, imaging and clinical examinations [69]. Conventional methods involve ex vivo genetic and molecular profiling and in vivo imaging to determine the type and stage of the tumor and thus the possible treatment [70]. In the ex vivo analysis, the expression of certain genes and proteins is quantified, and the biomarker profiling is investigated to determine the possible therapeutics and the probability of showing side effects. Whole genome profiling made it possible to lower the side effects by enhancing the diagnosis and the choice of the drug and its dose. The commonly used imaging techniques for the in vivo analysis are magnetic resonance imaging (MRI), positron emission tomography (PET), and computer tomography (CT), among others. However, the ex vivo genetic and molecular profiling and in vivo imaging subdivide the patients into three categories: those that respond to the treatment, those that do not and those that develop side effects to that particular treatment. Therefore, it is better to refer to it as "stratified medicine" rather than personalized medicine [70]. This is due to the absence of suitable sensitive contrast agents to visualize specific gene and protein expression profiles of patients. Instead of the separate administration of the diagnostic tools and therapeutics, theranostics form the basis of personalized and precision medicine. Theranostic refers to the combination of molecular imaging and therapy. Nanotechnology proved to be a key player in theranostics, especially for cancer therapy, helping in determining the stage of the disease, the appropriate treatment, treatment plan, and the adverse side effects. Nanotheranostics include nanosensors and nanomedicines; nanosensors detect the biomarkers and nanomedicines deliver the drug providing advantages over the free drug (Fig. 1.2). In addition to the previously mentioned properties of nanoparticles, their high surface area to volume ratio enabled them to have high loading capacity for the imaging probes, ligands, and drug molecules. The conventional imaging techniques detect the presence of significant tumor shrinkage; however, this is not always the case, especially in the short term where some drugs may induce cell cycle arrest. This necessitates the use of molecular imaging methods.

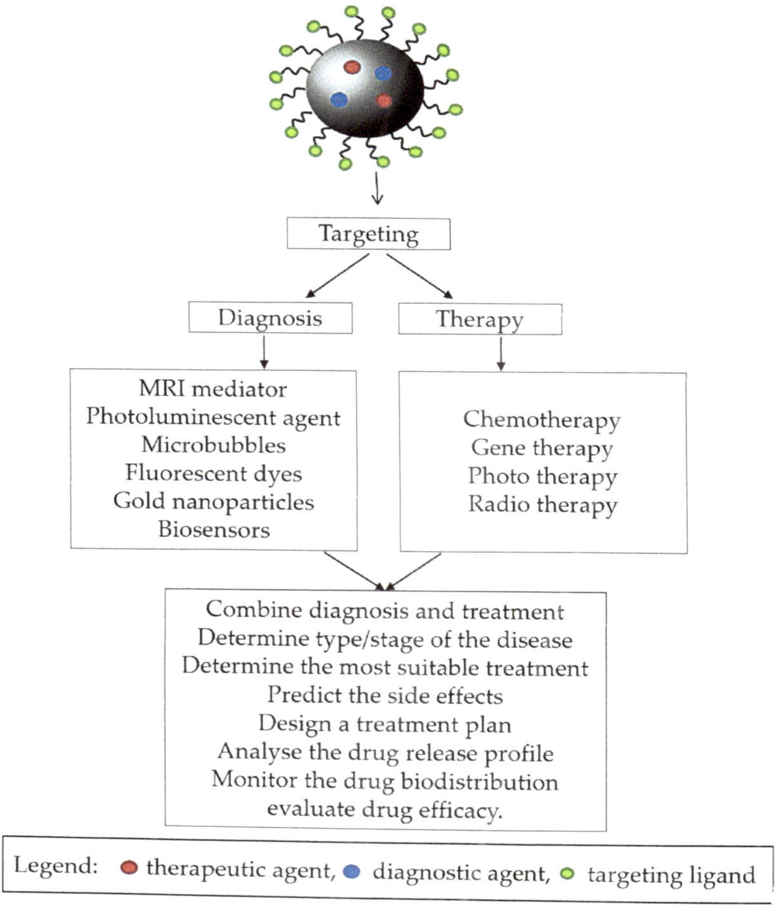

Figure 1.2 Nanotheranostics in personalized and precision medicine.

Nanoparticles can be incorporated with different targeting moieties, loaded with several particles having different therapeutic efficiency and contrasts, thus enhancing the image resolution and allowing the discovery of small lesions not visible by the traditional imaging techniques and providing an idea about the molecular changes that occur in the tumor after the treatment (e.g., shrinkage). Usually, each nanoparticle is loaded with thousands of a certain metal molecule that are important to enhance the contrast of the conventional MRI. An interesting example of the importance of nanotheranostics is the detection of small regions of angiogenesis associated with tumor development that are not yet detectable by the conventional MRI; where sparse angiogenic biomarkers are detected by ligands attached on the surface of the nanoparticle and visualized by the molecular MRI [69–71].

Depending on several conditions, the physician decides whether or not to adopt personalized medicine approach. If the conventional therapy showed side effects or if the tumor developed resistance to certain chemotherapeutics, precision and personalized medicine would follow. With the advent of nanotheranostics, it became possible to combine diagnosis and drug delivery together. The nanotheranostics, once introduced into the body, would locate, identify, and destroy the tumor.

References

. Buzea, C., Pacheco, I. I., and Robbie, K. (2010). Nanomaterials and nanoparticles: Sources and toxicity, *Biointerphases*, **2**, pp. MR17–MR71.

tken, R. J., Creely, K. S., and Tran, C. L. (2004). Nanoparticles: occupational hygiene review, *Heal. Saf. Exec. Res. Rep.*, **274**, -113.

it, D. B., Sayes, C. M., Reed, K. L., and Swain, K. A. (2008). effects related to nanoparticle exposures: Environmental, nd safety considerations for assessing hazards and risks, *l. Ther.*, **120**, pp. 35–42.

Fenniri, H., and Singh, B. (2007). Nanotechnology-based systems, *J. Occup. Med. Toxicol.*, **2**, p. 16.

5. Mudshinge, S. R., Deore, A. B., Patil, S., and Bhalgat, C. M. (2011). Nanoparticles: Emerging carriers for drug delivery, *Saudi Pharm. J.*, **19**, pp. 129–41.
6. Singh, S., Pandey, V. K., Tewari, R. P., and Agarwal, V. (2011). Nanoparticle based drug delivery system: Advantages and applications, *Indian J. Sci. Technol.*, **4**, pp. 177–180.
7. Aarti, N., Mukesh, P. R., and Shilpa, P. C. (2014). Nanoparticles - an overview, *Int. J. Res. Dev. Pharm. L. Sci.*, **3**, pp. 1121–1127.
8. Irving, B. (2007). Nanoparticle drug delivery systems, *Inno. Pharm. Biotechnol.*, **24**, pp. 58–62.
9. Wilczewska, A. Z., Niemirowicz, K., Markiewicz, K. H., and Car, H. (2012). Nanoparticles as drug delivery systems, *Pharmacol. Rep.*, **64**, pp. 1020–1037.
10. Nikam, A. P., Ratnaparkhiand, M. P., and Chaudhari, S. P. (2014). Nanoparticles - an overview, *Int. J. Res. Dev. Pharm. L. Sci.*, **3**, pp. 1121–1127.
11. Mekaru, H., Lu, J., and Tamanoi, F. (2015). Development of mesoporous silica-based nanoparticles with controlled release capability for cancer therapy, *Adv. Drug Deliv. Rev.*, **95**, p. 409.
12. Velavan, P., Karuppusamy, C., and Venkatesan, P. (2015). Nanoparticles as drug delivery systems, *J. Pharm. Sci. Res.*, **7**, pp. 1118–1122.
13. Akbarzadeh, A., Rezaei-Sadabady, R., Davaran, S., Joo, S. W., Zarghami, N., Hanifehpour, Y., Samiei, M., Kouhi, M., and Nejati-Koshki, K. (2013). Liposome: Classification, preparation, and applications, *Nanoscale Res. Lett.*, **8**, p. 102.
14. Kumari, A., Yadav, S. K., and Yadav, S. C. (2010). Biodegradable polymeric nanoparticles based drug delivery systems, *Colloids Surf. B Biointerfaces,* **75**, pp. 1–18.
15. Deb, S., Ghosh, K., and Shetty, S. D. (2015). Nanoimaging in cardiovascular diseases: Current state of the art, *Indian J. Med. Res.,* **141**, pp. 285–298.
16. Abbasi, E., Aval, S. F., Akbarzadeh, A., Milani, M., Nasrabadi, H. T., Joo, S. W., Hanifehpour, Y., Nejati-Koshki, K., and Pashaei-Asl, R. (2014). Dendrimers: Synthesis, applications, and properties, *Nanoscale Res. Lett.,* **9**, p. 247.
17. Thomas, S. C., Harshita, Mishra, P. K., and Talegaonkar, S. (2015). Ceramic Nanoparticles: Fabrication methods and applications in drug delivery, *Curr. Pharm. Des.,* **21**, pp. 6165–6188.

18. Akbarzadeh, A., Samiei, M., and Davaran, S. (2012). Magnetic nanoparticles: Preparation, physical properties, and applications in biomedicine, *Nanoscale Res. Lett.*, **7**, p. 144.

19. Omri, A., Suntres, Z. E., and Shek, P. N. (2002). Enhanced activity of liposomal polymyxin B against Pseudomonas aeruginosa in a rat model of lung infection, *Biochem. Pharmacol.*, **64**, pp. 1407–1413.

20. Sunderland, C. J., Steiert, M., Talmadge, J. E., Derfus, A. M., and Barry, S. E. (2006). Targeted nanoparticles for detecting and treating cancer, *Drug Dev. Res.*, **67**, pp. 70–93.

21. dos Santos Giuberti, C., de Oliveira Reis, E. C., Ribeiro Rocha, T. G., Leite, E. A., Lacerda, R. G., Ramaldes, G. A., and de Oliveira, M. C. (2011). Study of the pilot production process of long-circulating and pH-sensitive liposomes containing cisplatin, *J. Liposome Res.*, **21**, pp. 60–69.

22. Bai, J., Li, Y., Du, J., Wang, S., Zheng, J., Yang, O., and Chen, X. (2007). One-pot synthesis of polyacrylamide-gold nanocomposite, *Mater. Chem. Phys.*, **106**, pp. 412–415.

23. Saraog, G. K., Gupta, P., Gupta, U. D., Jain, N. K., and Agrawal, G. P. (2010). Gelatin nanocarriers as potential vectors for effective management of tuberculosis, *Int. J. Pharm.*, **385**, pp. 143–149.

24. Martinem, A., Iglesias, I., Lozano, R., Teijon, J. M., and Blanco, M. D. (2011). Synthesis and characterization of thiolated alginate-albumin nanoparticles stabilized by disulfide bonds. Evaluation as drug delivery systems, *Carbohydr. Polym.*, **83**, pp. 1311–1321.

25. des Rieux, A., Fievez, V., Garinot, M., Schneider, Y. J., and Préat, V. (2006). Nanoparticles as potential oral delivery systems of proteins and vaccines: A mechanistic approach, *J. Control. Release*, **116**, pp. 1–27.

26. Luo, G., Yu, X., Jin, C., Yang, F., Fu, D., Long, J., Xu, J., Zhan, C., and Lu, W. (2010). LyP-1-conjugated nanoparticles for targeting drug delivery to lymphatic metastatic tumors, *Int. J. Pharm.*, **385**, pp. 150–156.

27. Mora-Huertas, C. E., Fessi, H., and Elaissari, A. (2010). Polymer-based nanocapsules for drug delivery, *Int. J. Pharm.*, **385**, pp. 113–142.

28. Pandey, R., and Khuller, G. K. (2006). Oral nanoparticle-based antituberculosis drug delivery to the brain in an experimental model, *J. Antimicrob. Chemotherapy.*, **57**, pp. 1146–1152.

29. Svenson, S., and Tomalia, D. A. (2005). Dendrimers in biomedical applications – reflections on the field, *Adv. Drug Deliv. Rev.*, **57**, pp. 2106–2129.

30. Balogh, L., Swanson, D. R., Tomalia, D. A., Hagnauer, G. L., and McManus, A. T. (2001) Dendrimer-silver complexes and nanocomposites as antimicrobial agents, *Nano. Lett.*, **1**, pp. 18–21.

31. Devarakonda, B., Hill, R. A., Liebenberg, W., Brits, M., and de Villiers, M. M. (2005). Comparison of the aqueous solubilization of practically insoluble niclosamide by polyamidoamine (PAMAM) dendrimers and cyclodextrins, *Int. J. pharm.*, **304**, pp. 193–209.

32. D'Emanuele, A., and Attwood, D. (2005). Dendrimer-drug interactions. *Adv. Drug Deliv. Rev.*, **57**, pp. 2147–2162.

33. Menjoge, A. R., Kannan, R. M., and Tomalia, D. A. (2010). Dendrimer-based drug and imaging conjugates: Design considerations for nanomedical applications, *Drug Discov. Today*, **15**, pp. 171–187.

34. Yih, T. C., and Al-Fandi, M. (2006). Engineered nanoparticles as precise drug delivery systems. *J. Cell. Biochem.*, **97**, pp. 1184–1190.

35. Slowing, I., Trewyn, B. G., Giri, S., and Lin, V. S.-Y. (2007). Mesoporous silica nanoparticles for drug delivery and biosensing applications, *Adv. Funct. Mater*, **17**, pp. 1225–1236.

36. Foldvari, M., and Bagonluri, M. (2008). Carbon nanotubes as functional excipients for nanomedicines: I. Pharmaceutical properties, *Nanomedicine*, **4**, pp. 173–182.

37. Beg, S., Rizwan, M., Sheikh, A. M., Hasnain, M. S., Anwer, K., and Kohli, K. (2011). Advancement in carbon nanotubes: Basics, biomedical applications and toxicity, *J. Pharm. Pharmacol.*, **63**, pp. 141–163.

38. Arsawang, U., Saengsawang, O., Rungrotmongkol, T., Sornmee, P., Wittayanarakul, K., Remsungnen, T., and Hannongbua, S. (2011). How do carbon nanotubes serve as carriers for gemcitabine transport in a drug delivery system? *J. Mol. Graph. Model.*, **29**, pp. 591–596.

39. Perry, J. L., Martin, C. R., and Stewart, J. D. (2011). Drug-delivery strategies by using template-synthesized nanotubes, *Chemistry*, **17**, pp. 6296–6302.

40. Chen, Z., Pierre, D., He, H., Tan, S., Pham-Huy, C., Hong, H., and Huang, J. (2011). Adsorption behavior of epirubicin hydrochloride on carboxylated carbon nanotubes, *Int. J. Pharm.*, **28**, pp. 153–161.

41. Arruebo, M., Fernández-Pacheco, R., Ibarra, M. R., and Santamaría, J. (2007). Magnetic nanoparticles for drug delivery, *Nano Today*, **2**, pp. 22–32.

42. Couvreur, P., Barratt, G., Fattal, E., Legrand, P., and Vanthier, C. (2002). Nanocapsule technology; a review, *Crit. Res. Ther. Drug Carrier Syst.*, **19**, pp. 99–134.

43. Ealias, A. M., and Saravanakumar, M. P. (2017). A review on the classification, characterization, synthesis of nanoparticles and their application, *IOP Conf. Ser.: Mater. Sci. Eng.*, **263**, pp. 032019.

44. Champeau, R. (2006). Assessing safety health risks of nanomaterials, *Phys. Org.*, **15**, p. 2005.

45. Clogston, J. D., and Patri, A. K. (2011). Zeta potential measurement, *Methods Mol. Biol.*, **697**, pp. 63–70.

46. Dora, C. P., Singh, S. K., Kumar, S., Datusalia, A. K., Deep, A. (2010). Development and characterization of nanoparticles of glibenclamide by solvent displacement method, *Acta. Pol. Pharm.*, **67**, pp. 283–290.

47. Tiruwa, R. (2016). A review on nanoparticles - preparation and evaluation parameters, *Indian J. Pharm. Biol. Res.*, **4**, pp. 27–31.

48. Drasler, B., Vanhecke, D., Rodriguez-Lorenzo, L., Petri-Fink, A., and Rothen-Rutishauser, B. (2017). Quantifying nanoparticle cellular uptake: Which method is best? *Nanomedicine (Lond)*, **12**, pp. 1095–1099.

49. Parasuraman, S. (2011). Toxicological screening, *J. Pharmacol. Pharmacother.*, **2**, pp. 74–79.

50. Kommaleddy, S., Tiwari, S. B., and Amiji, M. M. (2005). Long circulating polymeric nanovectors for tumor selective gene delivery technol, *Cancer Res. Treat.*, **4**, pp. 615–625.

51. Alexis, F., Pridgen, E., Molnar, L. K., and Farokhzad, O. C. (2008). Factors affecting the clearance and biodistribution of polymeric nanoparticles, *Mol. Pharm.*, **5**, pp. 505–515.

52. Todescato, F., Fortunati, I., Minotto, A., Signorini, R., Jasieniak, J., and Bozio, R. (2016). Engineering of semiconductor nanocrystals for light emitting applications, *Materials.*, **9**, p. 672.

53. Laad, M., and Jatti, V. K. S. (2016). Titanium oxide nanoparticles as additives in engine oil, *J. King Saud Univ. Eng. Sci.*, **30**(2), pp. 0–6.

54. Mueller, N. C., and Nowack, B. (2008). Exposure modeling of engineered nanoparticles in the environment, *Environ. Sci. Technol.*, **42**, pp. 4447– 4453.

55. Alsenz, J., and Kansy, M. (2007). High throughput solubility measurement in drug discovery and development, *Adv. Drug Deliv. Rev.*, **59**, pp. 546–567.

56. Luo, Y., and Prestwich, G. D. (2002). Cancer-targeted polymeric drugs, *Curr. Cancer Drug Targets.*, **2**, pp. 209–226.

57. Kwon, G. S. (2003). Polymeric micelles for delivery of poorly water-soluble compounds, *Crit. Rev. Ther. Drug Carrier Syst.*, **20**, pp. 357–403.

58. Cheng, J., Teply, B. A., Sherifi, I., Sung, J., Luther, G., Gu, F. X., Levy-Nissenbaum, E., Radovic-Moreno, A. F., Langer, R., and Farokhzad, O. C. (2007). Formulation of functionalized PLGA-PEG nanoparticles for in vivo targeted drug delivery, *Biomaterials.*, **28**, pp. 869–876.

59. Bharali, D. J., Khalil, M., Gurbuz, M., Simone, T. M., and Mousa, S. A. (2009). Nanoparticles and cancer therapy: A concise review with emphasis on dendrimers, *Int. J. Nanomed.*, **4**, pp. 1–7.

60. Petschauer, J. S., Madden, A. J., Kirschbrown, W. P., Song, G., and Zamboni, W. C. (2015). The effects of nanoparticle drug loading on the pharmacokinetics of anticancer agents, *Nanomedicine (Lond)*, **10**, pp. 447–463.

61. Lehar, J., Krueger, A. S., Avery, W., Heilbut, A. M., Johansen, L. M., Price, E. R., Rickles, R. J, Short, G. F., 3rd, Staunton, J. E., Jin, X., Lee, M. S., Zimmermann, G. R., and Borisy, A. A. (2009). Synergistic drug combinations tend to improve therapeutically relevant selectivity, *Nat. Biotechnol.*, **27**, pp. 659–666.

62. Alam, M. F., Khan, G., Safhi, M. M., Alshahrani, S., Siddiqui, R., Sivagurunathan Moni, S., and Anwer, T. (2018). Thymoquinone ameliorates doxorubicin-induced cardiotoxicity in swiss albino mice by modulating oxidative damage and cellular inflammation, *Cardiol. Res. Pract.*, **2018**, 1483041.

63. Zidan, A. A., El-Ashmawy, N. E., Khedr, E. G., Ebeid, E. M., Salem, M. L., and Mosalam, E. M. (2018). Loading of doxorubicin and thymoquinone with F2 gel nanofibers improves the antitumor activity and ameliorates doxorubicin-associated nephrotoxicity, *Life Sci.*, **207**, pp. 461–470.

64. Siemann, D. W. (2011). The unique characteristics of tumor vasculature and preclinical evidence for its selective disruption by tumor-vascular disrupting agents, *Cancer Treat. Rev.*, **37**, pp. 63–74.

65. Viallard, C., and Larrivee, B. (2017). Tumor angiogenesis and vascular normalization: Alternative therapeutic targets, *Angiogenesis*, **20**, pp. 409–426.

66. Cho, K., Wang, X., Nie, S., Chen, Z. G., and Shin, D. M. (2008). Therapeutic nanoparticles for drug delivery in cancer, *Clin. Cancer Res.*, **14**, pp. 1310–1316.

67. Ventola, C. L. (2017). Progress in nanomedicine: Approved and investigational nanodrugs, *P. T.*, **42**, pp. 742–755.

68. Gu, F. X., Karnik, R., Wang, A. Z., Alexis, F., Levy-Nissenbaum, E., Hong, S., Langer, R. S., and Farokhzad, O. C. (2007). Targeted nanoparticles for cancer therapy, *Nanotoday.*, **2**, pp. 14–21.

69. Jain, K. K. (2005). Role of nanobiotechnology in developing personalized medicine for cancer, *Technol. Cancer Res. Treat.*, **4**, pp. 645–650.

70. Lammers, T., Rizzo, L. Y., Storm, G., and Kiessling, F. (2012). Personalized nanomedicine, *Clin. Cancer Res.*, **18**, pp. 4889–4894.

71. Mura, S., and Couvreur, P. (2012). Nanotheranostics for personalized medicine, *Adv. Drug Deliv. Rev.*, **64**, pp. 1394–1416.

Chapter 2

Nanoparticles in Cancer Treatment: Types and Preparation Methods

Jyoti Ahlawat,[a] Emmanuel Zubia,[b] and Mahesh Narayan[a]

[a]*Chemistry & Biochemistry Department,*
The University of Texas at El Paso, Texas, USA
[b]*Environment Science & Engineering Department,*
The University of Texas at EL Paso, Texas, USA

mnarayan@utep.edu

Nanoparticles are particles ranging from 1 to 100 nm in size. Such particles act as carrier moieties/vehicles for the delivery of chemotherapeutics for entry through fenestrations in tumor vasculature. These particles can be surface modified or engineered for binding to receptors on the cancer cell membrane or to nuclear receptor sites. Such modifications result in increased delivery of drugs at the target site with minimal toxicity to the surrounding healthy cells. Therefore, increased understanding of the interaction of nanoparticles, their preparation methods, and medicinal applications is mandatory so that clinical translation

Nanoparticle Drug Delivery Systems for Cancer Treatment
Edited by Hala Gali-Muhtasib and Racha Chouaib

ISBN 978-981-4800-54-9 (Hardcover), 978-0-429-34125-0 (eBook)
www.jennystanford.com

can be achieved in the near future. This chapter highlights different carrier platforms, their preparation methods, and the various nanoparticle-based medicines used in cancer chemotherapy.

2.1 Introduction

The field of nanoparticles in cancer therapy has focused on the efficient distribution of chemotherapeutics via entrapment, creating sustained drug release profile. The ideal nanotherapeutic will target specific cancer biomarker(s) and release the medication in a dose-dependent manner, having little repercussions in healthy cells. In this chapter, we provide a brief introduction on cancer and the interaction of nanoparticles with cancer cells, followed by types of nanoparticles used in research and some of the common methodologies to create such nanoparticles, culminating with a summary of nanoparticle-based medications used in cancer chemotherapy.

Cancer is the second leading cause of death in the United States [1]. In 2017, the most common incidence of cancer was breast and prostate cancer, for females and males, respectively, thereafter lung and bronchus, and colorectal cancers are common cancers for both genders [1]. Cancer is a complex ailment that manifests itself by the uncontrolled division of abnormal cells, its pathogenesis can stem from genetic dysregulation or mutations [2] to chronic or acute exposure to xenobiotic or environmental toxicants [3]. Cancer cells can spread beyond layers of tissue, metastasizing in different organs through a process called infiltration.

2.2 Nanoparticle Physiology and Their Effects on the Body

The definition of a nanoparticle, according to the ASTM E2456-06 standard, are particles having sizes that range from 1 to 100 nm. However, this definition has not been strictly followed by all authors, as nanoparticles might increase in geometric length due to their hydrodynamic diameter or by the interactions formed during intended functions. As such, certain publications refer to nanoparticles as particles that are larger than 0.1 micrometer in length.

In healthy tissues, nanoparticles are removed from circulation by the glomerular filtration system or by the mononuclear phagocyte system (MPS) [4, 5]. The glomerular filtration system is located in the nephrons of the kidneys and uses hydrostatic pressure to filter waste products from the blood. The MPS is part of the immune system and its main function is to remove pathogens from blood circulation; the MPS is composed of macrophages and monocytes located in reticular connective tissues. The primary mechanism of nanoparticle removal in the MPS is through endocytosis [6]. There is a size dependent correlation between nanoparticle size and their interaction with macrophages; in liposomes, nanoparticles smaller than 100 nm are removed at a slower rate from the blood in comparison to bigger nanoparticles [6]. In addition, a negative surface charge has been associated with rapid removal from the blood in comparison to neutral charges. One common way to address this issue is to graph a polymer coat (i.e., polyethylene glycol, phosphatidylinositol and phosphatidylserine) onto nanoparticles to protect them from immune recognition.

Depending on the composition of the nanoparticles, they can interact with proteins, lipids, and sugars. The resulting interaction often leads to the formation of a different bimolecular interphase known as coronas [7]. There are two types of coronas: hard and soft. The hard corona comprises biomolecules that have the greatest binding affinity to the nanoparticle, this association can last many hours and although disassociation is possible, it interferes with the molecular composition of the nanoparticle resulting in changes of uptake drug delivery and degradation [7]. The soft corona is considered to have weak interactions with biomolecules that result in transient binding [7].

Cancer nanotherapeutics have taken advantage of the enhanced permeability and retention (EPR) effect that is common in tumors. EPR arises as a result of the incomplete endothelial lining in the vasculature of cancer cells whereby the pericytes' coverage is diminished and basement membranes are loosely associated with endothelial cells. This leads to relatively large pores, ranging from 100 nm to 3 μm in diameter in the tumor vasculature [4, 5]. The EPR effect creates significantly higher vascular permeability and hydraulic conductivity in tumors, which allows for nano-size drug accumulation [5]. There are several methods to further enhance the EPR effect, thus increasing the efficacy of nanoparticle accumulation. These include modulating the tumor blood flow (i.e., vasoconstrictor agents, nitric oxide, etc.), modulating tumor vasculature and stroma (i.e., osnoporation, anti-angiogenic treatment and collagenase treatment) and killing cancer cells (i.e., radiation therapy) [5].

The delivery of therapeutic drugs to the central nervous system has always presented a special challenge due to the blood–brain barrier. The blood–brain barrier is an endothelial cell lining with tight junctions bounded to a basement membrane and astrocytes. It is hydrophobic in nature and effective at preventing the diffusion of hydrophilic molecules and pathogens from entering the brain parenchyma. One-way nanoparticles could pass this barrier is by coating them with surfactants such as polysorbate, and poly(butyl)cyanoacrylate [8].

There is limited data on the adverse toxicological effects of nanoparticle delivery systems. Most of the case studies on nanoparticle toxicity are via inhalation and damage to the respiratory track [8]. The toxic effect of nanoparticles is mainly studied in animal models where metal nanoparticles and carbon nanotubes have shown an increase in radical production and disruption of platelet function [8, 9]. An example of these toxic effects is cationic gold and polystyrene nanoparticles where they have been linked to hemolysis and blood clotting [8].

2.3 Nanoparticles with Medicinal Applications

The use of nanoparticles as drug delivery methods is advantageous as it increases the bioavailability of poorly soluble therapeutics (via encapsulation or cross-linking), and protects the drug from degradation and lower systemic toxicity [10]. Additionally, biomarkers can be incorporated in the formulation to increase the specificity of the target drug. Here, we review some of the common nanoparticles used in research.

2.3.1 Liposomes and Micelles

Liposomes are structural analogs of vesicles—amphipathic in nature with a lipid bilayer assembled into a closed colloidal structure with a polar surface area and a polar core having hydrophobic chains in between [4]. Depending on their composition (natural or synthetic lipids), they can be biocompatible and biodegradable and range in size from nano- to micrometers. Liposomes are similar to micelles; however, the main difference resides in that micelles are composed of monolayers having a hydrophobic core and a polar surface area [11]. Figure 2.1 provides an illustration of the different types of fatty acids that compose a micelle or a liposome. In order to enhance the lifetime of both micelles and liposomes, they can be coated with poly ethylene glycol (PEG) [4]. In 1995, the first liposomal anticancer drug (Doxil®), which is also the first FDA-approved nanomedicine, was marketed in the USA [12]. Pegylated liposomal doxorubicin and nab-paclitaxel are nanoparticle formulations of doxorubicin and paclitaxel that have demonstrated higher therapeutic efficacy against breast cancer in comparison to the parent compound [4].

Figure 2.1 Illustration of the shape of (A) micelle and (B) liposome nanoparticles.

2.3.2 Protein-Based Nanoparticles

Polypeptide-based nanoparticles present a unique advantage over other nanoparticles in regards to toxicity and unique functionalities that depend on the chemical composition of the protein and their surface scaffolds. Endogenous proteins are often used for the creation of nanoparticles as they elicit a low immune response, and some of their physicochemical properties (biodegradability, nonantigenicity, metabolism, surface charge, and structure) are already well defined. If the protein reagents are pure, these nanoparticles are relatively easy to prepare. One of the most ubiquitously used proteins in nanoparticle research is albumin. Albumin is a globular protein found in blood plasma and functions as a transporter of fatty acids, hormones and other compounds. In the context of nanoparticles, this protein makes use of its hydrophobic motif to reversibly bind hydrophobic chemotherapeutics and increase their bioavailability. One of the advantages of albumin is that it binds to glycoprotein 60, which mediates transcytosis, allowing it to be transported to the interior of the cell [13]. Figure 2.2A provides an illustration of a single monomeric protein assembled into a nanoparticle with a hollow cavity. An example of an albumin bound nanoparticle is Abraxane® (Celgene). This paclitaxel–albumin-bound nanoparticle increases the solubility and tumor delivery of paclitaxel, a drug that was approved by the FDA to treat breast cancer [12].

Other nanoparticles that comprise proteins are viruses. Due to their small size, these systems are considered to be naturally occurring nanoparticles. Without going into the classification and morphology of different types of viruses, we will consider the simplest structure which consists of a nucleic acid (DNA or RNA) and its capsid [14]. The capsid is a polymeric protein that functions as a barrier from nucleases. In addition, these capsids can have specific saccharides or proteins specific to cell receptors that attach and infect the host cell [14]. Figure 2.2B illustrates a simple structure of the tobacco mosaic virus that possesses a rod like structure and is used to deliver phenanthriplatin, a platinum (II)-based anticancer agent [15].

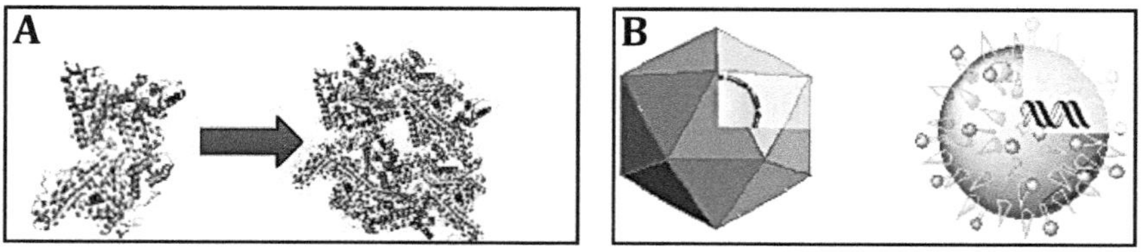

Figure 2.2 Protein-based nanoparticle assembly (A) albumin-based protein and (B) Virus structure nanoparticle.

2.3.3 Synthetic Polymers

These nanoparticles are usually self-assembled polymers that can be engineered to have various degrees of complexity. One important criterion for formulating these nanoparticles is biocompatibility and biodegradation. Some examples of these particles include Poly(D,L-lactic/glycolic acid) (PLGA), Polystyrene, and Poly-alkyl acrylate [6].

2.3.4 Dendrimers

Dendrimers are branched tree-like polymers that can be formulated with sugars, amino acids and nucleic acids. In essence, these nanoparticles are analogous to a step-by-step controlled synthesis, where their unique architectural motif can be carefully designed to accommodate the binding of the therapeutic agent via intermolecular forces [4]. The size of these nanoparticles can be easily controlled as they tend to increase linearly with every increasing dendrimer generation. The last dendrimer generation (the periphery) can have biomarkers that are tailored for specific cancer cell recognition sites. Since the synthesis of these molecules is carefully planned, there is a potential to create self-assembled dendrimers, where their functional groups recognize each other, leading to the spontaneous formation and encapsulation of any therapeutic agent. Figure 1.3 illustrates the branch-like structures of these polymers. These dendrimers can reach very complex structures; an example is the dendrimer nanoparticle polyethylene glycol-modified polyamidoamine dendrimer, which was conjugated with a peptide HAIYPRH (T7). This dendrimer was used to co-deliver a gene encoding human tumor necrosis factor-related apoptosis-inducing ligand (pORF-hTRAIL) and doxorubicin [16].

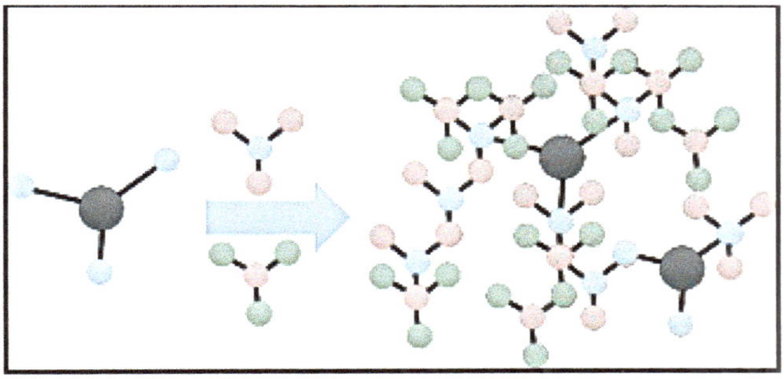

Figure 2.3 Theoretical representation of dendrimer synthesis and resulting aggregation.

2.3.5 Metal-Based and Inorganic Nanoparticles

As the name indicates they are metallic nanoparticles generally composed of gold, silver, or titanium and have been used for controlled drug release in chemotherapy [4]. It is generally desired that these metals are inert; however, when using them one must account for bioaccumulation and toxicity of the metal. An advantage of metallic nanoparticles is the ability of metals to conduct heat. Thus, some can be used in hyperthermic tumor therapy, where photo stimulation can provide the thermal energy to make this therapy highly specific.

Iron oxide nanoparticles are ferromagnetic and their bio-distribution can be controlled via magnetic fields; in addition, they can absorb in the visible and near infrared spectrum (680–1000 nm) [17]. Nanocube formulation of iron oxide has proven to provide efficient heat convection via magnetic field and photo radiation functioning in a cumulative highly specific thermic tumor therapy [17].

Inorganic silica core nanoparticles often provide a canvas to where a researcher can functionalize or coat them with various chemicals to achieve complex systems that have high specificity and drug delivery capabilities. An example of this versatility are the mesoporous silica nanoparticles; successful tumor reduction was noted with a 48 nm mesoporous silica nanoparticle [18]. The silica nanoparticle was first coated with PEG and then functionalized with 2-carboxyethyl-2-pyridyl disulfide to later load the pores with doxorubicin and finally encapsulated with beta cyclodextrin. To further add stability, the nanoparticle was PEG coated and further functionalized with an adamantane unit at one end and a folic acid unit at the other end [18].

2.4 Preparation Methods for Nanoparticles

Here, we review various preparation methods used in nanoparticle research.

2.4.1 Nanoprecipitation/Solvent Displacement Method

In this method (Fig. 2.4), a lipophilic drug or polymer is dissolved in an organic solvent like dichloromethane in the presence or absence of a surfactant and mixed with an aqueous solution-containing stabilizer. The rapid diffusion of the organic solvent into the aqueous phase sets up an interfacial turbulence causing the polymers to deposit on the interface. This emulsion is then passed through a high-pressure homogenizer to reduce the particle size to 100–200 nm [19].

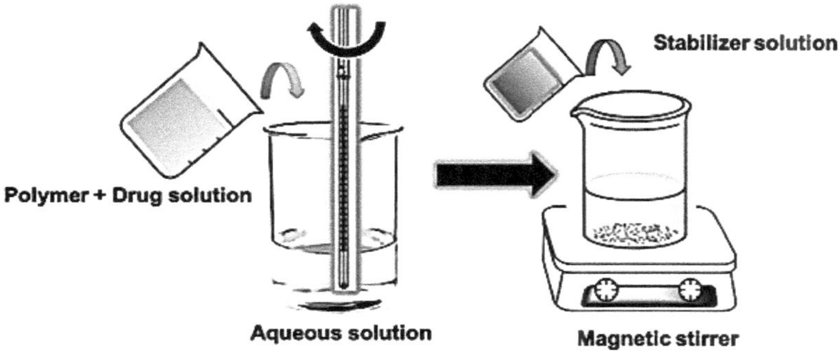

Figure 2.4 Schematic illustration of the nanoprecipitation method.

2.4.2 Solvent Emulsification-Diffusion Method

In this method, (Fig. 2.5) the lipophilic material is dissolved in an organic solvent such as cyclohexane, and ethyl acetate. The mixture is then emulsified in an aqueous phase containing surfactant like polysorbate-80 or poloxamer-188 to form oil in water emulsion. After stable emulsion is formed, the organic solvent is removed from the emulsion by evaporating either by continuous stirring or under reduced pressure (40–60 mbar). Although the process is energy extensive and the synthesized nanoparticles have polydisperse distribution, the process is continuous and commercially demonstrated [19–21].

2.4.3 Microemulsion

Water-in-oil microemulsion system involves stirring a mixture of low melting fatty acid such as stearic acid, an emulsifier such as polysorbate-20, and a co-emulsifier such as butanol and water at a temperature of 65–70°C. The hot mixture is then dispersed in cold water with temperatures between 2–3°C under constant stirring. High-temperature gradients prevent aggregation [20, 21].

In a study, Long Zhang Zhu et al. (2009) synthesized chitosan-coated magnetic nanoparticles using reverse microemulsion method. Particles with average sizes of 100–120 nm, low aggregation, and enhanced magnetic responsivity were synthesized as the carrier for 5-florouracil (CS-5-FuMNPs). Although the method is labor intensive, it requires low mechanical energy input for working [22].

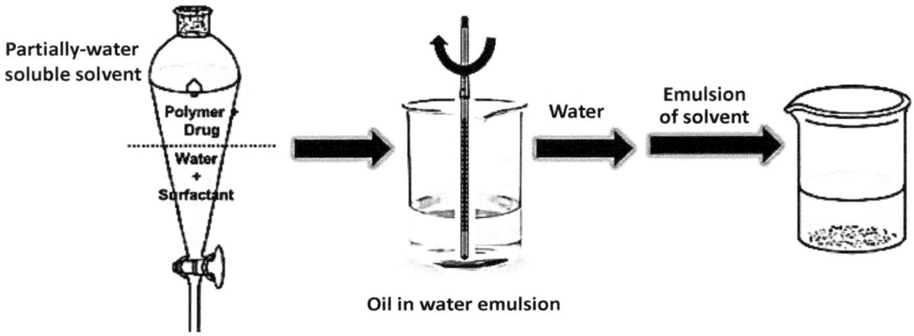

Figure 2.5 Schematic illustration of solvent emulsification-diffusion method.

2.4.4 Particle Replication in Non-Wetting Templates

One, emerging method in the preparation of nanoparticles is a technique called PRINT (Particle Replication in Non-wetting Templates), in which nanoparticles can be formulated using a nano template (Fig. 2.6). The advantage of this technique is that there is an increased control over the shape, size, composition and surface functionality resulting in future mass production of nanoparticles in industry [21].

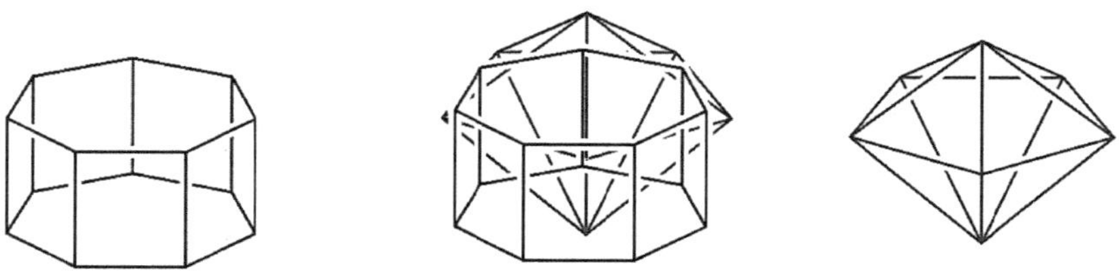

Figure 2.6 Illustration of the process of nanoparticle preparation using PRINT technique.

2.5 Nanomedicines in Drug Delivery

As previously discussed, using nanocarriers as drug delivery vehicles provides many advantages such as improved pharmacokinetics, enhanced therapeutic efficacy, reduced dose (dose-limiting), targeted delivery for increased specificity and lowered toxicity. An extensive list of examples regarding types of nanoparticles and different cross-links or formulations can be found in references [14] and [19], the following table (Table 2.1) is just a small list that attempts to illustrate the different variations and formulations of ligand-targeted particulate nanomedicines.

Table 2.1 The different variations and formulations of ligand-targeted particulate nanomedicines

Nanoparticle type	Active ingredient	Additional formulation/ component	Name	Cancer targeted
Liposome	Doxorubicin	PEG	Liposomal Doxorubicin (Doxil)	HIV-related Kaposi sarcoma, ovarian cancer, and multiple myeloma
	Doxorubicin	—	Liposomal doxorubicin (Myocet)	Metastatic Breast Cancer
	Melanoma antigens	Antigen	Lipovaxin-MM	Melanoma
Micelle	Rapamycin	—	Nab-rapa-mycin (ABI-Glioblastoma 009)	Perivascular epithelioid cell neoplasms
Polymeric micelle	Paclitaxel	Polymeric	Nab-paclitaxel (Abraxane)	Breast, lung and pancreatic cancer
Metal based	—	Iron oxide NP	NanoTherm	Glioblastoma
	Recombinant tumor necrosis factor alpha (TNF)	Colloidal gold (PEGylated)	CYT-6091	Inoperable ductal carcinoma of the breast
	—	Hafnium oxide NP	NBTXR3	Adult soft tissue sarcoma
Lipid NP	siRNA against PLK1	—	TKM-080301	Adrenocortical cancer

The finding that macromolecules accumulate inside solid tumors over time by virtue of their EPR effect has influenced our understanding, as well as the development and clinical translation of nanomedicines for cancer treatment. Such ligand-targeted nanomedicines allow targeted delivery of chemotherapeutics which can be applied to target sites where the EPR effect is not present. The aim of this next section is to discuss the various ligand-targeted particulate nanomedicines and does not focus on the stimuli-responsive nanomedicines and ligand-lacking particulate nanomedicines [23].

2.5.1 Lipid-Based Nanomedicines

(i) MM-302 is an anti-HER2 antibody attached to PEGylated liposomes loaded with doxorubicin (PLD) and has progressed to phase I clinical trials. Over the years, many parameters of this formulation have been optimized, which include antibody construct, composition of liposome and the method of conjugation [23]. Importantly, in vivo studies revealed that gold-labeled anti-HER2 PLD were localized within the tumor cells while the control formulation was present in the extracellular stroma of the tumor [24]. Moreover, in HER-2 negative xenografts, the biodistribution of anti-HER2 PLD and the control formulation was reported to be similar showing that both formulations were present inside the tumor but anti-HER2 PLD was associated directly with tumor cells [24]. Furthermore, comparable circulation time was observed for anti-HER2 PLD and control formulation indicating that the antibody fragment had no effect on the rate of clearance after multiple doses [25]. In another study, it was demonstrated that treatment of tumors using anti-HER2 PLD had more inhibitory effects on tumor growth compared to control formulations [25].

(ii) 2B3-101 is anti-GSH doxorubicin loaded liposome with an ability to cross the blood-brain barrier (BBB) for the treatment of central nervous system diseases such as glioma [23]. It can easily cross the BBB via glutathione transporters without disrupting the functionality of the BBB. Interestingly, in a study on rats, it was demonstrated that increasing the concentration of glutathione on the surface of pegylated liposomes loaded with ribavirin showed higher amounts of the drug inside the brain [26]. In another study, both 2B3-101 and PLD showed significant antitumor effects in a human breast cancer xenograft model in mice [27].

2.5.2 Polymer-Based Nanomedicines

(i) BIND-014 is a polymeric nanoparticle composed of PLA (poly (D,L-lactide)) and PEG (polyethylene glycol) copolymers loaded with docetaxel for the treatment of solid tumors [23, 28]. It is targeted to tumor cells expressing the prostate-specific membrane antigen using targeting ligand ACUPA (S,S-2-[3-[5-amino-1-carboxypentyl]-ureido]-pentanedioic acid) [28, 29]. In a study, targeted PLGA-based nanoparticles delivered fourfold more anticancer drug to the target site compared to control formulations in a mouse model bearing prostrate xenograft tumors that are positive for human prostate-specific membrane antigen [30]. In another xenograft study, targeted nanoparticles loaded with docetaxel showed significant inhibitory effects on tumor growth compared to control formulations [31].

(ii) CALAA-01 is a polymeric nanoparticle composed of a linear cyclodextrin-containing polymer (CDP) backbone, adamantane-conjugated PEG (AD-PEG), Tf-conjugated AD-PEG (Tf-PEG-ADD), and siRNA for the treatment of solid tumors [23, 32]. CALAA-01 functions by interfering with the catalytic activity of ribonucleotide reductase 2 enzyme [33]. Tf-nanoparticles were shown to inhibit tumor growth in a density-dependent manner [34]. In another study, significant inhibition in tumor growth was reported in human head and neck bearing mice cancer xenografts using CALAA-01 treatment compared to control formulation [35].

2.5.3 Bacteria-Derived Minicells as Nanomedicines

Erbitux®EDVsPAC are targeted minicells that can carry wide range of anticancer agents, (paclitaxel and cisplatin), siRNA and plasmid DNA encoding shRNA [36, 37]. These minicells can be engineered to carry target-specific moieties on their surface. The targeting antibodies for minicells are bispecific in nature which can recognize both lipopolysaccharide on minicells and overexpressed receptors such as EGFR on tumor cells [38]. Interestingly, EGFR-targeted minicells loaded with short hairpin RNA (shRNA) can be used to treat or reverse drug resistance in colon cancer cells by targeting the P glycoprotein MDR1 [37].

2.5.4 Retroviral Vectors

Rexin-G is a retroviral based nanomedicine for the treatment of cancers such as bone, soft tissue and pancreatic cancers. The main disadvantage of this murine leukemia virus-based nanomedicine is the lack of tissue specificity [39]. It is engineered to achieve targeted delivery by attaching high-affinity collagen-binding motif on its surface. It acts by interfering with mutant cyclin G1 gene of the cell cycle in tumor cells [40]. Rexin-G was observed to inhibit tumor growth and enhance survival compared to control formulation in human tumor xenografts [41]. Interestingly, improvement in physiological conditions such as wound healing and liver function were reported. This can be attributed to the targeting of collagen by the engineered retrovirus [42].

2.5.5 Virosomes

Immunopotentiating reconstituted influenza virosomes (IRIVs) are based on viral envelope protein integrated into the liposomes and are among the most advanced targeted delivery virosomal systems. Excellent immunogenicity and superior tolerability make them one of the best candidates for drug delivery applications. A couple of examples are Inflexal®V and Expaxal®. Inflexal®V is a trivalent influenza vaccine consisting of hemagglutinin and neuraminidase as the vaccine protecting agents, and Expaxal® is a hepatitis A vaccine and consists of inactivated hepatitis A virus as the virosome components. In a recent study, a virosomal chemotherapeutic formulation consisting of phosphatidylethanolamine-PEG-anchored antibodies as targeting agents and hemagglutinin for the cytoplasmic delivery of entrapped doxorubicin was synthesized and its effect on tumor progression was analyzed. It was reported that the formulation inhibited tumor progression in the mouse model. Moreover, virosomes comprising positively charged lipids (e.g., DOSPER) were formulated. The formulation had a plasmid encoding tumor-associated antigen protein. Mice immunized with this virosomal formulation showed a solid T-cell response [43].

2.6 Theranostic Nanoparticles: An Approach toward Personalized Medicine

Theranostics is a promising way to address the unmet medical needs by acting as a tailored treatment for individual patients. It involves the concept of personalized medicine, which has the potential to overcome limitations of conventional medicines by incorporation of therapy and imaging tools into a single system. The most commonly used nano-sized materials for theranostic applications include dendrimer, inorganic nanoparticles, lipid-based nanoparticles, and polymeric nanoparticles. These nano-sized particles offer huge advantages for the theranostic application as these nanoparticles have large surface-area-to-volume ratio, which allows for more drug loading. Moreover, due to their nano-size, such particles can easily extravasate into tumor sites due to the leaky vasculature and poor lymphatic drainage of the tumor (EPR effect) aiding in targeted imaging and therapy. Although, theranostic nanoparticles offer huge advantages in cancer therapy including chemotherapy (overcoming toxicity issues and drug-resistance), siRNA therapy (increasing bioavailability and enhancing stability), and photodynamic therapy (increased solubility in water and reduced systemic toxicity); the technology is comparatively new and this combination of imaging tools and therapy in one system has not been studied in full detail so far (Fig. 2.7). Thus, development of such single system (imaging and therapy) nanoparticles that are safe, simple, target-specific, and effective will continue to be the future focus as it holds great potential for clinical translation [43–46].

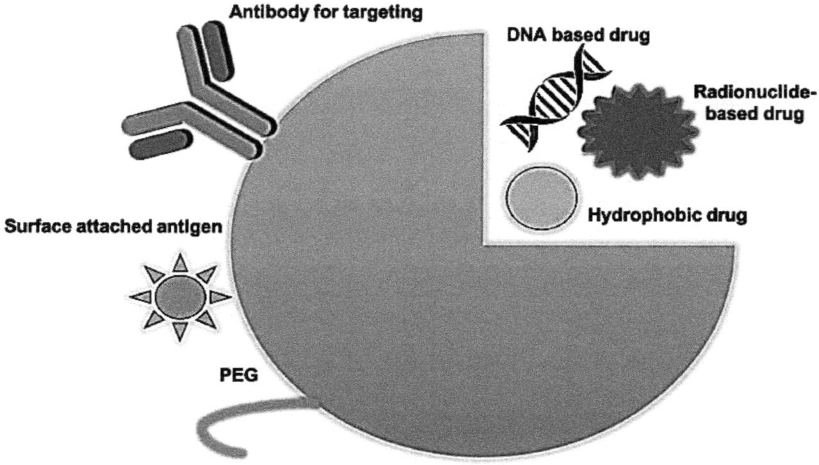

Figure 2.7 Schematic illustration of theranostic nanoparticles.

2.7 Conclusions

The use of nanoparticulate drug delivery systems is set to spread rapidly. Recent advances in drug delivery have led to designing customized systems that can achieve spatial and temporal control in addition to overcoming several physiological barriers associated with the use of traditional chemotherapeutic drugs. While older formulations rely more on the EPR effect, recent nanoparticle drug delivery systems employ ligand targeting toward cancer cells that overexpress specific antigens/biomarkers thus achieving extended release profile. Such nanoparticle design allows increased bioavailability, reduced drug dosage frequency, enhanced solubility of poorly soluble drugs, and increased patient compliance and safeguards the stability and structure of the drug by encapsulating it. There is a need to minimize the cost and technical difficulties in achieving such complex nanomedicines by establishing collaborations between researchers working in industry and particle toxicology to ensure exchange of ideas and knowhow so that these formulations are translated into clinically applicable therapeutics.

References

1. Siegel, R. L., Miller, K. D., and Jemal, A. (2015). Cancer statistics, 2015. *CA: A Cancer Journal for Clinicians*, **65**(1), pp. 5–29.
2. Whitfield, M. L., George, L. K., Grant, G. D., and Perou, C. M. (2006). Common markers of proliferation. *Nature Reviews Cancer*, **6**(2), p. 99.
3. Shimada, T. (2017). Inhibition of carcinogen-activating cytochrome P450 enzymes by xenobiotic chemicals in relation to antimutagenicity and anticarcinogenicity. *Toxicological Research*, **33**(2), p. 79.
4. Wang, A. Z., Langer, R., and Farokhzad, O. C. (2012). Nanoparticle delivery of cancer drugs. *Annual Review of Medicine*, **63**, pp. 185–198.
5. Nakamura, Y., Mochida, A., Choyke, P. L., and Kobayashi, H. (2016). Nanodrug delivery: Is the enhanced permeability and retention effect sufficient for curing cancer? *Bioconjugate Chemistry*, **27**(10), pp. 2225–2238.
6. Song, G., Petschauer, J. S., Madden, A. J., and Zamboni, W. C. (2014). Nanoparticles and the mononuclear phagocyte system: Pharmacokinetics and applications for inflammatory diseases. *Current Rheumatology Reviews*, **10**(1), pp. 22–34.
7. Burgum, M. J., Evans, S. J., Jenkins, G. J., Doak, S. H., and Clift, M. J. (2018). Considerations for the human health implications of nanotheranostics. *In Handbook of Nanomaterials for Cancer Theranostics*, (pp. 279–303). Elsevier.
8. De Jong, W. H., and Borm, P. J. (2008). Drug delivery and nanoparticles: Applications and hazards. *International Journal of Nanomedicine*, **3**(2), p. 133.
9. Karlsson, H. L., Cronholm, P., Gustafsson, J., and Moller, L. (2008). Copper oxide nanoparticles are highly toxic: A comparison between metal oxide nanoparticles and carbon nanotubes. *Chemical Research in Toxicology*, **21**(9), pp. 1726–1732.
10. Xu, X., Ho, W., Zhang, X., Bertrand, N., and Farokhzad, O. (2015). Cancer nanomedicine: From targeted delivery to combination therapy. *Trends in Molecular Medicine*, **21**(4), pp. 223–232.
11. Li, J., Wang, X., Zhang, T., Wang, C., Huang, Z., Luo, X., and Deng, Y. (2015). A review on phospholipids and their main applications in

drug delivery systems. *Asian Journal of Pharmaceutical Sciences*, **10**(2), pp. 81–98.

12. Bobo, D., Robinson, K. J., Islam, J., Thurecht, K. J., and Corrie, S. R. (2016). Nanoparticle-based medicines: A review of FDA-approved materials and clinical trials to date. *Pharmaceutical Research*, **33**(10), pp. 2373–2387.
13. Merlot, A. M., Kalinowski, D. S., and Richardson, D. R. (2014). Unraveling the mysteries of serum albumin—more than just a serum protein. *Frontiers in Physiology*, **5**, p. 299.
14. Baron, S. (1996). Structure and Classification of Viruses. Medical microbiology. Galveston, TX. *University of Texas Medical Branch at Galveston*, 41.
15. Czapar, A. E., Zheng, Y. R., Riddell, I. A., Shukla, S., Awuah, S. G., Lippard, S. J., and Steinmetz, N. F. (2016). Tobacco mosaic virus delivery of phenanthriplatin for cancer therapy. *ACS Nano*, **10**(4), pp. 4119–4126.
16. Han, L., Huang, R., Li, J., Liu, S., Huang, S., and Jiang, C. (2011). Plasmid pORF-hTRAIL and doxorubicin co-delivery targeting to tumor using peptide-conjugated polyamidoamine dendrimer. *Biomaterials*, **32**(4), pp. 1242–1252.
17. Espinosa, A., Di Corato, R., Kolosnjaj-Tabi, J., Flaud, P., Pellegrino, T., and Wilhelm, C. (2016). Duality of iron oxide nanoparticles in cancer therapy: Amplification of heating efficiency by magnetic hyperthermia and photothermal bimodal treatment. *ACS Nano*, **10**(2), pp. 2436–2446.
18. Zhang, Q., Wang, X., Li, P. Z., Nguyen, K. T., Wang, X. J., Luo, Z., Zhang, H., Tan, N. S., and Zhao, Y. (2014). Biocompatible, uniform, and redispersible mesoporous silica nanoparticles for cancer-targeted drug delivery in vivo. *Advanced Functional Materials*, **24**(17), pp. 2450–2461.
19. Ahlawat, J., Henriquez, G., and Narayan, M. (2018). Enhancing the delivery of chemotherapeutics: Role of biodegradable polymeric nanoparticles. *Molecules*, **23**(9), p. 2157.
20. Mohanraj, V. J., and Chen, Y. (2006). Nanoparticles-a review. *Tropical Journal of Pharmaceutical Research*, **5**(1), pp. 561–573.
21. Zhu, L., Ma, J., Jia, N., Zhao, Y., and Shen, H. (2009). Chitosan-coated magnetic nanoparticles as carriers of 5-fluorouracil: Preparation, characterization and cytotoxicity studies. *Colloids and Surfaces B: Biointerfaces*, **68**(1), pp. 1–6.

22. van der Meel, R., Vehmeijer, L. J., Kok, R. J., Storm, G., and van Gaal, E. V. (2016). Ligand-targeted particulate nanomedicines undergoing clinical evaluation: Current status. In *Intracellular Delivery III* (pp. 163–200). Springer, Cham.

23. Kirpotin, D. B., Drummond, D. C., Shao, Y., Shalaby, M. R., Hong, K., Nielsen, U. B., Marks, J. D., Benz, C. C., and Park, J. W. (2006). Antibody targeting of long-circulating lipidic nanoparticles does not increase tumor localization but does increase internalization in animal models. *Cancer Research*, **66**(13), pp. 6732–6740.

24. Park, J. W., Hong, K., Kirpotin, D. B., Colbern, G., Shalaby, R., Baselga, J., Shao, Y., Nielsen, U. B., Marks, J. D., Moore, D., and Papahadjopoulos, D. (2002). Anti-HER2 immunoliposomes: Enhanced efficacy attributable to targeted delivery. *Clinical Cancer Research*, **8**(4), pp. 1172–1181.

25. Rip, J., Appeldoorn, C. C. M., Manca, F. M., Dorland, R., Van Kregten, J. M. R., and Gaillard, P. J. (2010). Receptor-mediated delivery of drugs across the blood-brain barrier. *Pharmacology and Toxicology of the Blood-Brain Barrier: State of the Art, Needs for Future Research and Expected Benefits for the EU*. Brussels, Belgium.

26. Gaillard, P. J., Gladdines, W., Appeldoorn, C. C., Rip, J., Boogerd, W. J., Beijnen, J. H., Brandsma, D., and van Tellingen, O. (2012). Development of glutathione pegylated liposomal doxorubicin (2B3-101) for the treatment of brain cancer.

27. Hrkach, J., Von Hoff, D., Ali, M. M., Andrianova, E., Auer, J., Campbell, T., De Witt, D., Figa, M., Figueiredo, M., Horhota, A., and Low, S. (2012). Preclinical development and clinical translation of a PSMA-targeted docetaxel nanoparticle with a differentiated pharmacological profile. *Science Translational Medicine*, **4**(128), pp. 128ra39–128ra39.

28. Maresca, K. P., Hillier, S. M., Femia, F. J., Keith, D., Barone, C., Joyal, J. L., Zimmerman, C. N., Kozikowski, A. P., Barrett, J. A., Eckelman, W. C., and Babich, J. W. (2008). A series of halogenated heterodimeric inhibitors of prostate specific membrane antigen (PSMA) as radiolabeled probes for targeting prostate cancer. *Journal of Medicinal Chemistry*, **52**(2), pp. 347–357.

29. Cheng, J., Teply, B. A., Sherifi, I., Sung, J., Luther, G., Gu, F. X., Levy-Nissenbaum, E., Radovic-Moreno, A. F., Langer, R., and Farokhzad, O. C. (2007). Formulation of functionalized PLGA–PEG nanoparticles for in vivo targeted drug delivery. *Biomaterials*, **28**(5), pp. 869–876.

30. Farokhzad, O. C., Cheng, J., Teply, B. A., Sherifi, I., Jon, S., Kantoff, P. W., Richie, J. P., and Langer, R. (2006). Targeted nanoparticle-aptamer bioconjugates for cancer chemotherapy in vivo. *Proceedings of the National Academy of Sciences*, **103**(16), pp. 6315–6320.
31. Davis, M. E. (2009). The first targeted delivery of siRNA in humans via a self-assembling, cyclodextrin polymer-based nanoparticle: From concept to clinic. *Molecular Pharmaceutics*, **6**(3), pp. 659–668.
32. Cerqueira, N. M., Pereira, S., Fernandes, P. A., and Ramos, M. J. (2005). Overview of ribonucleotide reductase inhibitors: An appealing target in anti-tumour therapy. *Current Medicinal Chemistry*, **12**(11), pp. 1283–1294.
33. Bartlett, D. W., and Davis, M. E. (2007). Physicochemical and biological characterization of targeted, nucleic acid-containing nanoparticles. *Bioconjugate Chemistry*, **18**(2), pp. 456–468.
34. Rahman, M. A., Amin, A. R., Wang, X., Zuckerman, J. E., Choi, C. H. J., Zhou, B., Wang, D., Nannapaneni, S., Koenig, L., Chen, Z., and Yen, Y. (2012). Systemic delivery of siRNA nanoparticles targeting RRM2 suppresses head and neck tumor growth. *Journal of Controlled Release*, **159**(3), pp. 384–392.
35. MacDiarmid, J. A., Mugridge, N. B., Weiss, J. C., Phillips, L., Burn, A. L., Paulin, R. P., Haasdyk, J. E., Dickson, K. A., Brahmbhatt, V. N., Pattison, S. T., and James, A. C. (2007). Bacterially derived 400 nm particles for encapsulation and cancer cell targeting of chemotherapeutics. *Cancer Cell*, **11**(5), pp. 431–445.
36. MacDiarmid, J. A., Amaro-Mugridge, N. B., Madrid-Weiss, J., Sedliarou, I., Wetzel, S., Kochar, K., Brahmbhatt, V. N., Phillips, L., Pattison, S. T., Petti, C., and Stillman, B. (2009). Sequential treatment of drug-resistant tumors with targeted minicells containing siRNA or a cytotoxic drug. *Nature Biotechnology*, **27**(7), p. 643.
37. MacDiarmid, J. A., and Brahmbhatt, H. (2011). Minicells: Versatile vectors for targeted drug or si/shRNA cancer therapy. *Current Opinion in Biotechnology*, **22**(6), pp. 909–916.
38. Hall, F. L., Liu, L., Zhu, N. L., Stapfer, M., Anderson, W. F., Beart, R. W., and Gordon, E. M. (2000). Molecular engineering of matrix-targeted retroviral vectors incorporating a surveillance function inherent in von Willebrand factor. *Human Gene Therapy*, **11**(7), pp. 983–993.
39. Gordon, E. M., and Hall, F. L. (2010). Rexin-G, a targeted genetic medicine for cancer. *Expert Opinion on Biological Therapy*, **10**(5), pp. 819–832.

40. Gordon, E. M., Chen, Z. H., Liu, L., Whitley, M., Liu, L., Wei, D., Groshen, S., Hinton, D. R., Anderson, W. F., Beart Jr, R. W., and Hall, F. L. (2001). Systemic administration of a matrix-targeted retroviral vector is efficacious for cancer gene therapy in mice. *Human Gene Therapy*, **12**(2), pp. 193–204.

41. Gordon, E. M., and Hall, F. L. (2009). The 'timely' development of Rexin-G: First targeted injectable gene vector. *International Journal of Oncology*, **35**(2), pp. 229–238.

42. Felnerova, D., Viret, J. F., Glück, R., and Moser, C. (2004). Liposomes and virosomes as delivery systems for antigens, nucleic acids and drugs. *Current Opinion in Biotechnology*, **15**(6), pp. 518–529.

43. Ryu, J. H., Lee, S., Son, S., Kim, S. H., Leary, J. F., Choi, K., and Kwon, I. C. (2014). Theranostic nanoparticles for future personalized medicine. *Journal of Controlled Release*, **190**, pp. 477–484.

44. Ahlawat, J., Deemer, E. M., and Narayan, M. (2019). Chitosan nanoparticles rescue rotenone-mediated cell death. *Materials*, **12**(7), p. 1176.

45. Ahlawat, J., Kumar, V., and Gopinath, P. (2019). Carica papaya loaded poly(vinyl alcohol)-gelatin nanofibrous scaffold for potential application in wound dressing. *Materials Science and Engineering: C*, p. 109834.

46. Kumar, A., Ahlawat, J., and Narayan, M. (2019). 3D Bioprinting for Organs, Skin, and Engineered Tissues. *3D Printing Technology in Nanomedicine*, p. 115.

Chapter 3

The Role of Nanoparticles in Cancer Therapy through Apoptosis Induction

Marveh Rahmati,[a] Saeid Amanpour,[a] and Hadiseh Mohammadpour[a,b]

[a]*Cancer Biology Research Center, Cancer Institute of Iran, Imam Khomeini Hospital, Tehran University of Medical Sciences, Tehran, Iran*
[b]*Iran National Tumor Bank, Cancer Institute of Iran, Tehran University of Medical Sciences, Tehran, Iran*

m_rahmati@sina.tums.ac.ir

Nanoparticles (NPs) are currently being used for the diagnosis and treatment of many human diseases, including cancer. The characteristics of NPs, such as their size, shape, surface area, and distribution in the body, have led to their wide use in both in vitro and in vivo research. Despite the optimal properties of NPs, their cytotoxic effects on normal cells and living organs could lead to limited applications. Thus, there is an urgent need for more research to improve the efficacy of NPs and reduce their cytotoxicity. However, on the other side of the coin, the cytotoxic potential of

Nanoparticle Drug Delivery Systems for Cancer Treatment
Edited by Hala Gali-Muhtasib and Racha Chouaib

ISBN 978-981-4800-54-9 (Hardcover), 978-0-429-34125-0 (eBook)
www.jennystanford.com

NPs could be exploited for cancer therapy. In this chapter, we highlight the apoptotic effects of NPs and discuss nanomedicine approaches and nano-based drugs that are promising as therapeutic candidates for inducing apoptotic cell death in cancers.

3.1 Introduction

According to WHO reports, cancer is the second leading cause of mortality worldwide, and is responsible for an estimated 9.6 million deaths in 2018, meaning that about 1 in 6 deaths is due to cancer. The International Agency for Research on Cancer (IARC) estimates that one-in-five men and one-in-six women worldwide will develop cancer over the course of their lifetime, and those one-in-eight men and one-in-eleven women will die from their disease [1]. Cancer occurs through a complicated multistep carcinogenesis process such as genetic alteration in fundamental cell signaling pathways that control cell proliferation and cell death [2, 3]. Cancer is initiated in a local tissue site with the potential to spread or invade to other tissues in a process named metastasis [4]. The conventional cancer therapies include chemotherapy, radiation therapy, and surgery [5, 6]. These therapies are used routinely based on the clinical signs and pathological stages of the disease [3, 7]. Despite progressions in treatment protocols, long-term survival of patients is poor and the side effect of chemotherapy is painful for patients. The most clinical challenges are still the early detection, poor delivery of drugs, chemoresistance as well as non-specific effects of current treatments [5, 8, 9]. Therefore, finding new therapies with fewer side effects and more specific activities are always considered by scientists. In this regard, the molecular pathways involved in carcinogenesis and cancer cell survival are more studied. Apoptosis is the major type of programmed cell death known to remove unwanted cells and balance cell proliferation and cell death during growth and development of different pathways [10, 11]. Obviously, dysregulation of apoptosis may lead to a wide range of diseases, including neurodegenerative diseases, autoimmune disorders and cancers [10, 12]. In fact, mutations and genetic alterations in critical genes of apoptosis can be observed in cancer [11, 12]. By contrast, the excessive apoptosis is observed as a main pathological factor in many diseases such as immune deficiency syndrome (HIV/AIDS), neurodegenerative diseases, myelodysplastic syndromes and toxin-induced liver disease [10, 13]. Therefore, cell death pathways are considered as a therapeutic target for the treatment of a wide spectrum of human diseases. In addition, apoptosis serves as a cytotoxic mechanism of most current anti-cancer drugs. However, there is an urgent need to develop innovative technologies to design new apoptotic modulators for the treatment of cancer. In this chapter, we will discuss nanomedicine approaches and nano-based drugs that are promising as therapeutic candidates for apoptosis induction in cancer [14, 15]. To better understand the mechanisms of nanoparticles, we first discuss the molecular pathways of apoptosis and then the use of nano-based drugs as apoptosis inducers.

3.2 Apoptosis versus Necrosis/Necroptosis

Apoptosis is described as type-I programmed cell death, while necrosis typically is known as non-programmed cell death [2]. During apoptosis, the unwanted damaged cells are removed by phagocytosis without leaking their contents into the cytoplasm, and subsequently no inflammatory response occurs. By contrast, necrosis is known as the uncontrolled or unprogrammed type of cell death that is associated with the induction of inflammation [16]. Numerous external stimuli, including specific growth factors, hypoxia radiation and chemotherapeutic drugs, can trigger apoptosis and necrosis. However, not all cells respond similarly to the same stimulus. While certain types of cells may induce apoptosis in response to radiation, other types of cells may undergo necrosis. In addition, the duration and intensity of the stimulus may affect the cell fate. For instance, anti-cancer drugs at lower doses induce apoptosis, but may provoke necrosis at higher doses [16, 17].

Two important molecular events that are exclusively considered as apoptosis hallmarks are (1) the externalization of phosphatidylserine and (2) activation of the caspase family of proteins [16, 18]. The morphological features of apoptosis are also different from necrosis [19, 20]. Apoptotic cells typically manifest extensive membrane blebbing, shrinkage of the cell, and nuclear condensation and fragmentation; while necrotic cells show membrane disruption, cell swelling, and decondensation of nuclei [19, 20]. However, recent reports have described a form of regulated necrosis, which is initiated via the activation of apoptotic cell signaling, but terminated with necrosis features [19]. Necroptosis is activated in response to the members of the "death receptor" subset of the TNF superfamily, without caspase activation. Necroptosis occurs in cells that express the kinase, RIPK3 to RIPK1. In a normal situation, CASP-8 suppresses the activation of RIPK3 onto RIPK1, and necroptosis is activated, followed by an infection with some viruses having caspase inhibitors. The most important difference between necroptosis and other types of cell death is the involvement of RIPK3/RIPK1 and the absence of caspase activity [21, 22].

3.3 Major Signaling Pathways in Apoptosis

Apoptosis is an energy-dependent process. Two main apoptotic signaling pathways, including the extrinsic death receptor pathway and the intrinsic mitochondrial pathway, have been well-characterized. However, there are additional pathways that involve the T-cell-mediated cytotoxicity and the perforin-granzyme pathway. The perforin/granzyme pathway induces apoptosis through either granzyme B or granzyme A. Another recent pathway of apoptosis is mediated by the endoplasmic reticulum (ER) which plays important roles in cell fate and will be discussed in details later [17]. Apoptosis is mediated by chronological activation of protein superfamily of caspases [16]. Caspases are expressed in an inactive form of proenzymes in most cells. When activated, they activate other procaspases, leading to the initiation of the cascade of caspase-dependent apoptosis pathway. Caspases are highly conserved cysteine-dependent aspartate-specific proteases. There are different types of caspases: initiator caspases, including CASP-2, 8, 9, and 10; effector caspases, such as CASP-3, 6, and 7; and inflammatory caspases which are CASP-1, 4, and 5). The other caspases that have been recently studied are (i) caspase-11, which is involved in the regulation of apoptosis and cytokine maturation during septic shock, (ii) caspase-12, which mediates apoptosis through the endoplasmic reticulum, (iii) caspase-13, which is believed to be a bovine gene, and (iv) caspase-14, which is highly expressed just in embryonic tissues [23]. Initiator caspases are inactive until specific oligomeric activator protein binds to them. Subsequently, they bind to effector caspases. Effector caspases are then activated through proteolytic cleavage. The activated caspases then proteolytically degrade the intracellular proteins necessary for programmed cell death.

There also exists a caspase-independent apoptotic pathway that is mediated by AIF (Apoptosis-Inducing Factor) [24]. AIF is phylogenetically an old flavoprotein observed in the mitochondrial intermembrane. Upon lethal stimuli, AIF translocates from mitochondria to the nucleus. It binds to DNA and mediates caspase-independent chromatin condensation and large scale DNA fragmentation [25, 26].

3.3.1 The Intrinsic Pathway of Apoptosis

The intrinsic pathway of apoptosis, also known as mitochondrial pathway, is activated by a wide range of stimuli, including DNA-damaging factors, activators of oncogenes, high level of intracellular Ca^{2+}, and oxidative stress. Apoptotic agents, which target mitochondria, may cause mitochondrial swelling through the formation of membrane pores, or may increase the mitochondrial outer membrane permeabilization (MOMP), leading to the release of the apoptotic effectors by a two-step process (Fig. 3.1). This process is initiated by an increase in MOMP, resulting in the release of cytochrome *c* and other apoptogenic factors into the cytosol. In the cytosol, cytochrome *c* assembles to a multi-protein complex called "apoptosome," which consists of apoptotic protease activating factor 1 (apaf1), dATP, cytochrome *c*, and procaspase-9 (Fig. 3.1). Then, the apoptosome cleaves the procaspase-9 to caspase-9, which in turn activates CASP-3, CASP-6, and CASP-7, respectively [27–29]. The mitochondrial proteins SMACs (Second Mitochondria-derived Activator of Caspases) are also released into the cytosol upon the increase in MOMP. SMAC proteins bind to factors that inhibit apoptosis (IAPs) in order to deactivate them, and in turn, the caspase cascade proceeds to execute apoptosis and create apoptotic hallmarks, such as plasma membrane blebbing and DNA fragmentation. The mitochondrial pathway is strongly under the control of BCL-2 family members. These proteins, which may have either anti-apoptotic or pro-apoptotic functions, mostly regulate mitochondrial permeability [30].

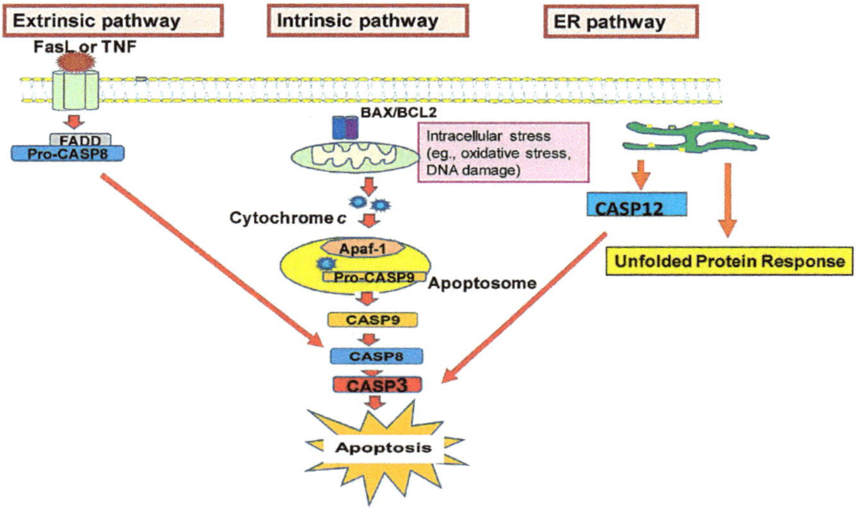

Figure 3.1 The extrinsic, intrinsic and ER stress mediated apoptotic pathways. Mammalian apoptosis occurs through the intrinsic and extrinsic pathways. In intrinsic pathway, cytochrome *c* is released into the cytosol, followed by increase of the mitochondrial membrane permeability and assembly to Apaf-1 and pro-caspase-9 to form the apoptosome. The apoptosome cleaves the pro-caspase 9 to caspase-9, which in turn activates the caspase-3. Extrinsic pathway is initiated by the binding of ligand to cell receptor, leading to caspase activation through the intermediate membrane proteins TRADD and/or FADD. ER stress mediated apoptosis is through CASP-12. FADD, Fas-associated death domain; Apaf-1, apoptotic protease activating factor 1.

3.3.1.1 BCL-2 Protein Family

The superfamily of BCl-2 family proteins located in the mitochondrial membrane determines cell fate decisions (i.e., cell survival or death) by controlling MOMP. Members of this family are classified into two groups, including pro-apoptotic or anti-apoptotic proteins, and are in the forms of homodimers or heterodimers. The pro-apoptotic proteins are further classified as multi-domain members, such as BCL-2 homology (BH) domain proteins such as BAX, BCLXs, and BAK, or single domain members, including BH3-only proteins BID, BIM, BAD, PUMA (p53 upregulated controller of apoptosis), and NOXA. The anti-apoptotic members contain multi-BH domain and include BCL-2, BCL-XL and MCL-1. Owing to the crucial role of BCL-2 family proteins, especially BH3-only proteins, in regulating and promoting apoptosis, they could be a potential target for therapeutic options [30]. Under normal physiological conditions, there is a balance between pro-apoptotic (BAX, BID, BAK or BAD) and anti-apoptotic (BCL-XL and BCL-2) members of the BCL-2 family. This balance can be disrupted by an internal stimulus, leading to an increase in the pro-apoptotic BCL-2 family members or the decrease in the anti-apoptotic BCL-2 family proteins. In most conditions, pro-apoptotic homodimers in the outer membrane of the mitochondrion are formed to make the mitochondrial membrane permeable to release cytochrome c and SMAC which, in turn, trigger the activation of apoptosis cascade [31].

3.3.2 The Extrinsic Pathway of Apoptosis

Two main signaling pathways, the TNF (Tumor Necrosis Factor)-TNF receptor (TNFR) and the Fas-Fas ligand (FasL), have been well-characterized in the extrinsic or cell receptor pathway of apoptotic mechanisms. Both involve receptors of the family coupled to extrinsic signals [32]. These pathways activate CASP-8. CASP-8 in turn activates CASP-3, which is responsible for proteolysis of cellular factors and finally creating the morphological signs of apoptosis (Fig. 3.1). All death receptors (DR) in cell surface are distinguished by their specific ligands. They translocate the death signal to the cytosol by their death domain. There are at least six members of DR family: TNF-R1, TRAILR 1 and 2 (TNF-Related Apoptosis-Inducing Ligand Receptor 1 and 2), CD95 (Fas), DR3, and DR6, which will be discussed below [33].

3.3.2.1 TNF signaling

Tumor necrosis factor (TNF), as a proinflammatory cytokine, plays an important role in various cellular events including septic shock, cell proliferation, differentiation, necrosis and apoptosis. Increased level of TNF is also involved in many diseases, including rheumatoid arthritis, malaria, AIDS and cancer.

TNFα is induced by cytokines which are released by inflammatory cells and particularly activated macrophages. TNFα is the main extrinsic stimulus of extrinsic apoptosis. TNFR1 and TNFR2 are two receptors for TNFα in most cells. After the binding of TNFα to TNFR1, CASP-8 is activated in DICS complex, which contains the intermediate membrane proteins TNF receptor-associated death domain (TRADD). TRADD subsequently recruits other effector proteins such as FADD/MORT1 (FAS-Associated Death Domain Protein) into DISC complex. Regardless of apoptosis, the TNFα-TNFR pathway can also indirectly activate some of the transcription factors involved in cell survival and inflammatory responses [34]. In response to TNF induction, the transcription factor NF-κB and MAP kinases, including ERK, p38, and JNK, are activated in most cell types. Apoptosis or necrosis could also be initiated.

3.3.2.2 Fas/FasL signaling

Upon interaction of FasL with Fas receptor (Apo-1 or CD95), the death-inducing signaling complex (DISC), which contains adaptor proteins such as FADD and procaspase-8, is formed. In some cell types, activated CASP-8 directly activates other members of the caspase family and subsequently leads to the execution of apoptosis. In other types of cells, the induction of Fas pathway is not sufficient to trigger an apoptotic response, rather this pathway cross-talk with mitochondrial pathway through proteolytic degradation of Bid producing a truncated form of Bid (tBid). Then, tBid acts through a heterodimer with BAX or BAK, leading to the activation of the mitochondrial pathway of apoptosis. In this situation, the extrinsic pathway functions as an amplifier to induce an apoptotic response [35].

3.3.3 The Perforin/Granzyme Pathway of Apoptosis

When cells are infected by viruses, the perforin/granzyme apoptosis pathway is initiated by cytotoxic lymphocytes to remove the infected cells. Perforin, also known as cytoplasmic granule toxin, is a kind of protein with a pore-forming ability in mitochondrial membrane. Granzyme is a serine protease protein, which contains cytotoxic granules of cytotoxic lymphocytes (CLs). Although, Granzyme is required for triggering apoptosis, it should be delivered appropriately by perforin. In humans, there are numerous granzymes including A, B, H, K, and M, but the Granzymes A and B are the most abundant enzymes that are involved in apoptosis. This pathway is initiated when granzymes could enter into target cells. This internalization is facilitated by perforin. Granzyme B activates pro-apoptotic BH3-interacting domain death agonist Bid, leading to the activation of CASP-3. Activated CASP-3 is subsequently able to proceed with executive apoptosis. Granzyme B can also inactivate MCL-1, which is a member of the anti-apoptotic BCL-2 family [36]. The granzyme A pathway is also involved in apoptosis via activating a parallel, caspase-independent cell death pathway through single-stranded DNA damage [37].

3.3.4 ER Stress Pathway

Endoplasmic reticulum (ER) is an organelle implicated in the secretion and correct folding of proteins, Ca^{2+} balance, as well as maintaining the quality control of proteins and cell homeostasis. In different situations, such as physiological or pathological conditions, the demand for folding proteins is increased, resulting in an elevation of unfolded or misfolded protein levels in the ER lumen. This burden on the ER is known as the ER stress. To cope with the stress, the unfolded protein response (UPR) is activated. If the URR cannot restore the stress, the UPR mediates apoptosis [38]. The studies have shown that two main pathways, a transcription factor- and a caspase-dependent signaling pathway, mediate ER stress-dependent apoptosis. The UPR mediators, such as transcription factor GADD153/CHOP, can disrupt the balance between BCL-2 and BAX, resulting in the induction of apoptosis. ER stress-induced apoptosis also occurs through the activation of CASP-12, independent of mitochondrial and death receptors pathways (Fig. 3.1) [39]. There is contradictory information about ER stress-mediated apoptosis. Some have demonstrated that UPR mediates apoptosis through CASP-12, while some claimed that CASP-12 is not related to UPR-mediated apoptosis, rather it is dependent on the mitochondrial apoptotic pathway [40, 41].

Interestingly, in most humans, CASP12 appears to be nonfunctional [42], and the ER-resident of CASP-4 is known to be implicated in ER stress-induced apoptosis [43–45].

3.4 Carcinogenesis and Apoptosis—Molecular Targeted Therapies

Due to the resistance of most tumor cells to apoptosis, induction of cell death leads to the promotion of healthy cells. Therefore, the targeting of apoptotic pathways is an interesting area for drug discovery in cancers. We will discuss nanomedicine, which has recently attracted fresh attention for cancer treatment.

3.5 Nanomedicine

Nanotechnology is a growing field with many promising applications in both drug discovery and drug delivery systems. Diverse physical and chemical characteristics of nanoparticles (NPs), including size, charge, shape, and surface properties, have a critical impact on the in vivo distribution, cellular internalization, stability in blood circulation, increased cellular uptake, elevated volume of distribution and half-life in tumor tissues with minimal damage to normal cells. Thereby, they could increase the therapeutic index and can enhance the efficacy of currently used chemotherapeutic drugs. Advances in nanotechnology could also help to create novel therapeutics and wide applications of diagnostic methods in cancer. Taken together, nanodrugs or nanocarriers are recently offering a unique opportunity to improve survival in cancer patients [46]. Recent advances suggest that nanotechnology will be a good tool used in disease prevention, diagnosis and treatment [47]. In this section, we will focus on the major physical parameters of nanoparticles that are used in medicine.

3.5.1 Size

Particle size is one of the primary factors used to determine the circulation time of the nanoparticles. After systemic administration, nanoparticles accumulate in the spleen due to mechanical filtration and are removed by the reticulo-endothelial system [46]. The size of 100–200 nm is accepted as optimal size for drug delivery systems, because nanocarriers with this size are large enough to avoid filtration in the spleen and the uptake in the liver [48]. Particles with a diameter smaller than 5 nm are rapidly cleared from the blood circulation by extravasation or renal clearance [49]. However, particles, with a diameter larger than 15 μm; accumulate in organs such as the liver, spleen and bone marrow [48].

3.5.2 Shape

Particle shape is important for nanoparticle degradation and cellular uptake. For example, spherical particles, are more susceptible to degradation than hemispherical particles [49, 50]. Besides, the ability of spherical nanoparticles to be deformed is very important to avoid spleen filtration since the spleen exhibits asymmetric filtering units.

3.5.3 Surface Characteristics

Surface properties of nanoparticles are important determinants of blood circulation time. After administration, nanoparticles may be associated with blood proteins which are known as "opsonins," such as immunoglobulins and complement proteins that contribute to the recognition of nanoparticles by macrophages. Therefore, opsonization is the key factor needed to determine whether the nanoparticles are sustained in blood circulation or engulfed by macrophages. Modifying the surface of nanoparticles is a potent strategy to enhance or reduce their circulation time in blood and tissues [51, 52].

3.5.4 Release Characteristics

The release characteristics of NP encapsulated drugs affect the presence of drugs in the plasma to reach the target site. Conventional drugs have different side effects, while, nanoparticles are used at a favorite concentration within the therapeutic range to reach the target site and exhibit fewer side effects. The plasma drug level of NPs is constant over a longer period of time compared to free drug [52].

3.6 Targeted Cancer Therapy by Nanoparticles

The desired anticancer agents should have some properties as follows: First, they can be targeted to a specific tumor site. Second, they should be able to kill the tumor cells selectively without affecting healthy cells. These two important aspects of anti-tumor drugs help increase the survival of patients. In this regard, nanoparticles have recently become considered for cancer targeted therapy due to their ability to deliver the anti-cancer drugs to the tumor tissues in a passive or active form [53]. These two models of targeting are discussed below. Despite the appropriate properties of NPs, the cytotoxic effects of NPs in vivo should be considered, as this cytotoxicity leads to limited applications of NPs. Consequently, there is an urgent need for more research on NP effects to improve their efficacy and reduce their cytotoxicity [54].

3.6.1 Passive Targeting

The size and blood circulation time of NPS as well as vasculature and micro-environment of tumors are the most important factors in passive targeting. This targeting approach could increase the drug bioavailability and efficacy. Indeed, nanodrugs use the differences between normal and tumor vasculature to reach the targeted site. These differences include angiogenic blood vessels which have large gaps in tumor tissues between adjacent endothelial cells. In addition, the vasculature of tumors is leaky and defective, so an increased level of some mediators, such as nitric oxide, vascular endothelial growth factor (VEGF), basic fibroblast growth factor (FGF) and prostaglandins, is secreted. These properties of tumor vasculature are named enhanced permeability and retention (EPR) effect. This effect enables nanodrugs or other macromolecules to accumulate in tumors [55]. The microenvironment surrounding tumor cells are also different from that of normal cells. The fast-growing cancer cells have a high metabolic rate, and the level of oxygen and nutrients is usually not sufficient for them. Therefore, tumor cells depend on glycolysis pathway for their nutrient and energy supply, thus creating an acidic environment. The liposomes are pH sensitive and release their contents into this microenvironment. In addition, cancer cells express and release some unique enzymes, such as matrix metalloproteinases, which are important for the movement and survival of tumor cells [56].

3.6.2 Active Targeting

Despite the advantages of passive targeting approaches, several limitations exist. Certain NPs are difficult to be delivered passively to tumors due to the lack of an EPR effect; hence, the permeability in blood vessels may not be identical throughout the tumor [57]. To overcome these limitations, NPs are designed to bind to specific targets (active targeting) in the target cells [58, 59]. In order to create drug delivery-based therapy, some factors must be considered. First, the receptor on the delivery system should be more abundantly expressed on tumor cells than on normal cells. Second, they should be expressed on all targeted tumor cells. Internalization of drug usually occurs through receptor-mediated endocytosis. The endosomes move to target organelles after which the drug is released to the cytoplasm [15].

3.7 The Role of Nanoparticles in Cancer Therapy via Apoptosis

Nanoparticles (NPs) can induce oxidative stress in cells through diverse mechanisms: first, by a direct generation of ROS; second, by an indirect generation of ROS and reactive nitrogen species (RNS) through stimulating inflammatory cells [51]; third, by the indirect changes in mitochondrial integrity through NADPH oxidase or cellular calcium homeostasis; and last, by ROS generation through releasing ions or soluble compounds [45]. Any mechanism that leads to ROS production ultimately leads to damage to DNA and proteins as well as destruction of organelles such as mitochondria. The damaged mitochondria leads to the activation of apoptosis [54, 60]. The main operation of ROS in various cellular mechanisms is involved in cell cycle regulation, proliferation, self-renewal, differentiation, and apoptosis. The primary function of p53 as a tumor suppressor, upon DNA damage, is to induce cell cycle arrest and to repair the damage. If the damage cannot be repaired, apoptosis is initiated [38].

3.8 Mechanisms of Nanoparticle-Targeted Therapeutic Strategies

3.8.1 NPs Induce Mitochondrial Apoptosis

The mitochondrial pathway of apoptosis is the main cytotoxic effect of NPs on the cells. Several NPs such as CuO, TiO_2, ZnO, and silica NPs could induce apoptosis through ROS-mediated oxidative stress and subsequently activate the mitochondrial pathways [61–64]. ROS production is the most well-known method to initiate apoptosis through the mitochondria [65]. The other examples are described in Table 3.1.

3.8.2 NPs Induce Extrinsic Pathway of Apoptosis

Numerous studies have shown the ability of NPs to induce apoptosis via the extrinsic pathways, or mitochondrial-independent pathway [65]. For example, copper and tungsten carbide-cobalt (WC-co) NPs can trigger extrinsic apoptotic pathways through an increased level of ROS and RNS in kidney tissues [82, 83]. Another study revealed that metallic nickel NPs could induce apoptotic in JB6 cells through a caspase-8/AIF-mediated cytochrome c-independent pathway [84]. The time, dose, and the size of NPs are important determinants in apoptosis induction. There are more examples in Table 3.1.

3.8.3 NPs Induce ER-Mediated Apoptosis

As mentioned before, various conditions such as physiological, pathological, and different environmental stimuli could induce ER stress and subsequently UPR to restore the homeostasis in cells. Recently, ER stress mediators have been more studied to find a new therapeutic target in cancers and autoimmune diseases [85, 86].

Table 3.1 Nanoparticle induced apoptosis through different pathways: intrinsic, extrinsic and ER stress pathway

Nanoparticle (NP)	Cell line	Intrinsic pathway	Extrinsic pathway	ER stress pathway	Ref.
Cadmium telluride quantum dots	Hepatocellular carcinoma HepG2 cells	ROS generation, induction of CASP-9 and -3	—	—	[66]
Cadmium sulfide quantum dots	HepG2 cells Human prostate cancer LNCaP cells	ROS generation, induction of CASP-9	—	—	[67, 68]
Cerium oxide NP	Hepatoma SMMC-7721 cells	ROS generation, induction of JNK and p38 MAPKs	—	—	[69]
Tungsten carbide-cobalt NP	Mouse epidermal JB6 cells	Cytochrome *c* release	—	—	[70]
Al_2O_3 NP	Human mesenchymal stem cells	Mitochondrial depolarization and CASP-9 activation	—	—	[71]
Cobalt NP	Hepatocellular carcinoma HepG2 cells	Activation of CASP-3, increase in oxidative stress markers, such as, lipid hydroperoxide, superoxide dismutase and catalase	—	—	[72]
Chitisan coated cobalt oxide NP	Breast cancer MCF7 cells	Oxidative stress, upregulation of BAX and CASP9 activation	TNF/Fas upregulation TNFα-mediated apoptosis	—	[73]
	Leukemia Jurkat cells	Oxidative stress and caspase activation			
Selenium NP modified with folic acid	Human lung epithelial cancer (A549)	ROS generation and upregulation of BCL-2	—	—	[74]

Table 3.1 (*Continued*)

Nanoparticle (NP)	Cell line	Intrinsic pathway	Extrinsic pathway	ER stress pathway	Ref.
Titanium dioxide (TiO_2) NP	HepG2 cells	DNA damage and upregulation of p53, BAX, Cyto c, Apaf-1 and caspases	—	—	[75–77]
		Activation of caspases			
	Leukemia U937 cells	Upregulation of BAX and CASP-3 and -9			
	A549 cells				
AgNP	Human THP-1 monocytes	—	—	Upregulation of ER stress markers	[78]
	Human breast MCF-7 and T-47D cancer cells				
AuNP	Human neutrophils	—	—	Upregulation of ER stress markers	[79]
ZnO NP	Human umbilical vein endothelial cells (HUVECs)	—	—	Upregulation of ER stress markers (GRP78, PERK and GADD34)	[80]
SiO_2 NP	Human hepatoma Huh7 cells	—	—	Upregulation of ER stress markers, MAPK and p53	[81]

Several NPs could trigger the ER stress and the upregulation of UPR mediators. Metal NPs, such as silver NPs (AgNPs) and gold NPs (AuNPs), are promising candidates drug delivery systems in different disorders and could induce apoptosis through upregulation of ER stress mediators [87, 88]. In addition, numerous NPs have a potential to be used in disease treatment. For example, AgNPs that induce cell death apoptosis through ER stress pathway in MCF-7 and T-47D breast cancer cell line highlight a new potential strategy for the treatment of breast cancers [87]. ZnO NPs are also reported, because they cause ER stress by increasing the level of the misfolded protein, subsequently leading to apoptosis through constant activation of the UPR pathway [89].

3.8.4 Other Mechanisms for Inducing Apoptosis via Nanoparticles

3.8.4.1 Hyperthermia

Hyperthermia is a technique that makes cancer cells more susceptible to treatment by raising the temperature in the tumor [90]. A smaller rise in temperature, usually to 40–45°C is appropriate for cancerous cells to induce apoptosis, while excessive temperature induces necrosis due to the coagulation of proteins and other biological macromolecules; this process is known as thermoablation [90]. The best control for hyperthermia is to use magnetic NPs. In this way, healthy tissues around the cancer are kept safe since the normal cells of the body are also affected upon increasing the temperature of the whole body. In this regard, the type and concentration of nanoparticles should be considered to give better results [91].

The two techniques of hyperthermia-mediated apoptosis are extrinsic and intrinsic heating (magnetic heating). Studies on the mechanism of nanoparticle-mediated hyperthermia in prostate cancer cells have demonstrated that magnetic heating promotes necrosis by triggering ROS formation and also reduces Ki-67, TOP2A, and TPX2 at the mRNA level while extrinsic heating induces apoptosis by less formation of ROS and upregulating the HSP70 (heat shock protein 70) expression. Despite the differences between both hyperthermia techniques, different cell lines and xenograft tumors have been also studied. Authors showed that the controlled magnetic heating in most situations is more efficient, and the tumor volume is decreased [92]. The combination of nanoparticles with other drugs in hyperthermia leads to an increase of the drug effect in real target tumors [90].

3.8.4.2 Other models for testing nanoparticles

In addition to common methods used for testing anti-cancer drugs, in silico and computational models have been widely used to calculate the amount and the distribution of NPs in tumor tissues. These models are useful to facilitate the computations and reduce the cost and the time of cell testing. In addition, they reduce the pain and suffering from laboratory animals. Recently, using a 3D printer, tumor tissues could be simulated and produced. In this way, the effect of the mechanical and real properties of the cancer tissue on the distribution and evaluation of the therapeutic effects of nanoparticles can be studied much better [93]. This method can provide a platform to fabricate a personalized in vitro cancer model which could simulate the most important features of the cancerous tissues (including shape and vascular network), which can be used to study the magnetic hyperthermia in a simulated media compatible to in vivo conditions [93].

3.9 Nanodrugs Used in vivo

Unlike cell culture experiments, in vivo experiments require that the NPs pass through some physical barriers (e.g., blood-brain) and physiological barriers (e.g., serum proteins) [94–96]. If NPs could pass through these barriers and reach the target cell, they must enter the cell via its membrane and escape from endosomes as well. They also must escape from the immune system [97]. In this regard, NPs should be tested one by one in suitable animal models [12–16]. Table 3.2 summarizes some NPs which have been tested in animal models to get a better response to treatment.

3.10 Clinical Trials Using Nanodrugs

To test nanoparticles in humans, the protocol should be approved by the Food and Drug Administration (FDA) in the United States, a process that is time-consuming and expensive. The drug approval process in the FDA is generally categorized into three parts. Preclinical, clinical, and post-marketing phases, each phase needs a precise process.

Table 3.2 The effects of nanoparticles in in vivo models

Nanoparticles (NPS)	Animal model	Mechanism of action	Ref.
Silver (Ag) NPs	Adult zebrafish	Oxidative stress and apoptosis (Bax, Noxa, and p21 upregulation)	[98, 99]
Ag NPs	Drosophila melanogaster	DNA damage-mediated apoptosis (up regulated of p53 and caspases)	
AgNPs	Swiss albino mice	DNA damage-mediated apoptosis	
Silver and gold NPs	Mice	ROS-mediated apoptosis	[100]
Gold-ursolic acid into poly (DL-lactide-co-glycolide) NPs	Nude mice	Apoptosis (upregulated p53 and downregulation of BCL-2	[101]
ZnO NP	Embryonic zebrafish	Oxidative stress, apoptosis	[102]
Dextran-coated lanthanum strontium manganese oxide NP	C57BL/6J mice	Hyperthermia-induced apoptosis via enhanced expression of HSP70 and 90	[100, 103]
Metal oxide NPs (CuO, Fe_3O_4, Fe_2O_3, TiO2)	Mice	Oxidative DNA damage	[100]
CUO NPS	Embryonic zebrafish	ROS and Apoptosis regulated cytotoxicity	[104]

The preclinical steps involve the tests from initial discovery, the study of efficacy, safety, appropriate dose and toxicity of NPs in both cell lines and animal models. These data should be confirmed by the Investigational New Drug (IND). Then the clinical phase is initiated.

This clinical trial is subdivided into three phases: I, II, and III. In this regard, following the successful completion of all steps, the new drug is registered by the New Drug Application (NDA) in the FDA. Upon NDA approval, the drug can be marketed according to specific labeling.

As part of the NDA approval, the FDA may request further studies for the post-marketing phase, or Phase IV. In fact, the entire FDA approval process, in addition to being lengthy and labor-intensive, is also stringent. It is estimated that it takes approximately 10–15 years to develop a new medicine with a cost of approximately $1 billion [105]. The FDA-approved nanodrugs used in clinical trials are introduced in Table 3.3. For more information, please see https://www.fda.gov/Drugs/default.htm.

Table 3.3 Some examples of nanoparticles approved by the FDA and which are undergoing clinical trials

No.	Nanoparticle (NP)	Status	Type of cancer
1	Polymeric NP loaded with cetuximab and decorated With somatostatin Analogue	Recruiting	Colon cancer, colorectal cancer
2	Carbon NP	Not recruiting	Rectal cancer
3	Albumin-bound paclitaxel followed by epirubicin in combination with cyclophosphamide	Recruiting	Triple negative breast cancer
4	Albumin-bound paclitaxel combined with carboplatin as neoadjuvant chemotherapy	Recruiting	HER2 negative breast cancer
5	Nanoparticle paclitaxel protein bound + Cisplatin + Gemcitabine	Recruiting	Pancreatic cancer
6	Combining CRLX101, a nanoparticle camptothecin, with enzalutamide	Recruiting	Prostate cancer
7	Paclitaxel albumin-stabilized nanoparticle	Active, not recruiting	Breast cancer
8	TUSC2-nanoparticles and erlotinib	Active, not recruiting	Lung cancer
9	Paclitaxel albumin-stabilized nanoparticle formulation	Active, not recruiting	Breast cancer

3.11 Conclusion

Nanoparticles (NPs) are currently being used by many scientists for the diagnosis and treatment of many human diseases, such as cancers. The size, shape, surface area, and distribution of NPs in the body are the most important characteristics that determine their use in research. Despite the appropriate properties of NPs, their cytotoxic effects on normal cells and living organs could lead to their limited applications in the clinic. Consequently, there is an urgent need for conducting more research on NP effects to improve their efficacy and reduce cytotoxicity. Interestingly, the cytotoxic potential of NPs and their apoptosis-inducing effects have been exploited for cancer therapy.

Acknowledgments

M. R. and S. A. acknowledge support from the Cancer Biology Research Center, Tehran University of Medical Sciences.

References

1. Bray F., Ferlay J., Soerjomataram I., Siegel R. L., Torre L. A., Jemal A. (2018). Global cancer statistics 2018: GLOBOCAN estimates of incidence and mortality worldwide for 36 cancers in 185 countries, *CA: A Cancer J. Clin.*, **68**, pp. 394–424.
2. Taylor R. C., Cullen S. P., Martin S. J. (2008). Apoptosis: Controlled demolition at the cellular level, *Nat. Rev. Mol. Cell Biol.*, **9**, p. 231.
3. Labi V., Erlacher M. (2015). How cell death shapes cancer, *Cell Death Disease*, **6**, p. e1675.
4. Quail D. F., Joyce J. A. (2013). Microenvironmental regulation of tumor progression and metastasis, *Nat. Med.*, **19**, p. 1423.
5. Palumbo M. O., Kavan P., Miller W., Panasci L., Assouline S., Johnson N., Cohen V., Patenaude F., Pollak M., Jagoe R. T. (2013). Systemic cancer therapy: Achievements and challenges that lie ahead, *Front. Pharm.*, **4**, p. 57.
6. Arruebo M., Vilaboa N., Sáez-Gutierrez B., Lambea J., Tres A., Valladares M., González-Fernández A. (2011). Assessment of the evolution of cancer treatment therapies, *Cancers*, **3**, pp. 3279–3330.

7. Reichert J. M., Wenger J. B. (2008). Development trends for new cancer therapeutics and vaccines, *Drug Discov. Today*, **13**, pp. 30–37.

8. Housman G., Byler S., Heerboth S., Lapinska K., Longacre M., Snyder N., Sarkar S. (2014). Drug resistance in cancer: An overview, *Cancers*, **6**, pp. 1769–1792.

9. Gottesman M. M., Fojo T., Bates S. E. (2002). Multidrug resistance in cancer: Role of ATP–dependent transporters, *Nat. Rev. Cancer*, **2**, p. 48.

10. Thompson C. B. (1995). Apoptosis in the pathogenesis and treatment of disease, *Science*, **267**, pp. 1456–1462.

11. Kerr J. F., Winterford C. M., Harmon B. V. (1994). Apoptosis. Its significance in cancer and cancer therapy, *Cancer*, **73**, pp. 2013–2026.

12. Tatton W., Olanow C. (1999). Apoptosis in neurodegenerative diseases: The role of mitochondria, *Biochim. Biophys. Acta (BBA)-Bioenergetics*, **1410**, pp. 195–213.

13. Moosavi M. A., Rahmati M., Ashtari N., Alizadeh J., Hashemi M., Bathaei S. Z., Ghavami S. (2017). Apoptosis, autophagy, and unfolded protein response and cerebellar development, in: *Development of the Cerebellum from Molecular Aspects to Diseases, Springer*, pp. 153–178.

14. Wicki A., Witzigmann D., Balasubramanian V., Huwyler J. (2015). Nanomedicine in cancer therapy: Challenges, opportunities, and clinical applications, *J. Control. Release*, **200**, pp. 138–157.

15. Misra R., Acharya S., Sahoo S. K. (2010). Cancer nanotechnology: Application of nanotechnology in cancer therapy, *Drug Discov. Today*, **15**, pp. 842–850.

16. Elmore S. (2007). Apoptosis: A review of programmed cell death, *Toxicol. Pathol.*, **35**, pp. 495–516.

17. Baig S., Seevasant I., Mohamad J., Mukheem A., Huri H., Kamarul T. (2017). Potential of apoptotic pathway-targeted cancer therapeutic research: Where do we stand? *Cell Death Disease*, **7**, p. e2058.

18. Compton M. M. (1992). A biochemical hallmark of apoptosis: Internucleosomal degradation of the genome, *Cancer Metastasis Rev.*, **11**, pp. 105–119.

19. Fink S. L., Cookson B. T. (2005). Apoptosis, pyroptosis, and necrosis: Mechanistic description of dead and dying eukaryotic cells, *Infect. Immun.*, **73**, pp. 1907–1916.

20. Ziegler U., Groscurth P. (2004). Morphological features of cell death, *Physiology*, **19**, pp. 124–128.

21. Kaczmarek A., Vandenabeele P., Krysko D. V. (2013). Necroptosis: The release of damage-associated molecular patterns and its physiological relevance, *Immunity*, **38**, pp. 209–223.

22. Festjens N., Berghe T. V., Vandenabeele P. (2006). Necrosis, a well-orchestrated form of cell demise: Signalling cascades, important mediators and concomitant immune response, *Biochim. Biophys. Acta (BBA)-Bioenergetics*, **1757**, pp. 1371–1387.

23. Elmore S. (2007). Apoptosis: A review of programmed cell death, *Toxicol. Pathol.*, **35**, pp. 495–516.

24. Wang J., Lenardo M. J. (2000). Roles of caspases in apoptosis, development, and cytokine maturation revealed by homozygous gene deficiencies, *J. Cell Sci.*, **113**, pp. 753–757.

25. Candé C., Cecconi F., Dessen P., Kroemer G. (2002). Apoptosis-inducing factor (AIF): Key to the conserved caspase-independent pathways of cell death?, *J. Cell Sci.*, **115**, pp. 4727–4734.

26. Lorenzo H. K., Susin S. A., Penninger J., Kroemer C. J. C. D. (1999). Apoptosis inducing factor (AIF): A phylogenetically old, caspase-independent effector of cell death, *Cell Death Differ.*, **6**, p. 516.

27. Ghobrial I. M., Witzig T. E., Adjei A. A. (2005). Targeting apoptosis pathways in cancer therapy, *CA Cancer J. Clin.*, **55**, pp. 178–194.

28. Scorrano L., Ashiya M., Buttle K., Weiler S., Oakes S. A., Mannella C. A., Korsmeyer S. J. (2002). A distinct pathway remodels mitochondrial cristae and mobilizes cytochrome c during apoptosis, *Dev. Cell*, **2**, pp. 55–67.

29. Kim J. S., He L., Lemasters J. J. (2003). Mitochondrial permeability transition: A common pathway to necrosis and apoptosis, *Biochem. Biophys. Res. Commun.*, **304**, pp. 463–470.

30. Gross A., McDonnell J. M., Korsmeyer S. J. (1999). BCL-2 family members and the mitochondria in apoptosis, *Genes Dev.*, **13**, pp. 1899–1911.

31. Murphy K. M., Ranganathan V., Farnsworth M. L., Kavallaris M., Lock R. B. (2000). BCL-2 inhibits Bax translocation from cytosol to mitochondria during drug-induced apoptosis of human tumor cells, *Cell Death Differ.*, **7**, p. 102.

32. Wajant H. (2002). The Fas signaling pathway: More than a paradigm, *Science*, **296**, pp. 1635–1636.

33. Susin S. A., Daugas E., Ravagnan L., Samejima K., Zamzami N., Loeffler M., Costantini P., Ferri K. F., Irinopoulou T., Prévost M. C., Brothers

G., Mak T. W., Penninger J., Earnshaw W. C., Kroemer G. (2000). Two distinct pathways leading to nuclear apoptosis, *J. Exp. Med.*, **192**, pp. 571–580.

34. Chen, G., and Goeddel, D. V. (2002). TNF-R1 signaling: a beautiful pathway. *Science.*, **296**(5573), pp. 1634–1635.
35. Wajant, H. (2002). The Fas signaling pathway: more than a paradigm. *Science.*, **296**(5573), pp. 1635–1636.
36. Boivin W. A., Cooper D. M., Hiebert P. R., Granville D. J. (2009). Intracellular versus extracellular granzyme B in immunity and disease: Challenging the dogma, *Lab Invest.*, **89**, p. 1195.
37. Martinvalet D., Zhu P., Lieberman J. (2005). Granzyme A induces caspase-independent mitochondrial damage, a required first step for apoptosis, *Immunity*, **22**, pp. 355–370.
38. Rutkowski, D. T., Kaufman, R. J. (2004). A trip to the ER: Coping with stress, *Trends Cell Biol.*, **14**, pp. 20–28.
39. Szegezdi E., Fitzgerald U., Samali A. (2003). Caspase-12 and ER-stress-mediated apoptosis, *Ann. N. Y. Acad. Sci.*, **1010**, pp. 186–194.
40. Ruiz-Vela A., Opferman J. T., Cheng E. H., Korsmeyer S. J. (2005). Proapoptotic BAX and BAK control multiple initiator caspases, *EMBO Rep.*, **6**, pp. 379–385.
41. Obeng E. A., Boise L. H. (2005). Caspase-12 and caspase-4 are not required for caspase-dependent endoplasmic reticulum stress-induced apoptosis, *J. Biol. Chem.*, **280**, pp. 29578–29587.
42. Fischer H., Koenig U., Eckhart L., Tschachler E. (2002). Human caspase 12 has acquired deleterious mutations, *Biochem. Biophys. Res. Commun.*, **293**, pp. 722–726.
43. Iurlaro R., Muñoz-Pinedo C. (2016). Cell death induced by endoplasmic reticulum stress, *FEBS J.*, **283**, pp. 2640–2652.
44. Upton J. P., Wang L., Han D., Wang E. S., Huskey N. E., Lim L., Truitt M., McManus M. T., Ruggero D., Goga A., Papa F. R., Oakes S. A. (2012). IRE1α cleaves select microRNAs during ER stress to derepress translation of proapoptotic Caspase-2, *Science*, **338**, pp. 818–822.
45. Sandow J. J., Dorstyn L., O'Reilly L. A., Tailler M., Kumar S., Strasser A., Ekert P. G. (2014). differentiation, ER stress does not cause upregulation and activation of caspase-2 to initiate apoptosis, *Cell Death Differ.*, **21**, p. 475.
46. Aslan B., Ozpolat B., Sood A. K., Lopez-Berestein G. (2013). Nanotechnology in cancer therapy, *J. Drug Target*, **21**, pp. 904–913.

47. Misra R., Acharya S., Sahoo S. K. J. D. D. T. (2010). Cancer nanotechnology: Application of nanotechnology in cancer therapy, *Drug Discov. Today*, **15**, pp. 842–850.

48. Petros R. A., DeSimone J. M. (2010). Strategies in the design of nanoparticles for therapeutic applications, *Nat. Rev. Drug Discov.*, **9**, p. 615.

49. Wang J., Tian S., Petros R. A., Napier M. E., Desimone J. M. (2010). The complex role of multivalency in nanoparticles targeting the transferrin receptor for cancer therapies, *J. Am. Chem. Soc.*, **132**, pp. 11306–11313.

50. Champion J. A., Katare Y. K., Mitragotri S. (2007). Particle shape: A new design parameter for micro-and nanoscale drug delivery carriers, *J. Control. Release*, **121**, pp. 3–9.

51. Owens III D. E., Peppas N. A. (2006). Opsonization, biodistribution, and pharmacokinetics of polymeric nanoparticles, *Int. J. Pharm.*, **307**, pp. 93–102.

52. Sakamoto J. H., van de Ven A. L., Godin B., Blanco E., Serda R. E., Grattoni A., Ziemys A., Bouamrani A., Hu T., Ranganathan S. I. J. P. R. (2010). Enabling individualized therapy through nanotechnology, *Pharmacol. Res.*, **62**, pp. 57–89.

53. Bertrand N., Wu J., Xu X., Kamaly N., Farokhzad O. C. (2014). Cancer nanotechnology: The impact of passive and active targeting in the era of modern cancer biology, *Adv. Drug Deliv. Rev.*, **66**, pp. 2–25.

54. Ajdary M., Moosavi M., Rahmati M., Falahati M., Mahboubi M., Mandegary A., Jangjoo S., Mohammadinejad R., Varma R. (2018). Health concerns of various nanoparticles: A review of their in vitro and in vivo toxicity, *Nanomaterials*, **8**, p. 634.

55. Wang X., Yang L., Chen Z., Shin D. M. (2008). Application of nanotechnology in cancer therapy and imaging, *CA: A Cancer J. Clin.*, **58**, pp. 97–110.

56. Cho K., Wang X., Nie S., Shin D. M. (2008). Therapeutic nanoparticles for drug delivery in cancer, *Clin. Cancer Res.*, **14**, pp. 1310–1316.

57. Fang J., Nakamura H., Maeda H. (2011). The EPR effect: Unique features of tumor blood vessels for drug delivery, factors involved, and limitations and augmentation of the effect, *Adv. Drug Deliv. Rev.*, **63**, pp. 136–151.

58. Aslan B., Ozpolat B., Sood A. K., Lopez-Berestein G. (2013). Nanotechnology in cancer therapy, *J. Drug Target.*, **21**, pp. 904–913.

59. Brigger I., Dubernet C., Couvreur P. (2012). Nanoparticles in cancer therapy and diagnosis, *Adv. Drug Deliv. Rev.*, **64**, pp. 24–36.

60. Wason M. S., Colon J., Das S., Seal S., Turkson J., Zhao J., Baker C. H. (2013). Medicine, Sensitization of pancreatic cancer cells to radiation by cerium oxide nanoparticle-induced ROS production, *Nanomedicine*, **9**, pp. 558–569.

61. Shafagh M., Rahmani F., Delirezh N. (2015). CuO nanoparticles induce cytotoxicity and apoptosis in human K562 cancer cell line via mitochondrial pathway, through reactive oxygen species and P53, *Iranian J. Basic Med. Sci.*, **18**, p. 993.

62. Meena R., Rani M., Pal R., Rajamani P. (2012). Nano-TiO_2-induced apoptosis by oxidative stress-mediated DNA damage and activation of p53 in human embryonic kidney cells, *Appl. Biochem. Biotechnol.*, **167**, pp. 791–808.

63. Bai D.-P., Zhang X.-F., Zhang G.-L., Huang Y.-F., Gurunathan S. (2017). Zinc oxide nanoparticles induce apoptosis and autophagy in human ovarian cancer cells, *Int. J. Nanomed.*, **12**, p. 6521.

64. Ye Y., Liu J., Xu J., Sun L., Chen M., Lan M. (2010). Nano-SiO_2 induces apoptosis via activation of p53 and Bax mediated by oxidative stress in human hepatic cell line, *Toxicol. In Vitro*, **24**, pp. 751–758.

65. R. Mohammadinejad, M. A. Moosavi, S. Tavakol, D. Ö. Vardar, A. Hosseini, Rahmati M., Dini L., Hussain S., Mandegary A., Klionsky D. J. (2019). Necrotic, apoptotic and autophagic cell fates triggered by nanoparticles, *Autophagy*, **15**, pp. 4–33.

66. Nguyen, K. C., Willmore, W. G., and Tayabali, A. F. (2013). Cadmium telluride quantum dots cause oxidative stress leading to extrinsic and intrinsic apoptosis in hepatocellular carcinoma HepG2 cells. *Toxicology.*, **306**, pp. 114–123.

67. Paesano L., Perotti A., Buschini A., Carubbi C., Marmiroli M., Maestri E., Iannotta S., Marmiroli N. (2017). Data on HepG2 cells changes following exposure to cadmium sulphide quantum dots (CdS QDs), *Data Brief.*, **11**, pp. 72–97.

68. Al-Ali A., Singh N., Manshian B., Wilkinson T., Wills J., Jenkins G. J., Doak S. H. (2015). Quantum dot induced cellular perturbations involving varying toxicity pathways, *Toxicol. Res.*, **4**, pp. 623–633.

69. Cheng G., Guo W., Han L., Chen E., Kong L., Wang L., Ai W., Song N., Li H., Chen H. (2013). Cerium oxide nanoparticles induce cytotoxicity in human hepatoma SMMC-7721 cells via oxidative stress and the activation of MAPK signaling pathways, *Toxicol. In Vitro*, **27**, pp. 1082–1088.

70. Ding M., Kisin E., Zhao J., Bowman L., Lu Y., Jiang B., Leonard S., Vallyathan V., Castranova V., Murray A. (2009). Size-dependent effects of tungsten carbide–cobalt particles on oxygen radical production and activation of cell signaling pathways in murine epidermal cells, *Toxicol. Appl. Pharmacol.*, **241**, pp. 260–268.

71. Manke, A., Wang, L., and Rojanasakul, Y. (2013). Mechanisms of nanoparticle-induced oxidative stress and toxicity. *BioMed Research International.*, **2013**, pp. 1–15.

72. Alarifi S., Ali D., Al Omar Suliman Y., Ahamed M., Siddiqui M. A., Al-Khedhairy A. A. (2013). Oxidative stress contributes to cobalt oxide nanoparticles-induced cytotoxicity and DNA damage in human hepatocarcinoma cells, *Int. J. Nanomed.*, **8**, pp. 189.

73. Chattopadhyay S., Dash S. K., Mahapatra S. K., Tripathy S., Ghosh T., Das B., Das D., Pramanik P., Roy S. (2014). Chitosan-modified cobalt oxide nanoparticles stimulate TNF-α-mediated apoptosis in human leukemic cells, *JBIC J. Biol. Inorg. Chem.*, **19**, pp. 399–414.

74. Wu H., Zhu H., Li X., Liu Z., Zheng W., Chen T., Yu B., Wong K.-H. (2013). Induction of apoptosis and cell cycle arrest in A549 human lung adenocarcinoma cells by surface-capping selenium nanoparticles: An effect enhanced by polysaccharide–protein complexes from *Polyporus rhinocerus*, *J. Agric. Food Chem.*, **61**, pp. 9859–9866.

75. Wang Y., Cui H., Zhou J., Li F., Wang J., Chen M., Liu Q. (2015). Cytotoxicity, DNA damage, and apoptosis induced by titanium dioxide nanoparticles in human non-small cell lung cancer A549 cells, *Environ. Sci. Pollut. Res.*, **22**, pp. 5519–5530.

76. Vamanu C. I., Cimpan M. R., Høl P. J., Sørnes S., Lie S. A., Gjerdet N. R. (2008). Induction of cell death by TiO_2 nanoparticles: Studies on a human monoblastoid cell line, *Toxicol. In Vitro*, **22**, pp. 1689–1696.

77. Shukla R. K., Kumar A., Gurbani D., Pandey A. K., Singh S., Dhawan A. (2013). TiO_2 nanoparticles induce oxidative DNA damage and apoptosis in human liver cells, *Nanotoxicology*, **7**, pp. 48–60.

78. Simard J.-C., Durocher I., Girard D. (2016). Silver nanoparticles induce irremediable endoplasmic reticulum stress leading to unfolded protein response dependent apoptosis in breast cancer cells, *Apoptosis*, **21**, pp. 1279–1290.

79. Noël C., Simard J.-C., Girard D. (2016). Gold nanoparticles induce apoptosis, endoplasmic reticulum stress events and cleavage of cytoskeletal proteins in human neutrophils, *Toxicol. In Vitro*, **31**, pp. 12–22.

80. Chen R., Huo L., Shi X., Bai R., Zhang Z., Zhao Y., Chang Y., Chen C. (2014). Endoplasmic reticulum stress induced by zinc oxide nanoparticles is an earlier biomarker for nanotoxicological evaluation, *ACS Nano*, **8**, pp. 2562–2574.

81. Christen V., Camenzind M., Fent K. (2014). Silica nanoparticles induce endoplasmic reticulum stress response, oxidative stress and activate the mitogen-activated protein kinase (MAPK) signaling pathway, *Toxicol. Rep.*, **1**, pp. 1143–1151.

82. Sarkar A., Das J., Manna P., Sil P. C. (2011). Nano-copper induces oxidative stress and apoptosis in kidney via both extrinsic and intrinsic pathways, *Toxicology*, **290**, pp. 208–217.

83. Zhao J., Bowman L., Magaye R., Leonard S. S., Castranova V., Ding M. (2013). Apoptosis induced by tungsten carbide-cobalt nanoparticles in JB6 cells involves ROS generation through both extrinsic and intrinsic apoptosis pathways, *Int. J. Oncol.*, **42**, pp. 1349–1359.

84. Zhao J., Bowman L., Zhang X., Shi X., Jiang B., Castranova V., Ding M. (2009). Metallic nickel nano-and fine particles induce JB6 cell apoptosis through a caspase-8/AIF mediated cytochrome c-independent pathway, *J. Nanobiotechnol.*, **7**, p. 2.

85. Minamino T., Komuro I., Kitakaze M. (2010). Endoplasmic reticulum stress as a therapeutic target in cardiovascular disease, *Circ. Res.*, **107**, pp. 1071–1082.

86. Rahmati, M., Moosavi, M. A., and McDermott, M. F. (2018). ER stress: a therapeutic target in rheumatoid arthritis? *Trends in Pharmacological Sciences*, **39**(7), pp. 610–623.

87. Simard J. C., Durocher I., Girard D. (2016). Silver nanoparticles induce irremediable endoplasmic reticulum stress leading to unfolded protein response dependent apoptosis in breast cancer cells, *Apoptosis*, **21**, pp. 1279–1290.

88. Noël C., Simard J. C., Girard D. (2016). Gold nanoparticles induce apoptosis, endoplasmic reticulum stress events and cleavage of cytoskeletal proteins in human neutrophils, *Toxicol. In Vitro*, **31**, pp. 12–22.

89. Kuang H., Yang P., Yang L., Aguilar Z. P., Xu H. (2016). Size dependent effect of ZnO nanoparticles on endoplasmic reticulum stress signaling pathway in murine liver, *J. Hazard. Mater.*, **317**, pp. 119–126.

90. Chatterjee D. K., Diagaradjane P., Krishnan S. (2011). Nanoparticle-mediated hyperthermia in cancer therapy, *Ther. Deliv.*, **2**, pp. 1001–1014.

91. Khoulenjani S. B., Amanpour S., Mirzadeh H., Tabar Z. Z. (2019). A review on application of in vitro and in vivo models of cancerous tumors for studying hyperthermia effect, *Basic Clin. Cancer Res.*, **11**(1), pp. 71–81.

92. Ludwig R., Teran F. J., Teichgraeber U., Hilger I. (2017). Nanoparticle-based hyperthermia distinctly impacts production of ROS, expression of Ki-67, TOP2A, and TPX2, and induction of apoptosis in pancreatic cancer, *Int. J. Nanomed.*, **12**, pp. 1009–1018.

93. Zahedi-Tabar Z., Bagheri-Khoulenjani S., Mirzadeh H., Amanpour S. (2019). 3D in vitro Cancerous tumor models: Using 3D printers, *Med. Hypotheses*, **124**, pp. 91–94.

94. Paunovska K., Sago C. D., Monaco C. M., Hudson W. H., Castro M. G., Rudoltz T. G., Kalathoor S., Vanover D. A., Santangelo P. J., Ahmed R. (2018). A direct comparison of in vitro and in vivo nucleic acid delivery mediated by hundreds of nanoparticles reveals a weak correlation, *Nano Lett.*, **18**, pp. 2148–2157.

95. Zuckerman J. E., Choi C. H. J., Han H., Davis M. E. (2012). Polycation-siRNA nanoparticles can disassemble at the kidney glomerular basement membrane, *Proc. Natl. Acad. Sci. U. S. A.*, **109**, pp. 3137–3142.

96. Tsoi K. M., MacParland S. A., Ma X.-Z., Spetzler V. N., Echeverri J., Ouyang B., Fadel S. M., Sykes E. A., Goldaracena N., Kaths J. M., et al. (2016). Mechanism of hard-nanomaterial clearance by the liver, *Nat. Mater.*, **15**, p. 1212.

97. Zhang Y., Sloan S. A., Clarke L. E., Caneda C., Plaza C. A., Blumenthal P. D., Vogel H., Steinberg G. K., Edwards M. S., Li G., et al. (2016). Purification and characterization of progenitor and mature human astrocytes reveals transcriptional and functional differences with mouse, *Neuron*, **89**, pp. 37–53.

98. Choi J. E., Kim S., Ahn J. H., Youn P., Kang J. S., Park K., Yi J., Ryu D.-Y. (2010). Induction of oxidative stress and apoptosis by silver nanoparticles in the liver of adult zebrafish, *Aquat. Toxicol.*, **100**, pp. 151–159.

99. Alaraby M., Romero S., Hernández A., Marcos R. (2019). Toxic and genotoxic effects of silver nanoparticles in drosophila, *Environ. Mol. Mutagen.*, **60**, pp. 277–285.

100. Seabra A., Durán N. (2015). Nanotoxicology of metal oxide nanoparticles, *Metals*, **5**, pp. 934–975.

101. Wang S., Meng X., Dong Y. (2017). Ursolic acid nanoparticles inhibit cervical cancer growth in vitro and in vivo via apoptosis induction, *Int. J. Oncol.*, **50**, pp. 1330–1340.

102. Verma S. K., Panda P. K., Jha E., Suar M., Parashar S. (2017). Altered physiochemical properties in industrially synthesized ZnO nanoparticles regulate oxidative stress; induce in vivo cytotoxicity in embryonic zebrafish by apoptosis, *Sci. Rep.*, **7**, p. 13909.

103. Haghniaz R., Bhayani K. R., Umrani R. D., Paknikar K. M. (2013). Dextran stabilized lanthanum strontium manganese oxide nanoparticles for magnetic resonance imaging, *RSC Adv.*, **3**, pp. 18489–18497.

104. Sun Y., Zhang G., He Z., Wang Y., Cui J., Li Y. (2016). Effects of copper oxide nanoparticles on developing zebrafish embryos and larvae, *Int. J. Nanomed.*, **11**, p. 905.

105. Eifler A. C., Thaxton C. S. (2011). Nanoparticle therapeutics: FDA approval, clinical trials, regulatory pathways, and case study, in: *Biomedical Nanotechnology*, *Springer*, pp. 325–338.

Chapter 4

Cell Membrane-Mimetic Nanoparticles for Cancer Treatment: What Has Been Done So Far?

João Basso,[a,b,*] Jessica Silva,[a,b,*] Maria Mendes,[a,b,*] João Sousa,[a,b,c] Alberto Pais,[c] and Carla Vitorino[a,b,c]

[a]*Faculty of Pharmacy, University of Coimbra, Pólo das Ciências da Saúde, Azinhaga de Santa Comba, 3000-548 Coimbra, Portugal*
[b]*Centre for Neurosciences and Cell Biology, University of Coimbra, Rua Larga, Faculty of Medicine, Pólo I, 1st floor, 3004-504 Coimbra, Portugal*
[c]*Coimbra Chemistry Centre, Department of Chemistry, University of Coimbra, Rua Larga, 3004-535 Coimbra, Portugal*

csvitorino@ff.uc.pt

Targeted drug delivery systems hold great promise as potential therapeutic strategies against cancer, as they preferentially accumulate at tumor sites and reduce drug-related adverse effects. Among nano-sized systems, cell membrane-coated nanoparticles

*These authors contributed equally to this work and should be regarded as co-first authors.

Nanoparticle Drug Delivery Systems for Cancer Treatment
Edited by Hala Gali-Muhtasib and Racha Chouaib

ISBN 978-981-4800-54-9 (Hardcover), 978-0-429-34125-0 (eBook)
www.jennystanford.com

have emerged as a new class of biomimetic strategies, which combine the technological advantages of nanomaterials with the stealth and biocompatibility characteristics of cellular membranes. Several sources of cell membranes (including erythrocytes, leukocytes, mesenchymal stem cells, platelet, and cancer cells) and their association with targeting ligands have been explored to shield nanoparticles, assessing the high versatility of cell membrane-mimetic nanoparticles. This chapter addresses the most recent advances in cell membrane based nanosystems, highlighting their ability to increase nanoparticle circulation, promote an active targeting strategy, or trigger a desired immunity reaction against tumor cells.

4.1 Introduction

Nanoparticle (NP) drug delivery systems are far from being easily transposable to clinical practice although they are one of the most explored approaches in preclinical research [1]. This is mainly due to biological and technological barriers. As a consequence, biomimetic strategies for drug delivery, including surface modification with small molecules, carbohydrates, peptides and proteins, pathogen-derived particle functionalization or cellular membrane coating, are being explored for several bio-applications [2]. This chapter focuses on the application of cell membrane-coated nanoparticles (CMNPs) to cancer therapy, as they are able to overcome some limitations associated with conventional nanoparticles that are pinpointed as the cause of their frequent unsuccessful outcome [3].

The surface of CMNPs preserves the natural structure of cell membranes, taking advantages of their ability to interact with the body environment without being recognized by proteins and/or phagocytic cells. This new approach allows the use of a direct cell membrane coating to prolong their circulation time, to reduce immunogenicity and to enhance their targeting ability, thus increasing biocompatibility and biodegradability [4]. Aspects like the core and shell composition, particle size, shape, and coating of NPs are crucial for activating defense mechanisms (e.g., the innate or adaptive immune systems, the recognition by the reticuloendothelial system, RES, or opsonization). Thus, if one of them is activated, the human body can easily perceive such nanosystems as intruders and may promote severe immune responses, leading to the elimination of NPs from the blood circulation and compromising the efficiency of drug delivery to target sites.

So, CMNPs with a cell membrane specific coating provide a complete surface antigenic diversity and various biological benefits, which include longer blood circulation time and active targeting capabilities. Moreover, in theory, they should be nontoxic and invisible to the immune system, due to an outer surface possessing an inherent affinity to the disease features. The main biomimetic strategies consist of using erythrocytes, leukocytes, mesenchymal stem cells, platelet and cancer cells membranes to cloak and shield NPs [5–11]. These functional coatings provide an opportunity of specific targeting and drug delivery, increasing the therapeutic efficacy and reducing side effects. It is known that CMNPs can interact with cells by multiple pathways, taking part in various physiological and pathological processes [12]. For these reasons, they have been studied in connection to several pathologies, such as arterial diseases, bacterial infections, immune thrombocytopenias and cancer [13].

In the development of CMNPs for cancer treatment, the major problem associated with the respective translation to the clinic is the lack of knowledge of the nanoparticle–cell interactions, as well as the unpredictability of pharmacokinetic and pharmacodynamics processes. However, it is known that CMNPs have the ability to interact with the complex tumor microenvironment, to associate different tumor-targeting strategies and enhance the anti-tumor drug delivery therapy. In addition, numerous cell membrane coatings are related or involved in multiple mechanisms, including angiogenesis, pro-differentiation, growth factor pathways, and inhibition of tumor invasion. For these reasons, NPs cloaked with cell membranes have been explored to overcome the hurdles of conventional tumor-targeting drug delivery.

Cell membrane-coated nanoparticles will have a behavior similar to that of cells in the body or in the tumor micro-environment, with the advantage of prolonging the circulation time (erythrocytes), as mentioned, or enhancing the targeting ability to inflammation (leukocytes), circulating tumor cells (platelets) or native cancer cells (cancer cells or mesenchymal stem cells) [14]. All these advantages allow the development of CMNPs for tumor targeting, immune counter-response, drug delivery, imaging, and photothermal therapy activation. However, more investigation on CMNPs is necessary, due to the limitations in cell membrane extraction, the several steps of isolation processes, and the preparation methods (extrusion or/and sonication), which are complex and associated with a low yield (Table 4.1).

Table 4.1 Advantages and disadvantages of CMNPs

Advantages	Disadvantages
• The natural structures of cell membranes are preserved, making them a combination of both natural and synthetic components • Cell membranes express specific proteins as a "self" signal or present surface antigens to host dendritic cells, which is ideal to generate a robust immune response • Increase in circulation time • Reduced immunogenicity • Improved targeting • Higher biocompatibility and biodegradability • Efficient drug delivery systems • Increased therapeutic efficacy • Reduction of side effects • Multiple pathways of the NPs physiological and pathological processes are explored	• The availability of cell membranes is quite limited • Cell membrane isolation processes involve several steps • Preparation process is complex • Low yield • Scale-up process is difficult

In this chapter, the main advantages of CMNPs, considering different cell-membrane coatings, their capabilities as drug delivery and cancer diagnostic platforms will be discussed. Also, research involving both in vitro and in vivo studies, showing targeting properties to specific tumor tissues will be addressed. Although

CMNPs can be broadly used in several diseases, this review is essentially focused on cancer. Note that the presented tables (Tables 4.2–4.4) do not aim at exhaustively listing the relevant NP systems developed. The importance of CMNPs in personalized medicine and the difficulties associated with the respective clinical implementation will be considered.

4.2 Cell Membrane Biomimetic Strategies

4.2.1 Erythrocyte Membrane-Based Nanoparticles

Several strategies have been developed to improve the performance of drug delivery systems, as they are quickly recognized as foreign bodies that trigger an immunological reaction leading to their quick removal from the bloodstream. The leading standard surface engineering approach is to pegylate these platforms, thus forming a hydration shell that is able to evade immune responses. Yet, it has been shown that after multiple administrations, pegylated nanosystems trigger anti-PEG responses, hence the need to explore new stealth moieties [15]. Erythrocytes are the major cellular component of blood, being characterized by their biconcave shape and absence of nucleus and other organelles. Moreover, erythrocytes have a life span of 100 to 120 days before undergoing senescence and clearance from the bloodstream by splenic macrophages, and this may be due to the presence of the integrin-associated transmembrane protein CD47 [16, 17]. In fact, erythrocytes lacking CD47 receptors on their surface are rapidly cleared from the blood circulation following a macrophage response [17].

Considering the intrinsic stealth properties of erythrocytes and the delivery abilities of NPs, Hu et al. [8] designed a novel system in which polymeric NPs were superficially cloaked by these cell membranes. As a result, these NPs presented a higher in vivo residence time (up to 72 h vs. 48 h for pegylated NPs), with their polymeric cores retaining the respective function.

The higher complexity of biological membranes over simpler phospholipid arrangements and the need to retain the structure and functionality of the red blood cell membranes led to the development of top-down production methods to tailor polymeric and metallic NPs (Table 4.2). The process of NP cloaking by erythrocyte-derived membranes is relatively simple, consisting of two mechanical extrusion steps after the purification of the membranes (Fig. 4.1). However, there is a need to preserve the correct orientation and surface density of CD47 receptors, as these are key characteristics for a successful functionalization of erythrocyte membrane-cloaked drug delivery systems [18].

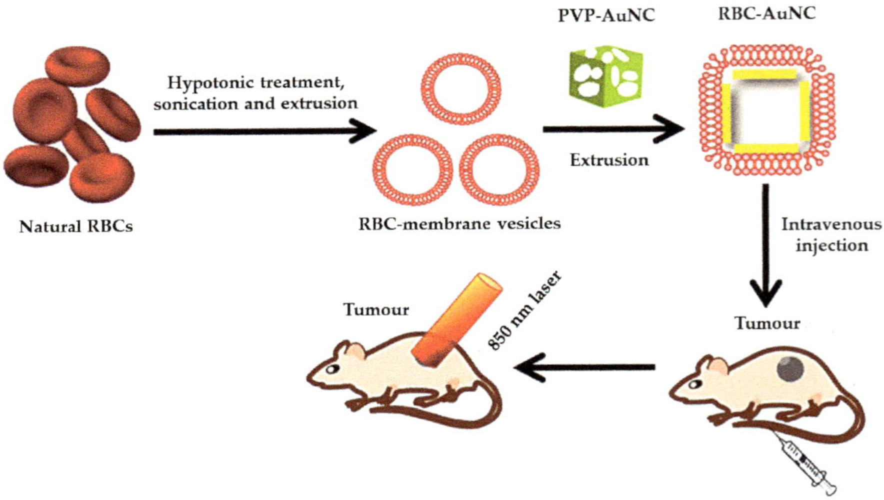

Figure 4.1 Common top-down approach to produce red blood cell membrane-cloaked NPs: cell membrane purification followed by two steps of mechanical extrusion. Adapted from ref. [19] with permission from the American Chemical Society.

Red blood cell membranes also include an abundant asymmetric distribution of glycans, the glycocalyx. The extracellular and hydrophilic coating with the negatively charged sialyl residues of the glycocalyx is critical for the stabilization and immune-evasive properties of these cells [20–22]. Moreover, glycocalyx plays a significant role on the correct electrostatic binding and orientation of the membrane onto the surface of the negatively charged solid NP cores [23].

Table 4.2 Non-exhaustive list of erythrocyte membrane coated NPs with proved anticancer activity

NP Core	Polymeric					Polymeric
Core Composition	PLGA	PLGA	PLGA	PLGA	PLGA	Chitosan
Payload	—	DOX	DOX	hgp100$_{25-33}$ antigenic peptide, Monophosphoryl lipid	DOX	DOX, PTX, Fe_3O_4 nanocrystals
Cancer	Lung/breast cancer	Glioma	Acute myeloid leukemia	Melanoma	Lymphoma	Lung
Major outcomes	Surface functionalization with folate and aptamer A549 of the RBC membrane led to an 8- and 2-fold increase in cell uptake by KB and MCF-7 cells, respectively	Following a surface functionalization with the DCDX peptide, the NPs efficiently target the brain, significantly increasing the overall survival of tumor bearing mice up to 34 days (vs. 26 days for saline)	Erythrocyte membrane coating plays a role in the controlled release of DOX, with a release of 20% following 72 h (vs. 40 h in pegylated NPs); also, it shows higher toxicity over AML Kasumi-1 cells, when compared to free DOX	In vivo results show that this functionalized biomimetic vaccine was found in 4.7–22.4% of different dendritic cells and macrophages; mannose functionalized platforms presented a preferential uptake in macrophages (31.4% vs. 22.4% in unmodified coated NPs) and increased tumor-occurring time, inhibited tumor growth and suppressed tumor metastasis in prophylactic, therapeutic, and metastatic melanoma models, respectively, following an intradermal administration	Although presenting a higher IC_{50} value (5.6 ng/mL vs. 1.4 ng/mL for free DOX) due to an incomplete drug release at 72 h, cloaked NPs decreased tumor growth and increase the overall survival of tumor bearing mice up to 47 days (vs. 24 days in control group), while not showing any immune response (IgM/IgG) on day 30 after multiple administrations	RBC cloaking decreased macrophage uptake (20%), when compared to pegylated (40%) and chitosan NPs (>80%) at 24 h; additionally, loaded nanosystems reduce tumor cell viability up to 20%, when submitted to an external magnetic field; RGD tailored NPs effectively target tumor cells to co-deliver DOX and PTX, in a superior performance than the naked NPs by a 17-fold factor, thus significantly increasing the overall survival of all tumor bearing mice at day 21 vs. 12 days for control
Ref.	[26]	[27]	[58]	[30]	[25]	[59]

Table 4.2 (*continued*)

NP Core	Metallic		Metallic		
Core Composition	Gold nanocages	Fe_3O_4 nanoclusters	Silica-Titanium dioxide (SiO_2-TiO_2)	Zirconium based metal-organic framework	Pegylated-Selenium NPs
Payload	—	—	DTX	IDG, Oxygen	Bevacizumab
Cancer	Hepatocellular carcinoma	Breast cancer	Breast cancer	Breast cancer	Melanoma
Major outcomes	RBC membrane showed no significant interference in the temperature rising ability of the nanocages after irradiation, while drastically increasing cell uptake (10^{10} more particles in tumor tissue, comparing to the base NP) and leading to a 100% survival at day 45 (vs. 80% with naked NPs and 20% with control)	RBC cloaking improved circulation time up to 48 h (vs. 8 h for naked clusters), by a shielding effect from macrophage uptake (2.4-fold difference in extracellular Fe concentration); thus, an enhanced photothermal effect was accomplished in tumor bearing mice, increasing tumor inhibition rates up to 94.24% (vs. 21.43% for uncoated particles)	This UV light–triggered drug delivery system efficiently releases DTX following irradiation, with the RBC coating degradation being responsible for the controlled release of the drug (100% with UV irradiation vs. ~12% without UV irradiation); additionally, the production of oxygen radicals promotes an additional photodynamic effect on MCF-7 cells, decreasing cell viability to ~25%(vs ~75% with coated NPs not submitted to UV irradiation)	Following a photothermal stimulus, this strategy efficiently disrupts the RBC cloaking and delivers oxygen to hypoxic tumors, causing an almost complete ablation of the tumor via photodynamic therapy in tumor bearing mice models, without adverse effects	This smart biomimetic platform effectively bypasses autoimmune reactions, displaying a strong anticancer, radiosensitizing and antiangiogenic effects in athymic nude mice models with X-ray irradiation, decreasing the relative tumor growth ratio to 17.4%. Tumor relative volume was found to be 2.5% at day 21, vs. 4.7%, 8.7% and ~15% for X-ray, NP and control groups, respectively
Ref.	[19]	[32]	[60]	[61]	[62]

Note: PLGA, Poly(lactic-co-glycolic acid); RBC, Red blood cell; DOX, Doxorubicin; NPs, Nanoparticles; IC_{50}, Half maximal inhibitory concentration; PTX, Paclitaxel; Fe_3O_4, Iron oxide; RGD, Arginyl-glycyl-aspartic acid peptide; UV, Ultraviolet; DTX- Docetaxel; IDG, Indocyanine green.

Poly(lactic-co-glycolic acid) (PLGA) is a biodegradable and biocompatible polymer extensively used to produce organic NPs [24]. Considering the extensive knowledge of this polymer and its intrinsic negative surface charge, a wide range of PLGA nanometric cores have been efficiently extruded and completely cloaked with extracted and purified erythrocyte membranes, thus forming novel biomimetic core-shell drug delivery vehicles, with a correct CD47/glycocalyx external orientation and a prolonged circulation time in vivo [8, 18, 23]. However, when the surface charge of the cores is reversed, for example, with the incorporation of polyethyleneimine, the extrusion process is not completed due to the formation of aggregates [23].

The combination of a loaded nanoparticulate system with a biomimetic erythrocyte membrane shell is a promising strategy for application in many diseases, including cancer. However, in spite of efficiently improving the pharmacokinetic profile of the therapeutic agent, it lacks an active targeting strategy, with the nude erythrocyte membrane-cloaked NPs only taking advantage of the leaky blood vessels to target the tumor (enhanced permeability and retention, EPR, effect) [25]. The presence of a heterogeneous and delicate biomembrane with lipids, proteins, and sugar moieties on the surface of the NPs requires a non-disruptive functionalization technique of these systems in order to retain the biomimetic properties of the shell membrane. Bearing this in mind, Fang et al. [26] developed a lipid insertion method and functionalized erythrocyte membranes with two ligands, folic acid, and a nucleolin-targeting aptamer AS1411, without damaging the pre-existing surface proteins. After incubating blocks of lipid-linker-ligand with empty red blood cells, the lipid spontaneously inserted itself into the membrane. These modified ghost cells were then used to cloak PLGA NPs via extrusion. Cellular uptake studies of folate and aptamer AS1411 functionalized NPs were conducted in three different cancer cell lines: the overexpressing folate receptors cervical cancer derived KB carcinoma cells, the non-expressing folate receptors A549 lung carcinoma cells, and the breast cancer cell line MCF-7, respectively. While not showing a preferential uptake in A549 cells, folate modified cloaked NPs showed an 8-fold increased uptake in KB cells compared to unmodified cloaked NPs. Likewise, the aptamer modified system showed a 2-fold increase in cellular uptake by MCF-7 cells, thus proving the applicability and efficacy of tailoring the erythrocyte membranes as an active anticancer targeting strategy. Interestingly, this lipid insertion technique failed when the positively charged ^{D}CDX peptide ligand was used, probably due to an electrostatic adsorption onto the surface of the membrane, compromising its targeting ability [27]. Considering the high affinity between streptavidin and biotin, Chai et al. designed two different blocks, one of streptavidin-linker-lipid and the second containing biotin-linker-^{D}CDX. The first was incubated and spontaneously inserted into the RBC membrane, with the latter being conjugated with the streptavidin modified RBC membrane-cloaked NPs, thus forming a ^{D}CDX functionalized carrier system. When loaded with doxorubicin (DOX), this brain targeting system crossed the blood brain barrier via nicotinic acetylcholine receptors to efficiently deliver its payload to the tumor site, increasing the overall survival of mice bearing brain tumors.

Although popular and increasing in number, biomimetic approaches with erythrocyte membrane coating are extremely versatile and not limited to drug delivery. Senescent and damaged erythrocytes have their membranes modified and are easily recognized by macrophages and dendritic cells, thus showing potential to deliver antigens to antigen presenting cells [28, 29]. This vaccine-like property was explored by Guo et al. [30] who developed a NP able to trigger an immune response against melanoma cells after an exposure to the antigenic peptide hgp100_{25-33}. In order to target mannose receptors of dendritic cells, mannose residues were inserted in the erythrocyte membranes, thus increasing the IFN-γ and T cell response, when compared to bare antigen carrying NPs.

Other approaches with non-polymeric cores have been designed, namely with metallic NPs, considering the unique magnetic, thermal, optical, catalytic and electrical properties of metals [31]. Recently, researchers took advantage of the benefits of erythrocyte mimetic NPs and used gold and iron NPs to effectively kill tumor cells through a heating process promoted by local light irradiation [19, 32]. Other examples regarding the use of metallic NPs within erythrocyte membrane structures are presented in Table 4.2.

Overall, the technological ease in membrane extraction and purification combined with a high residence time within the bloodstream has prompted the use of such membranes as a potential biomimetic strategy to increase the circulation time of artificial NPs, while reducing toxic effects and maintaining the drug loading and protective properties of such nanocarrier systems [33].

4.2.2 Leukocyte Membrane-Based Nanoparticles

Another strategy focuses on the coating of NPs with leukocyte membranes (also referred to as white blood cells, WBC) as potential drug delivery systems in cancer therapy. Leukocytes are intrinsically related with tissue injury, thus having the ability to interact with the inflammatory microenvironment of tumors [34]. In order to provide an inflammatory response, leukocytes are activated through the triggering of chemical signals mediated by cytokines (including chemokines and soluble mediators of cell killing), and other leukocyte-activating molecules (such as cytotoxic mediators, including reactive oxygen species, serine-, cysteine-, and metalloproteases, membrane perforating agents, and growth factors) released into the extracellular matrix by resident cells of inflamed tissue. Thereafter, leukocyte vascular extravasation from blood vessels to the site of initial tissue damage occurs via endothelial adhesion, transmigration and chemotaxis [34–36]. Since tumor-related leukocytes (macrophages, dendritic cells, neutrophils, eosinophils, mast cells, and lymphocytes), which are involved in neoplasia development and metastasis, have the ability to produce various cytokines and chemokines, the recruitment of leukocytes can also take place. They actively recognize and bind to tumor endothelium cells through their membrane surface biomarkers (e.g., CD45, CD3z, CD11a) [9, 34]. Thus, leukocyte membranes comprise a promising approach for anticancer drug delivery.

Different types of NPs, including those based on lipids, polymers or metals, have been explored and biofunctionalized with leukocyte membranes. Several examples of these cell membrane-based biomimetic NPs are depicted in Table 4.3. The major outcomes of these studies have shown high in vitro and in vivo stability and biocompatibility, prolonged systemic circulation, avoided opsonization and clearance by the RES, as well as an enhanced active targeting ability and accumulation in tumor sites, which are not so widely achieved when using erythrocyte membranes [37]. The first developed leukocyte membrane-based NP was a macrophage membrane-coated nanoporous silica particle, known as leuko-like vector (LLV), prepared by electrostatic and hydrophobic interactions between negatively charged cell membranes and positively charged NPs [9]. The authors have shown that leukocyte membranes on the NP surface display an important role in reducing NP opsonization, degradation and phagocytic uptake, with a preferential interaction with the inflamed endothelium. The therapeutic potential of this system was also evaluated using in vitro studies and considering free DOX, DOX-loaded NPs and DOX-loaded LLVs. It was observed that LLVs enhance the transport of DOX across the endothelium, improving their tumor therapeutic efficacy. Furthermore, in vivo studies were performed, which revealed a delay in liver clearance and LLVs integrity change, enhancing their circulation time and tumoritropic accumulation when compared to conventional NPs. In another study, a macrophage cloaked emtansine-loaded pH-sensitive liposome was developed by extrusion to target metastatic cancer cells [38]. This biomimetic nanoplatform displayed a specific and effective targeting ability to metastatic nodules and suppressed lung metastasis of breast cancer. When the $\alpha4\beta1$ integrins, expressed on macrophage membrane, were blocked by their specific anti-integrin $\alpha4\beta1$ monoclonal antibody, the nanosystem uptake by tumor cells was reduced by 25% in comparison with the respective cellular uptake of macrophage-coated emtansine-loaded liposomes. Thus, blocking prevents the interactions between $\alpha4\beta1$ integrins and vascular cell adhesion molecule-1 (VCAM-1) expressed on cancer cells membrane and, consequently, inhibits the recruitment of the platform by the tumor cells, compromising its therapeutic efficiency. Apart from the drug delivery function, leukocyte membrane-coated NPs improved the efficacy of photothermal therapy, and provided excellent optical properties extremely useful for tumor imaging [39, 40].

Table 4.3 Non-exhaustive list of leukocyte membrane coated NPs with proved anticancer activity

Membrane Coating	Macrophages	Plasma membrane of leukocytes	Monocytes	Neutrophils
NP core	Lipid		Polymeric	Polymeric
Core composition	DSPE-PEG and DOPE liposome	Grapefruit-derived nanovector NV	PLGA NP	PLGA NP
Payload	Emtansine	DOX	DOX	CFZ
Type of cancer	Lung metastasis of breast cancer	Colon and breast cancer	Metastasis of breast cancer	CTCs
Major outcomes	IC_{50} (free emtansine) = 0.57 μg/mL; IC_{50} (emtansine-loaded liposome) = 0.68 μg/mL; IC_{50} ($\alpha_4\beta_1$ integrins blocked macrophage-coated emtansine-loaded liposome) = 0.75 μg/mL; IC_{50} (macrophage-coated emtansine-loaded liposome) = 0.25 μg/mL Emtansine-loaded macrophage-coated liposome treatment leads to an 87.1% inhibition rate on lung metastasis of breast cancer, with an average number of metastatic nodules per lung equal to 3.8±2.6, which was only 12.9%, 17.6% and 24.0% of the negative control, free emtansine and emtansine-loaded liposome group treatments	Higher concentrations of DOX were detected in the tumor bearing mice injected with plasma membrane-coated grapefruit-derived DOX-loaded NVs when compared to mice treated with free DOX or DOX-loaded grapefruit-derived NVs; also, the extrusion of grapefruit-derived NVs with the activated leukocyte membranes demonstrated to be essential for higher efficiency delivery of DOX to tumor site	IC_{50} (free DOX) = 0.5 μM; IC_{50} (DOX-loaded NP) = 12 μM; IC_{50} (monocytes-coated DOX-loaded NPs) = 4 μM The uptake by tumor cells was significantly higher in the monocytes-coated DOX-loaded NPs in comparison to NP control samples	$IC5_0$ (free CFZ) = 58.2 ng/mL; IC_{50} (CFZ-loaded NPs) = 178 ng/mL, IC_{50} (neutrophils-coated CFZ-loaded NP) = 69.4 ng/mL Neutrophils-coated CFZ-loaded NPs ensure therapeutic potential for both preventing early metastasis and inhibiting the progress of already-formed metastasis
Ref.	[38]	[63]	[64]	[65]

(*Continued*)

Table 4.3 (*continued*)

Membrane Coating	hCTL	Macrophages	Macrophages	Macrophages	Macrophages
NP core		Metallic	Metallic		
Core composition	PLGA NP	AuNS	β-$NaYF_4$:Er^{3+},Yb^{3+} UCNP	NSNP (LLV)	MSNC
Payload	PTX	Cy7 or FITC-Dextran	—	DOX	DOX
Type of cancer	Gastric cancer	Breast cancer	Breast cancer	Melanoma	Breast cancer
Major outcomes	IC_{50} (free PTX) = 6.130 μg/mL; IC_{50} (PTX-loaded NPs) = 2.620 μg/mL; IC_{50} (hCTL membrane-coated PTX-loaded NPs) = 2.822 μg/mL hCTL membrane-coated PTX-loaded NPs promoted a decreased uptake rate of 23.99% by macrophages, prolonging their circulation time; also, an inhibition of tumor growth rate of 56.68% by hCTL membrane-coated PTX-loaded NPs was verified and this rate was increased to 88.50% with the concomitant application of local LDI	Macrophage membrane-coated AuNSs induced an uptake rate by tumor cells of 83.18% over the 42.15% of the uncoated AuNSs; furthermore, considering the injection of cell-membrane-based AuNSs, the content of Au element in the tumor was over 7.48% ID/g, while the correspondent for bare AuNSs was only 1.61% ID/g, suggesting a higher in vivo tumor accumulation and targeting efficacy, as well as an improving the efficacy of in vivo photothermal therapy	Excellent optical properties for fluorescent tumor imaging were reached with these UCNPs, since they emitted bright green fluorescence under the excitation of an external 980 nm NIR laser	For DOX-loaded LLVs, the in vitro tumor cell viability was reduced by 57% and endothelial cell viability was only reduced by 20%, suggesting the higher potential therapeutic effect of this leukocyte membrane-based nanosystem, which is able to successfully transport DOX through the tumor endothelium	After the injection of macrophage membrane-coated DOX-loaded MSNCs, the content of Si element in tumor was over 6% ID/g, whereas MSNCs only reached to 1.3% ID/g, suggesting the enhanced tumor targeting efficacy of cell membrane-coated MSNCs; also, the growth of tumor of mice was completely inhibited by these nanoplatforms
Ref.	[66]	[39]	[40]	[9]	[67]

Note: DSPE-PEG, Distearoylphosphatidylethanolamine-poly(ethylene glycol); DOPE, 1,2-dioleoyl-sn-glycero-3-phoshoethanolamine; IC50, Half maximal inhibitory concentration; DOX, Doxorubicin; NV, Nanovector; PLGA, Poly(lactic-co-glycolic acid); NP, Nanoparticle; CFZ, Carfilzomib; CTC, Circulating tumor cell; hCTL, Human cytotoxic T-lymphocytes; PTX, Paclitaxel; LDI, Low-dose irradiation; AuNS, Gold nanoshell; Cy7, Cyanine 7; FITC, Fluorescein isothiocyanate; ID, Injected dose; UCNP, Upconversion nanoparticle; NIR, Near infrared; NSNP, Nanoporous silica nanoparticle; LLV, Leuko-like vector; MSNC, Mesoporous silica nanocapsule.

4.2.3 Mesenchymal Stem Cell Membrane-Based Nanoparticles

Mesenchymal stem cells (MSC), categorized as adult stem cells, consist of multipotent cells capable of differentiating into various cell lineages such as osteoblast, chondrocyte, adipocyte, and myoblast [41, 42]. As in the case of cancer cells, MSC membranes contain several functional surface molecules able to recognize and interact with tumor cells, inducing a high tumor targeting ability [43]. Moreover, MSCs offer the advantage of easy in vitro expansion, since they can be isolated from various tissues and are available in large amounts. Also, they display low immunogenic and exhibit immunomodulatory activity in vivo [41].

Although the application of MSC membranes to coat NP surfaces has not been explored much, a MSC membrane-coated DOX-loaded gelatin nanogel was prepared by extrusion and directed at cervical cancer treatment [11]. An excellent tumor targeting capacity in vitro and higher accumulation in the tumor site in vivo and ex vivo were shown when compared with uncoated nanogels, indicating that this system offers a promising targeting drug delivery platform. Also, the in vivo antitumor therapeutic efficacy was evaluated, where the inhibition of tumor growth and the stabilization of tumor volume were confirmed for MSC membrane-coated DOX-loaded gelatin nanogels treated group. Additionally, a MSC membrane-cloaked superparamagnetic iron oxide nanoparticle (SPION) was developed using a simple and mild sonication method [41]. Despite the fast in vivo elimination of SPIONs, they present favorable magnetic properties to magnetic resonance imaging (MRI) and magnetic hyperthermia therapy. In vitro studies conducted on mice macrophages showed a preferential phagocytic activity on naked SPIONs (16 ppm of iron content of SPIONs in cells vs. 0.08 and 6.55 ppm in control and on MSC membrane-cloaked SPIONs, respectively). Moreover, MSC membrane-cloaked SPIONs displayed a preferential uptake by prostate cancer cells, as a stronger iron staining was observed. MSC membrane-cloaked SPIONs induced cancer cell death via a magnetic hyperthermia mechanism under alternating magnetic field applications, reducing cell viability up to ~87.3%. Also, they were demonstrated to have good colloidal stability and potential in vivo theranostic applications.

4.2.4 Platelet Membrane-Based Nanoparticles

Platelets (PLTs), the smallest circulating blood cells, are produced in bone marrow and are a blood component that is essential for maintaining hemostasis [6]. Their most important function is to stop bleeding by initiation and propagation of the coagulation process. PLTs have been used and associated with several phases of diseases, due to their function in maintaining hemostasis within the body [44].

The importance of PLT coating is emphasized by the significant crosstalk between PLTs and inflammatory processes and/or tumor cells through the receptors and signaling molecules. PLTs are able to bind to immune cells, recruiting them to inflammation sites, thus demonstrating an important interaction and displaying a naturally targeting behavior. At the same time, PLTs could accumulate in tumor microenvironments, modulating functions of both tumor and immune cells [45–47].

As mentioned before, PLTs are circulation sentinels in the bloodstream, having the capacity to recognize and interact with circulating tumor cells (CTCs). CTCs have received considerable attention for cancer treatment, due to their outstanding role in tumor progression and metastasis [5, 6]. In addition, PLTs are instinctively recruited to the sites of injured vessels, triggering a cascade that leads to clot formation that starts the healing process. Thus, recent improvements in biomaterials have demonstrated that PLT membranes can be masked onto organic and/or inorganic NPs, taking advantages of both systems [5, 48]. The advent of this new strategy results in decreased macrophage cell uptake, failure in nanoparticle-induced activation of the complement system and additional mimetic PLT properties, such as a specific adhesion to damaged tumor tissues [49].

In addition, it is possible to conjugate on the surface of biomimetic NPs cytokines or the tumor necrosis factor, producing a targeted therapeutic effect. The influence of PLT-coated NPs in the inhibition of growth and metastasis of drug-resistant tumors and tumor vasculature, as a dual-targeting strategy was addressed [5]. For this, a nanoplatform was developed by encapsulating melanin nanoparticles (MNPs) and DOX inside RGD peptide (c(RGDyC))-modified nanoscale platelet vesicles (RGD-NPVs). Following in vitro and in vivo studies, the results highlighted a versatile platelet-camouflaged nanoparticle, designed for an effective inhibitory behavior over the evolution and metastasis of drug-resistant breast cancer. Another study has shown that platelet membrane-coating of a photodynamic nanoparticle reduced the irradiation dose, which minimized skin damage without impacting the therapeutic outcome [50]. Additionally, it was possible to provide a longer circulation and active targeting effects, with the PLT coated-NPs demonstrating a higher tumor uptake when compared to the erythrocyte membrane-coated counterpart.

The development of platelet membrane-based nanoparticles for theranostics is also a reality. In fact, Rao et al. [51] developed a nanoplatform with increased system circulation and targeting activity against MCF-7 breast cancer cells. This was achieved by means of a PLT-derived coating onto Fe_3O_4 magnetic nanoparticles (MNs). The PLT-derived membrane was obtained from mice and injected into the same mouse donor, demonstrating a personalized medicine strategy with improved immunocompatibility.

In spite of this progress, there are some limitations that impair the fast evolution of platelet membrane-based nanoparticles, including platelet variability and complex isolation and preparation processes, associated with scale-up difficulties. These limit the translation of these basic scientific ideas to practice and subsequent clinical application.

4.2.5 Cancer Cell Membrane-Based Nanoparticles

Cancer cells display surface antigens with homologous adhesion domains responsible for the cell-to-cell specific adhesion [52, 53]. Intrinsic homotypic binding properties of cancer cells are responsible for formation of multicellular malignant cell aggregates in tumors, which in turn are related to the mechanical trapping theory of cancer metastasis [54].

The biomimetic functionalization of lipid, polymeric and metallic NPs with cancer cell membrane coating has been shown to be a potential approach for cancer therapy and diagnosis (Table 4.4), taking advantage of their effective escape from immune recognition and natural homotypic targeting abilities to the tumor site [37]. As an example, a cancer cell membrane protein–based biomimetic liposome was developed to enhance anti-glioma chemotherapeutic efficacy [10]. The membrane proteins of C6 glioma cells were extracted and embedded into the lipid bilayer of liposomes by successive extrusion, resulting in a stable biomimetic liposome. The DOX-loaded biomimetic nanosystem exhibited a higher in vitro homologous targeting ability, uptake efficiency and cytotoxicity, and a 2.25-fold deeper in vitro penetration in 3D tumor spheroid models, when compared to the conventional DOX-loaded liposome. Moreover, using dual-modal imaging techniques by co-loading indocyanine green (ICG), an FDA approved dye with near infrared fluorescence imaging and photoacoustic imaging features, into the DOX-loaded biomimetic liposome, the authors identified an improvement in tumor accumulation in C6 glioma tumor-bearing nude mice, when compared to free DOX and conventional DOX-loaded liposomes. Also, Chen et al. [53] developed, by extrusion, a breast cancer cell membrane-coated ICG-loaded PLGA NP for homologous-targeting dual-modal imaging and photothermal therapy. This system demonstrated to be an excellent fluorescence/photoacoustic imaging probe to real-time monitoring, possessing high spatial resolution and deep penetration, as well as an enhanced photothermal efficiency to eradicate xenografted tumors. Thus, cancer cell membrane coatings on NPs take the advantage of the homologous targeting to cancer cells for highly specific in vivo tumor imaging, due to the higher tumor accumulation via active homologous targeting and passive targeting of the enhanced EPR effect [53, 55]. Cancer cell membrane-based NPs have also been explored in cancer immunotherapy as cancer vaccines. Fang et al. explored the use of a melanoma cell membrane-coated PLGA NP [56] as a nanovaccine for cancer immunotherapy purposes, where the introduction of multiple cancer cell membrane antigens together with immunological adjuvants contributed to tumor-specific immune responses.

Table 4.4 Non-exhaustive list of cancer cell membrane coated NPs with proved anticancer activity

NP core	Lipid		Polymeric	Polymeric	
Core composition	DPPC, Chol and DSPE-PEG 2000 liposome	Egg lecithin (PC-98T) and Chol liposome	PLGA NP	PLGA NP	PLGA nanovaccine
Payload	DOX or DOX/ICG	ICG	ICG	—	Imiquimod
Type of cancer	Glioma	Glioma and lung metastasis of glioma	Breast cancer	Melanoma	Melanoma
Major outcomes	The DOX-loaded biomimetic liposome shown a higher uptake efficiency and cytotoxicity (cell viability was 29.63±3.34%) of C6 glioma cells when compared with free DOX (cell viability was 80.56±2.54%) and DOX-loaded liposome (cell viability was 61.95±1.59%). In vivo studies evidenced that the ICG fluorescence intensity of tumor tissue for DOX/ICG-loaded biomimetic liposome was 5.89- and 3.23-fold higher than free DOX/ICG and DOX/ICG-loaded liposome, respectively; the data demonstrated a more accentuated homotypic targeting ability and tumor accumulation by DOX-loaded biomimetic liposome	ICG-loaded biomimetic liposome exhibited the homologous targeting effect at the cellular level, as well as the specific-targeting ability to homologous xenografts glioma at the in vivo level, resulting in a promising nanoplatform for homologous-targeting phototherapy; moreover, the survival rate of ICG-loaded biomimetic liposome plus NIR irradiation group was 100% on day 21 post-treatment, which is higher than other groups	Cancer cell membrane-cloaked ICG-loaded NPs exhibit a cellular uptake 1.8-fold and 3.2-fold higher when compared with that of ICG-loaded NPs and free ICG, respectively; ICG-loaded biomimetic NP plus NIR irradiation induced up 93% of cell death and an in vivo survival rate of 100% on day 18 post-treatment with complete remission of tumor	Cancer cell membrane-coated NPs showed significantly increased particle-to-cell adhesion to source cancer cells when compared to bare NPs and structurally analogous erythrocyte membrane-coated NPs, demonstrating high potential for drug delivery applications	The modification of melanoma cell membrane-coated nanovaccine with mannose (by a lipid-anchoring method) enhanced uptake by APCs and demonstrated a higher antitumor effect, and the integration imiquimod, a TLR7 agonist, into nanovaccine increased tumor growth inhibition efficacy; also, with the combination of an anti-PD-1 checkpoint blockade, tumor progression was effectively inhibited and 3 out of 6 mice became tumor free and survived at day 45 post tumor inoculation
Ref.	[10]	[68]	[53]	[56]	[69]

Table 4.4 (*Continued*)

NP core	Metallic		Metallic
Core composition	β-NaYF 4 :Er^{3+},Yb^{3+} UCNP	AuN	MSNP
Payload	—	DOX	RNase A
Type of cancer	Breast, prostate, squamous and colorectal cancers	Metastasis of breast cancer	Cervical cancer
Major outcomes	It was shown that when cell membrane coating of UCNPs matched those of the host cancer cells, notable targeting ability was verified, while mismatch between the donor and host cells leads to almost no targeting property; also, cancer cell membrane-coated UCNPs demonstrated a higher accumulation in the tumor and lower accumulation in liver and spleen, when compared with the corresponding uncoated UCNPs	IC_{50} (cancer cell membrane-coated DOX-loaded AuNs) = 1.1 μg/mL; IC_{50} (cancer cell membrane-coated DOX-loaded AuNs + NIR irradiation) = 0.3 μg/mL The concomitant application of chemo and photothermal therapy accelerated the release of DOX at target points in vitro due to the laser-induced hyperthermia and the triggered phototoxicity of AuNs, enhancing therapy efficacy; about 98.9% and 98.5% inhibiting rates of the tumor volume and metastatic modules were achieved using the combinatorial therapy	The IC_{50} of cancer cell membrane-coated RNase A-loaded MSNP was 3-fold lower than that of bare RNase A-loaded MSNP; cell membrane cloaking enhanced the antitumor therapeutic effect due to homotypic tumor-targeting
Ref.	[55]	[70]	[71]

Note: DPPC, Dipalmitoylphosphatidylcholine; Chol, Cholesterol; DSPE-PEG, Distearoylphosphatidylethanolamine-poly(ethylene glycol); DOX, Doxorubicin; ICG, Indocyanine green; NIR, Near infrared; PLGA, Poly(lactic-co-glycolic acid); NP, Nanoparticle; TLR7, Toll-like receptor 7; UCNP, Upconversion nanoparticle; AuN, Gold nanocage; IC50, Half maximal inhibitory concentration; MSNP, Mesoporous silica nanoparticle.

4.2.6 Other Cell Membrane-Based Nanoparticles

Recently, hybrid cell-membrane-coated nanoparticles that gather the function of different cell types have been studied, exhibiting long circulation times and self-targeting properties [7, 52, 53]. It was observed that it is feasible to fuse platelet membranes and erythrocyte or leukocyte membranes, which may represent a cross-functional approach that retains the exclusive properties of each cell type. Dehaini et al. [52] explored the fusion of erythrocyte- and platelet-membranes to coat NPs for photothermal therapy, as a strategy to improve tumor distribution, achieve large photothermal efficacy, without compromising healthy tissues. A hybrid cell membrane nanoparticle to target CTCs was also developed, with the fusion of PLTs and leukocytes being used to coat magnetic beads for theranostic therapy [7]. This hybrid cell membrane system possesses the advantages of both cells: they can recognize and target CTCs (CTC–PLT interactions) but do not form clusters in circulation (CTC leukocyte interactions). Both studies showed that these dual membrane coated-NPs exhibited in vivo properties consistent with expectations. Using liposomes, He et al. [14] developed a biomimetic nanosystem that combines liposomes with plasma membrane components derived from both leukocytes (murine J774A.1 cells) and tumor cells (head and neck tumor cells HN12). The so-called "Leutusome" displayed original surface biomarkers from both cell types, allowing a superior targeting capability to tumor cells.

The same generic purposes of cell membrane coating can be applied to other cell types, such as cancer-associated fibroblasts. In fact, Liu et al. [57] designed an organic phototheranostics system, coated by cancer-associated fibroblasts. These NPs preferentially target cancer-associated fibroblasts, resulting in a higher accumulation in tumor tissues, when compared to the uncoated nanosystem. This approach demonstrated the ability of this kind of NPs in targeting components of the tumor microenvironment.

4.3 Conclusions

Despite the proven advantages of nanoparticle-based systems over conventional therapies, a successful translation to clinic still remains limited, due to the fast identification of these foreign bodies by immune defense mechanisms. Cell membrane coating of nanoparticles has been shown to outperform simpler systems, circumventing their limitations. These smarter biomimetic approaches are being explored due to their intrinsic biocompatibility and stealth or targeting properties. In fact, they may be naturally confused with regular cells, improving the biodistribution of nanoparticles or, when applied, promote a cancer cell-like stimulus, activating a desirable immune response or delivering their payload to tumor cells.

This kind of surface functionalization may take advantage of the biological characteristics of the patient, namely, surface markers and proteins, thus showing a high potential use in personalized medicine, with promising results regarding the progression and efficacy of the anticancer treatment.

Acknowledgments

The authors acknowledge Fundação para a Ciência e a Tecnologia (FCT), the Portuguese Agency for Scientific Research, for financial support through the Research Project no. 016648 (Ref. POCI-01-0145-FEDER-016648), the project PEst-UID/NEU/04539/2013, and COMPETE (Ref. POCI-01-0145-FEDER-007440). The Coimbra Chemistry Centre is supported by FCT, through the Project PEst-OE/QUI/UI0313/2014 and POCI-01-0145-FEDER-007630. Maria Mendes acknowledges the PhD research grant SFRH/BD/133996/2017 assigned by FCT.

References

1. Anselmo, A. C., and Mitragotri, S. (2016). Nanoparticles in the clinic. *Bioeng. Transl. Med.,* **1**, pp. 10–29.
2. Dehaini, D., Fang, R. H., and Zhang, L. (2016). Biomimetic strategies for targeted nanoparticle delivery. *Bioeng. Transl. Med.,* **1**, pp. 30–46.
3. Mendes, M., Sousa, J., Pais, A., and Vitorino, C. (2018). Clinical applications of nanostructured drug delivery systems: From basic research to translational medicine. In *Core-Shell Nanostructures for Drug Delivery and Theranostics*, pp. 43–116. Elsevier.

4. Narain, A., Asawa, S., Chhabria, V., and Patil-Sen, Y. (2017). Cell membrane coated nanoparticles: Next-generation therapeutics. *Nanomedicine,* **12**, pp. 2677–2692.

5. Jing, L., Qu, H., Wu, D., Zhu, C., Yang, Y., Jin, X., Zheng, J., Shi, X., Yan, X., and Wang, Y. (2018). Platelet-camouflaged nanococktail: Simultaneous inhibition of drug-resistant tumor growth and metastasis via a cancer cells and tumor vasculature dual-targeting strategy. *Theranostics,* **8**, p. 2683.

6. Hu, Q., Sun, W., Qian, C., Wang, C., Bomba, H. N., and Gu, Z. (2015). Anticancer platelet-mimicking nanovehicles. *Adv. Mater.,* **27**, pp. 7043–7050.

7. Rao, L., Meng, Q., Huang, Q., Wang, Z., Yu, G., Li, A., Ma, W., Zhang, N., Guo, S., and Zhao, X. (2018). Platelet–leukocyte hybrid membrane-coated immunomagnetic beads for highly efficient and highly specific isolation of circulating tumor cells. *Adv. Funct. Mater.,* **28**, p. 1803531.

8. Hu, C.-M. J., Zhang, L., Aryal, S., Cheung, C., Fang, R. H., and Zhang, L. (2011). Erythrocyte membrane-camouflaged polymeric nanoparticles as a biomimetic delivery platform. *Proc. Natl. Acad. Sci.,* **108**, pp. 10980–10985.

9. Parodi, A., Quattrocchi, N., van de Ven, A. L., Chiappini, C., Evangelopoulos, M., Martinez, J. O., Brown, B. S., Khaled, S. Z., Yazdi, I. K., Enzo, M. V., Isenhart, L., Ferrari, M., and Tasciotti, E. (2013). Synthetic nanoparticles functionalized with biomimetic leukocyte membranes possess cell-like functions. *Nat. Nanotechnol.,* **8**, pp. 61–68.

10. Jia, Y., Sheng, Z., Hu, D., Yan, F., Zhu, M., Gao, G., Wang, P., Liu, X., Wang, X., and Zheng, H. (2018). Highly penetrative liposome nanomedicine generated by a biomimetic strategy for enhanced cancer chemotherapy. *Biomater. Sci.,* **6**, pp. 1546–1555.

11. Gao, C., Lin, Z., Jurado-Sánchez, B., Lin, X., Wu, Z., and He, Q. (2016). Stem cell membrane-coated nanogels for highly efficient in vivo tumor targeted drug delivery. *Small,* **12**, pp. 4056–4062.

12. Luk, B. T., and Zhang, L. (2015). Cell membrane-camouflaged nanoparticles for drug delivery. *J. Control. Release,* **220**, pp. 600–607.

13. Bose, R. J. C., Lee, S.-H., and Park, H. (2016). Biofunctionalized nanoparticles: An emerging drug delivery platform for various disease treatments. *Drug Discov. Today,* **21**, pp. 1303–1312.

14. He, H., Guo, C., Wang, J., Korzun, W. J., Wang, X.-Y., Ghosh, S., and Yang, H. (2018). Leutusome: A biomimetic nanoplatform integrating plasma membrane components of leukocytes and tumor cells for remarkably enhanced solid tumor homing. *Nano Lett.,* **18,** pp. 6164–6174.

15. Knop, K., Hoogenboom, R., Fischer, D., and Schubert, U. S. (2010). Poly (ethylene glycol) in drug delivery: Pros and cons as well as potential alternatives. *Angew. chem. Int. Ed.,* **49,** pp. 6288–6308.

16. Föller, M., Huber, S. M., and Lang, F. (2008). Erythrocyte programmed cell death. *IUBMB Life,* **60,** pp. 661–668.

17. Oldenborg, P.-A., Zheleznyak, A., Fang, Y.-F., Lagenaur, C. F., Gresham, H. D., and Lindberg, F. P. (2000). Role of CD47 as a marker of self on red blood cells. *Science,* **288,** pp. 2051-2054.

18. Hu, C.-M. J., Fang, R. H., Luk, B. T., Chen, K. N. H., Carpenter, C., Gao, W., Zhang, K., and Zhang, L. (2013). 'Marker-of-self' functionalization of nanoscale particles through a top-down cellular membrane coating approach. *Nanoscale,* **5,** pp. 2664–2668.

19. Piao, J.-G., Wang, L., Gao, F., You, Y.-Z., Xiong, Y., and Yang, L. (2014). Erythrocyte membrane is an alternative coating to polyethylene glycol for prolonging the circulation lifetime of gold nanocages for photothermal therapy. *ACS Nano,* **8,** pp. 10414–10425.

20. Eylar, E. H., Madoff, M. A., Brody, O. V and Oncley, J. L. (1962). The contribution of sialic acid to the surface charge of the erythrocyte. *J. Biol. Chem.,* **237,** pp. 1992–2000.

21. Schauer, R. (2009). Sialic acids as regulators of molecular and cellular interactions. *Curr. Opin. Struct. Biol.,* **19,** pp. 507–514.

22. Kumar, D., and Rizvi, S. I. (2013). Erythrocyte membrane bound and plasma sialic acid during aging. *Biologia,* **68,** pp. 762–765.

23. Luk, B. T., Hu, C.-M. J., Fang, R. H., Dehaini, D., Carpenter, C., Gao, W., and Zhang, L. (2014). Interfacial interactions between natural RBC membranes and synthetic polymeric nanoparticles. *Nanoscale,* **6,** pp. 2730–2737.

24. Danhier, F., Ansorena, E., Silva, J. M., Coco, R., Le Breton, A., and Préat, V. (2012). PLGA-based nanoparticles: An overview of biomedical applications. *J. Control. Release,* **161,** pp. 505–522.

25. Luk, B. T., Fang, R. H., Hu, C.-M. J., Copp, J. A., Thamphiwatana, S., Dehaini, D., Gao, W., Zhang, K., Li, S., and Zhang, L. (2016). Safe and immunocompatible nanocarriers cloaked in RBC membranes for drug delivery to treat solid tumors. *Theranostics,* **6,** p. 1004.

26. Fang, R. H., Hu, C.-M. J., Chen, K. N. H., Luk, B. T., Carpenter, C. W., Gao, W., Li, S., Zhang, D.-E., Lu, W., and Zhang, L. (2013). Lipid-insertion enables targeting functionalization of erythrocyte membrane-cloaked nanoparticles. *Nanoscale,* **5**, pp. 8884–8888.

27. Chai, Z., Hu, X., Wei, X., Zhan, C., Lu, L., Jiang, K., Su, B., Ruan, H., Ran, D., Fang, R. H., Zhang, L., and Lu, W. (2017). A facile approach to functionalizing cell membrane-coated nanoparticles with neurotoxin-derived peptide for brain-targeted drug delivery. *J. Control. Release,* **264**, pp. 102–111.

28. Cremel, M., Guérin, N., Horand, F., Banz, A., and Godfrin, Y. (2013). Red blood cells as innovative antigen carrier to induce specific immune tolerance. *Int. J. Pharm.,* **443**, pp. 39–49.

29. Banz, A., Cremel, M., Rembert, A., and Godfrin, Y. (2010). In situ targeting of dendritic cells by antigen-loaded red blood cells: A novel approach to cancer immunotherapy. *Vaccine,* **28**, pp. 2965–2972.

30. Guo, Y., Wang, D., Song, Q., Wu, T., Zhuang, X., Bao, Y., Kong, M., Qi, Y., Tan, S., and Zhang, Z. (2015). Erythrocyte membrane-enveloped polymeric nanoparticles as nanovaccine for induction of antitumor immunity against melanoma. *ACS Nano,* **9**, pp. 6918–6933.

31. Sharma, A., Goyal, A. K., and Rath, G. (2018). Recent advances in metal nanoparticles in cancer therapy. *J. Drug Target.,* **26**, pp. 617–632.

32. Ren, X., Zheng, R., Fang, X., Wang, X., Zhang, X., Yang, W., and Sha, X. (2016). Red blood cell membrane camouflaged magnetic nanoclusters for imaging-guided photothermal therapy. *Biomaterials,* **92**, pp. 13–24.

33. Pierigè, F., Serafini, S., Rossi, L., and Magnani, M. (2008). Cell-based drug delivery. *Adv. Drug Deliv. Rev.,* **60**, pp. 286–95.

34. Coussens, L. M., and Werb, Z. (2002). Inflammation and cancer. *Nature,* **420**, pp. 860–867.

35. Firestein, G. S. (2012). Mechanisms of inflammation and tissue repair. In *Goldman's Cecil Medicine,* pp. 230–235. Elsevier.

36. Coussens, L. M., and Werb, Z. (2001). Inflammatory cells and cancer. *J. Exp. Med.,* **193**, pp. F23–F26.

37. Li, R., He, Y., Zhang, S., Qin, J., and Wang, J. (2018). Cell membrane-based nanoparticles: A new biomimetic platform for tumor diagnosis and treatment. *Acta Pharm. Sin. B,* **8**, pp. 14–22.

38. Cao, H., Dan, Z., He, X., Zhang, Z., Yu, H., Yin, Q., and Li, Y. (2016). Liposomes coated with isolated macrophage membrane can target lung metastasis of breast cancer. *ACS Nano,* **10**, pp. 7738–7748.

39. Xuan, M., Shao, J., Dai, L., Li, J., and He, Q. (2016). Macrophage cell membrane camouflaged Au nanoshells for in vivo prolonged circulation life and enhanced cancer photothermal therapy. *ACS Appl. Mater. Interfaces,* **8**, pp. 9610–9618.

40. Rao, L., He, Z., Meng, Q.-F., Zhou, Z., Bu, L.-L., Guo, S.-S., Liu, W., and Zhao, X.-Z. (2017). Effective cancer targeting and imaging using macrophage membrane-camouflaged upconversion nanoparticles. *J. Biomed. Mater. Res. Part A,* **105**, pp. 521–530.

41. Lai, P.-Y., Huang, R.-Y., Lin, S.-Y., Lin, Y.-H., and Chang, C.-W. (2015). Biomimetic stem cell membrane-camouflaged iron oxide nanoparticles for theranostic applications. *RSC Adv.,* **5**, pp. 98222–98230.

42. Ullah, I., Subbarao, R. B., and Rho, G. J. (2015). Human mesenchymal stem cells–current trends and future prospective. *Biosci. Rep.,* **35**, p. e00191.

43. Majumdar, M. K., Keane-Moore, M., Buyaner, D., Hardy, W. B., Moorman, M. A., McIntosh, K. R., and Mosca, J. D. (2003). Characterization and functionality of cell surface molecules on human mesenchymal stem cells. *J. Biomed. Sci.,* **10**, pp. 228–241.

44. Harker, L. A., Roskos, L. K., Marzec, U. M., Carter, R. A., Cherry, J. K., Sundell, B., Cheung, E. N., Terry, D., and Sheridan, W. (2000). Effects of megakaryocyte growth and development factor on platelet production, platelet life span, and platelet function in healthy human volunteers. *Blood,* **95**, pp. 2514–2522.

45. Yang, J., Wang, S., Liu, P., Dai, L., Chen, B., Luan, J., and Zhou, J. (2017). Platelet-inspired medicine for tumor therapy. *Oncotarget,* **8**, p. 115748.

46. Nash, G. F., Turner, L. F., Scully, M. F., and Kakkar, A. K. (2002). Platelets and cancer. *Lancet Oncol.,* **3**, pp. 425–430.

47. Yan, M., and Jurasz, P. (2016). The role of platelets in the tumor microenvironment: From solid tumors to leukemia. *Biochim. Biophys. Acta (BBA)-Molecular Cell Res.,* **1863**, pp. 392–400.

48. Li, J., Ai, Y., Wang, L., Bu, P., Sharkey, C. C., Wu, Q., Wun, B., Roy, S., Shen, X., and King, M. R. (2016). Targeted drug delivery to circulating tumor cells via platelet membrane-functionalized particles. *Biomaterials,* **76**, pp. 52–65.

49. Patel, S. R., Hartwig, J. H., and Italiano, J. E. (2005). The biogenesis of platelets from megakaryocyte proplatelets. *J. Clin. Invest.,* **115**, pp. 3348–3354.

50. Xu, L., Gao, F., Fan, F., and Yang, L. (2018). Platelet membrane coating coupled with solar irradiation endows a photodynamic

nanosystem with both improved antitumor efficacy and undetectable skin damage. *Biomaterials,* **159**, pp. 59–67.

51. Rao, L., Bu, L., Meng, Q., Cai, B., Deng, W., Li, A., Li, K., Guo, S., Zhang, W., and Liu, W. (2017). Antitumor platelet-mimicking magnetic nanoparticles. *Adv. Funct. Mater.,* **27**, p. 1604774.

52. Khaldoyanidi, S. K., Glinsky, V. V., Sikora, L., Glinskii, A. B., Mossine, V. V., Quinn, T. P., Glinsky, G. V., and Sriramarao, P. (2003). MDA-MB-435 human breast carcinoma cell homo- and heterotypic adhesion under flow conditions is mediated in part by Thomsen-Friedenreich antigen-galectin-3 interactions. *J. Biol. Chem.,* **278**, pp. 4127–4134.

53. Chen, Z., Zhao, P., Luo, Z., Zheng, M., Tian, H., Gong, P., Gao, G., Pan, H., Liu, L., Ma, A., Cui, H., Ma, Y., and Cai, L. (2016). Cancer cell membrane–biomimetic nanoparticles for homologous-targeting dual-modal imaging and photothermal therapy. *ACS Nano,* **10**, pp. 10049–10057.

54. Glinsky, V. V., Glinsky, G. V., Glinskii, O. V., Huxley, V. H., Turk, J. R., Mossine, V. V., Deutscher, S. L., Pienta, K. J., and Quinn, T. P. (2003). Intravascular metastatic cancer cell homotypic aggregation at the sites of primary attachment to the endothelium. *Cancer Res.,* **63**, pp. 3805–3811.

55. Rao, L., Bu, L.-L., Cai, B., Xu, J.-H., Li, A., Zhang, W.-F., Sun, Z.-J., Guo, S.-S., Liu, W., Wang, T.-H., and Zhao, X.-Z. (2016). Cancer cell membrane-coated upconversion nanoprobes for highly specific tumor imaging. *Adv. Mater.,* **28**, pp. 3460–3466.

56. Fang, R. H., Hu, C.-M. J., Luk, B. T., Gao, W., Copp, J. A., Tai, Y., O'Connor, D. E., and Zhang, L. (2014). Cancer cell membrane-coated nanoparticles for anticancer vaccination and drug delivery. *Nano Lett.,* **14**, pp. 2181–2188.

57. Li, J., Zhen, X., Lyu, Y., Jiang, Y., Huang, J., and Pu, K. (2018). Cell membrane coated semiconducting polymer nanoparticles for enhanced multimodal cancer phototheranostics. *ACS Nano,* **12**, pp. 8520–8530.

58. Aryal, S., Hu, C.-M. J., Fang, R. H., Dehaini, D., Carpenter, C., Zhang, D.-E., and Zhang, L. (2013). Erythrocyte membrane-cloaked polymeric nanoparticles for controlled drug loading and release. *Nanomedicine,* **8**, pp. 1271–1280.

59. Fu, Q., Lv, P., Chen, Z., Ni, D., Zhang, L., Yue, H., Yue, Z., Wei, W., and Ma, G. (2015). Programmed co-delivery of paclitaxel and doxorubicin boosted by camouflaging with erythrocyte membrane. *Nanoscale,* **7**, pp. 4020–4030.

60. Rao, L., Meng, Q., Huang, Q., Liu, P., Bu, L., Kondamareddy, K. K., Guo, S., Liu, W., and Zhao, X. (2016). Photocatalytic degradation of cell membrane coatings for controlled drug release. *Adv. Healthc. Mater.,* **5,** pp. 1420–1427.

61. Gao, S., Zheng, P., Li, Z., Feng, X., Yan, W., Chen, S., Guo, W., Liu, D., Yang, X., Wang, S., Liang, X.-J., and Zhang, J. (2018). Biomimetic O_2-Evolving metal-organic framework nanoplatform for highly efficient photodynamic therapy against hypoxic tumor. *Biomaterials,* **178,** pp. 83–94.

62. Liu, T., Shi, C., Duan, L., Zhang, Z., Luo, L., Goel, S., Cai, W., and Chen, T. (2018). A highly hemocompatible erythrocyte membrane-coated ultrasmall selenium nanosystem for simultaneous cancer radiosensitization and precise antiangiogenesis. *J. Mater. Chem. B,* **6,** pp. 4756–4764.

63. Wang, Q., Ren, Y., Mu, J., Egilmez, N. K., Zhuang, X., Deng, Z., Zhang, L., Yan, J., Miller, D., and Zhang, H.-G. (2015). Grapefruit-derived nanovectors use an activated leukocyte trafficking pathway to deliver therapeutic agents to inflammatory tumor sites. *Cancer Res.,* **75,** pp. 2520–2529.

64. Krishnamurthy, S., Gnanasammandhan, M. K., Xie, C., Huang, K., Cui, M. Y., and Chan, J. M. (2016). Monocyte cell membrane-derived nanoghosts for targeted cancer therapy. *Nanoscale,* **8,** pp. 6981–6985.

65. Kang, T., Zhu, Q., Wei, D., Feng, J., Yao, J., Jiang, T., Song, Q., Wei, X., Chen, H., Gao, X., and Chen, J. (2017). Nanoparticles coated with neutrophil membranes can effectively treat cancer metastasis. *ACS Nano,* **11,** pp. 1397–1411.

66. Zhang, L., Li, R., Chen, H., Wei, J., Qian, H., Su, S., Shao, J., Wang, L., Qian, X.-P., and Liu, B. (2017). Human cytotoxic T-lymphocyte membrane-camouflaged nanoparticles combined with low-dose irradiation: A new approach to enhance drug targeting in gastric cancer. *Int. J. Nanomed.,* **12,** pp. 2129–2142.

67. Xuan, M., Shao, J., Dai, L., He, Q., and Li, J. (2015). Macrophage cell membrane camouflaged mesoporous silica nanocapsules for in vivo cancer therapy. *Adv. Healthc. Mater.,* **4,** pp. 1645–1652.

68. Xu, H.-L., Shen, B.-X., Lin, M.-T., Tong, M.-Q., Zheng, Y.-W., Jiang, X., Yang, W.-G., Yuan, J.-D., Yao, Q., and Zhao, Y.-Z. (2018). Homing of ICG-loaded liposome inlaid with tumor cellular membrane to the homologous xenografts glioma eradicates the primary focus and prevents lung metastases through phototherapy. *Biomater. Sci.,* **6,** pp. 2410–2425.

69. Yang, R., Xu, J., Xu, L., Sun, X., Chen, Q., Zhao, Y., Peng, R., and Liu, Z. (2018). Cancer cell membrane-coated adjuvant nanoparticles with mannose modification for effective anticancer vaccination. *ACS Nano,* **12**, pp. 5121–5129.

70. Sun, H., Su, J., Meng, Q., Yin, Q., Chen, L., Gu, W., Zhang, Z., Yu, H., Zhang, P., Wang, S., and Li, Y. (2017). Cancer cell membrane-coated gold nanocages with hyperthermia-triggered drug release and homotypic target inhibit growth and metastasis of breast cancer. *Adv. Funct. Mater.,* **27**, p. 1604300.

71. Shao, D., Li, M., Wang, Z., Zheng, X., Lao, Y.-H., Chang, Z., Zhang, F., Lu, M., Yue, J., Hu, H., Yan, H., Chen, L., Dong, W., and Leong, K. W. (2018). Bioinspired diselenide-bridged mesoporous silica nanoparticles for dual-responsive protein delivery. *Adv. Mater.,* **30**, p. 1801198.

Chapter 5

Delivery of Immune Checkpoint Inhibitors Using Nanoparticles

Abdullah Shaito and Houssein Hajj Hassan

Department of Biological and Chemical Sciences,
Faculty of Arts and Sciences, Lebanese International University,
Beirut 1105, Lebanon

abdallah.shaito@liu.edu.lb, houssein.hajjhassan@liu.edu.lb

Currently immune checkpoint inhibition is one of the major targets of anticancer therapy. However, immune checkpoint inhibition therapy has the limitations of low efficacy, systemic adverse side effects and high cost. Several strategies are being followed to try to address these limitations. One way to minimize the limitations of immune checkpoint blockade is to enhance drug delivery. Nanoparticles represent a delivery method that can enhance the efficacy of immune checkpoint inhibitors and decrease their side effects. The aim of this chapter is to provide an overview of different immunotherapies, and the different drug delivery

Nanoparticle Drug Delivery Systems for Cancer Treatment
Edited by Hala Gali-Muhtasib and Racha Chouaib

ISBN 978-981-4800-54-9 (Hardcover), 978-0-429-34125-0 (eBook)
www.jennystanford.com

systems with focus on nanoparticles. Finally, an update on the use of nanoparticles as delivery vehicles and the potential solutions they offer for immune checkpoint inhibition therapy will be presented.

5.1 Cancer Immunotherapy

Methods of cancer treatment include surgical removal of the cancer, radiotherapy, stem cell transplantation or anticancer drug therapy that include chemotherapy, targeted therapy, hormone therapy, or immune therapy [1].

Currently, immunotherapy is one of the most promising cancer treatments [2] and the development of new immunotherapies has become a necessity [3, 4]. In recent years, immunotherapy has become widespread and has been used to treat both hematological and solid cancers [2]. Immunotherapy is a biological therapy that involves activation of the immune system to target and kill cancer cells through different approaches. Promising immunotherapy approaches include adoptive cell transfer, therapeutic monoclonal antibodies (mAbs), treatment vaccines, cytokine treatment using interferons and interleukins, Bacillus Calmette Guérin (BCG), which is a weakened bacterium used in the treatment of bladder cancer, and immune checkpoint inhibitors. Chimeric antigen receptor therapy also known as CAR T-cell therapy has stood out as a clinically effective type of adoptive cell transfer therapy. Immune checkpoint inhibitors, in particular, have shown potential in the treatment of several cancers and have been FDA approved for the treatment of melanoma (recurrent and/or metastatic), non-small cell lung cancers (NSCLCs), genitourinary cancers (GUCs), head and neck cancers (HNCs), renal cell carcinomas, urothelial carcinomas, non-Hodgkin lymphomas and other cancers [5].

In essence, immunotherapy against cancer entails the use of naturally or synthetically derived components to activate or enhance an immune response against cancer cells. [6]. Historically, there are many reports on tumor clearance following pathogen infection. In fact, there were several trials, dating back to the late 19th century, that used infections as therapy for cancer. The basic concept behind such trials was the possibility that the infection would stimulate an immune response, leading to an immune rejection of the tumor. In 1891, the first known immunotherapeutic trial in humans was performed by an American surgeon, William B. Coley, who inoculated "erysipelas," i.e., live cultured Streptococci, into a sarcoma patient stimulating an antitumor immune response [7]. Also, German clinicians, 150 years ago, had seriously attempted this strategy of infectious therapy [8, 9]. The clinical outcome of such an infectious therapy was varied, until finally it resulted in the use of BCG to treat patients with bladder tumors [10]. Most importantly, Coley's trials founded the basis for investigating the interaction between the immune system and tumor cells and raised the possibility of cancer therapy by utilizing this interaction.

The immune system is continuously under a complex and delicate balance between the identification and then elimination of the "non-self" foreign antigens as well as the necessity of suppressing an excessive uncontrolled immune response that may attack "self" antigens. Immune system eradication of cancer cells requires regulation of T cell-mediated responses by foreign or endogenous antigen presenting cells (APCs) such as dendritic cells (DCs) [11, 12]. Cancer cells continuously produce neo-antigens due to genetic mutations and other irregular cellular processes in cancer cells [13]. These neo-antigens are not recognized as "self" by the normal host defenses. The cancer cell neo-antigens lead to production of cytotoxic T-cells (CTLs). CTLs can identify neo-antigens presented by cancer cells, and then eliminate the cancer cells. In the case of cancer-initiated immune responses, DCs process and present the tumor antigen (usually peptides from tumor digested proteins) bound to major histocompatibility class I (MHCI) complexes [14]. The "self" and "non-self" antigens are bound to MHCI complexes and presented on APCs and are recognized by CD8+ effector T cells, or CTLs. But, an immune response is rarely initiated against a cancer cell. This is due, at least partly, to the immunosuppressive mechanisms that are usually acting to inhibit over-activated and damaging immune reactions [15].

A single signal is not enough for T cell activation; a T cell is not activated when the TCR (T cell receptor) binds an antigen-bound MHCI complex. T cell activation requires an additional signal viz. the binding of co-receptors. This second signal has been termed co-stimulation. In T cells, the second signal takes place when CD28 present on T cells binds to CD86 (B7-2) or CD80 (B7-1) expressed by APCs [16–18]. These co-receptors or co-stimulants, when active, act as accelerators that promote the T cell activation and, hence, the immune response. However, it was later discovered that some co-receptors act, in fact, as brakes that reduce the activation of T cells. These brake co-receptors are required, as mentioned above, to avoid an exaggerated and damaging immune response. The brakes inhibitory signaling pathways, are termed the immune checkpoints, and constitute any signal that inhibit priming of T cell activation or any signal that reduce an already initiated T cell response [19].

In addition, cancer cells can evade surveillance by the immune system. Evasion of immune system surveillance has been recently classified as a hallmark of cancer cells by Hanahan and Weinberg in their updated description of hallmarks of cancer [20]. Overall, mounting the immune response against cancer cells will require decreased evasion of immune surveillance, as well as, enhanced T cell stimulatory processes and reduced T cell inhibitory processes in such a way that a robust anticancer immunity is achieved with minimal potentially harmful side effects [5, 21].

The cancer-immunity cycle was proposed by Chen and Mellman as a cohesive model of processes that lead to promotion of anticancer immunity. The cancer-immunity cycle model is a sequential association of the events that should repeatedly occur in order to mount an efficient immune reaction to cancer cells [21]. The steps of a cancer-immunity cycle are illustrated in Fig. 5.1.

As shown in Fig. 5.1, this cycle does not usually proceed due to the fact that T cell activation is usually dampened by the immune checkpoints acting as brakes. As a result, immunosuppressive inhibitory processes such as activation of immune checkpoint inhibitors that reduce cancer cell recognition by CTLs, must be inhibited in order for the cycle to progress and last. The discovery of these immune inhibitory processes in the cancer-immunity cycle has resulted in the design of new therapeutic tools that aim to overcome the immune inhibitory processes and fight cancer cells as will be detailed in the section about immune checkpoint blockade.

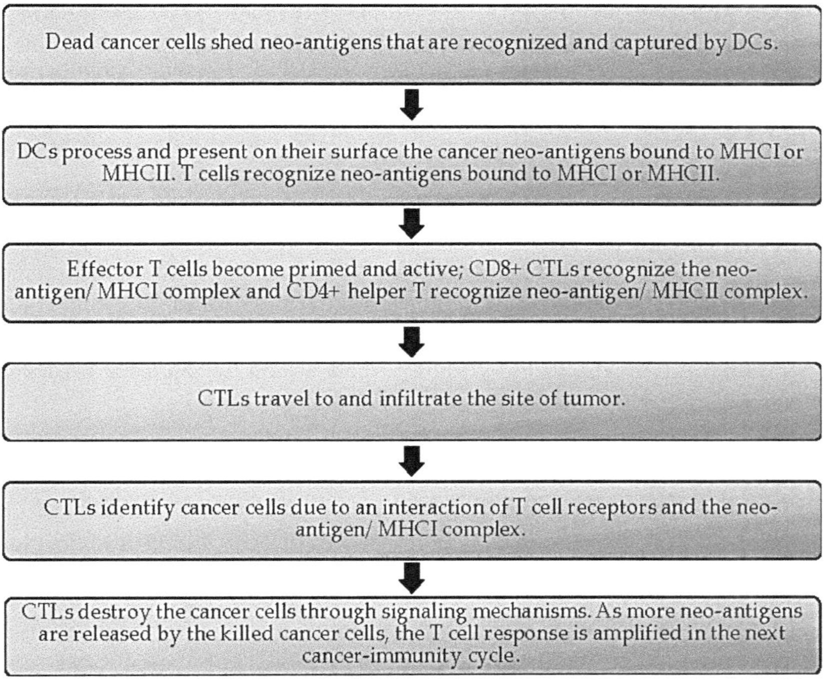

Figure 5.1 Cancer-immunity cycle.

5.2 Immune Checkpoint Inhibition

The first protein to be identified as an immune checkpoint molecule is CTLA-4 (cytotoxic T lymphocyte antigen-4). In 1987, a cDNA expressed in T cells, CTLA-4 (CD152), was cloned in the laboratory of Pierre Golstein [22]. CD28 and CTLA-4 were found to be structurally related and belong to the immunoglobulin superfamily. The function of CTLA-4 was unknown, but later it was found that CTLA-4 protein resides inside inactive T cells and rapidly translocates to the cell plasma membrane following T cell stimulation [23, 24]. Furthermore, CTLA-4 was found to recognize and bind CD80, in a similar way as CD28 but with higher affinity [25]. As a result, it was believed that CTLA-4 had a costimulatory effect on T cells that is similar to CD28. Subsequently, it was shown that CTLA-4 was in reality a negative regulator of T cell stimulation [26, 27]. Later on, the *ctla-4* gene was knocked-out in mice, the knock-out mice developed very severe autoimmune symptoms due to excessive T cell proliferation [28, 29]. Consequently, it became clear that CTLA-4 acts as a co-inhibitory molecule and that the response of T cells to stimulation is regulated not only by CD28 costimulatory signals, but also by CTLA-4 inhibitory signals.

Another immune checkpoint regulation is through PD-1 (CD279) and its ligand molecules PDL-1 (B7-H1, CD274) and PDL-2 (B7-DC, CD273). In the early 1990s, PD-1 was cloned by the group of Tasuku Honjo at Kyoto University in Japan [30]. This discovery took place at the peak of the identification of co-stimulation, but before identifying CTLA-4 as a co-inhibitor rather than a co-stimulator. Back then, Honjo and his group assumed that PD-1 is involved in pathways regulating apoptotic cell death, and they called it Programmed Cell Death-1, PD-1. PD-1 expression was demonstrated by stimulated T and B-cells as well as by immature T cells [31]. However, it took Honjo and his group around 10 years to uncover the real PD-1 function. *pd-1* was knocked-out in mice of several genetic backgrounds and it was determined that *pd-1* knock-out mice had different forms of autoimmune syndromes due to T cell over-proliferation. This phenotype is comparable to that seen in the *ctla-4* knock-out mice. Honjo concluded that PD-1 negatively regulated T cell activation [32–34]. In parallel, PDL-1 was identified as the ligand for PD-1 in both mice and humans [35, 36]. Soon after, PDL-2 was discovered as another PD-1 ligand that also prevents T-cell activation [37]. PDL-1 is present on macrophages, DCs and other immune cells, and on certain cancer cells as well. This leads to the conclusion that PD-1/PDL-1 or PD-1/PDL-2 signaling could be used by cancer cells to prevent an anticancer immune response [36]. Recently, it was confirmed that interactions between PD-1 on T cells and PDL-1 or PDL-2 play important roles in provoking T cell exhaustion, loss of effector functions, and dysfunction. In activated T cells, PD-1 can bind its ligand PDL-1, and the PD-1/PDL-1 complex can dampen or prevent the immune response via inhibition of CTL responses [38–40]. T cell exhaustion is a characteristic of chronic viral infections and cancer. Interestingly, in different animal models of viral infection, T cell function could be restored through blockade of PD-1, thus supporting the significance of PD-1 inhibitory pathway during a viral infection [41–44].

So far, more than twenty different checkpoint molecule pairs have been discovered, other than CTLA-4 and PD-1, including LAG-3/MHCII, TIGIT/CD155, and TIM3/Gal-9, which are expressed, with varying levels, not only on T-cells but by other immune cells as well [45–47].

Both of the CTLA-4 and the PD-1/PDL-1 immune checkpoints negatively regulate T cell activation yet in different ways. CTLA-4 engages with CD28 and dampens the costimulatory signal already initiated by CD28. Effectively, CTLA-4 puts a brake on activation of T cells already induced by CD28 co-stimulation or, in other words, it can inhibit the already primed T cells. PD-1, on the other hand, acts independent of CD28, and has the ability to directly inhibit activation of T cells. PD-1 can act even on non-primed T cells. Cancer cells can exploit the PD-1/PDL-1 pathway resulting in immunosuppression and T cell exhaustion, and thereby can evade immune anticancer responses. PD-1 or PDL-1 mAbs can disturb this T cell inhibitory signal, leading to reactivation of the anticancer responses of specific CTLs. Moreover, it was demonstrated that targeting CTLA-4, PD-1, or other immune checkpoint molecules can reverse T cell exhaustion and restore T cell functions and activity in chronic viral infections and during cancer [41, 48–53]. The mechanism of action of checkpoint inhibition during cancer immunotherapy is illustrated in Fig. 5.2.

Due to such discoveries, the 2018 Nobel Prize in Physiology or Medicine was jointly presented to Tasuku Honjo and James Allison for their discoveries of CTLA-4 and PD-1 immune checkpoints and for "their discovery of cancer therapy by inhibition of negative immune regulation." Subsequently, several immune checkpoints inhibitors have been developed and approved for therapy of several cancers as outlined in the time line of Fig. 5.3.

As shown in Fig. 5.1 different immune checkpoint antibodies were developed by several pharmaceutical companies, and those mainly target PD-1, or PDL-1 or CTLA-4. Ipilimumab, also known as Yervoy, is a CTLA-4 antibody that was first developed by James Allison and is now manufactured by Bristol-Myers Squibb (BMS). Nivolumab, also known as Opdivo, is a PD-1 mAb also manufactured by BMS. Pembrolizumab, known as Keytruda, is another PD-1 mAb produced by Merck. Genentech, on the other hand, developed an antibody against PDL-1, Atezolizumab (Tecentriq). Pfizer also developed Avelumab (Bavencio) as a PDL-1 antibody.

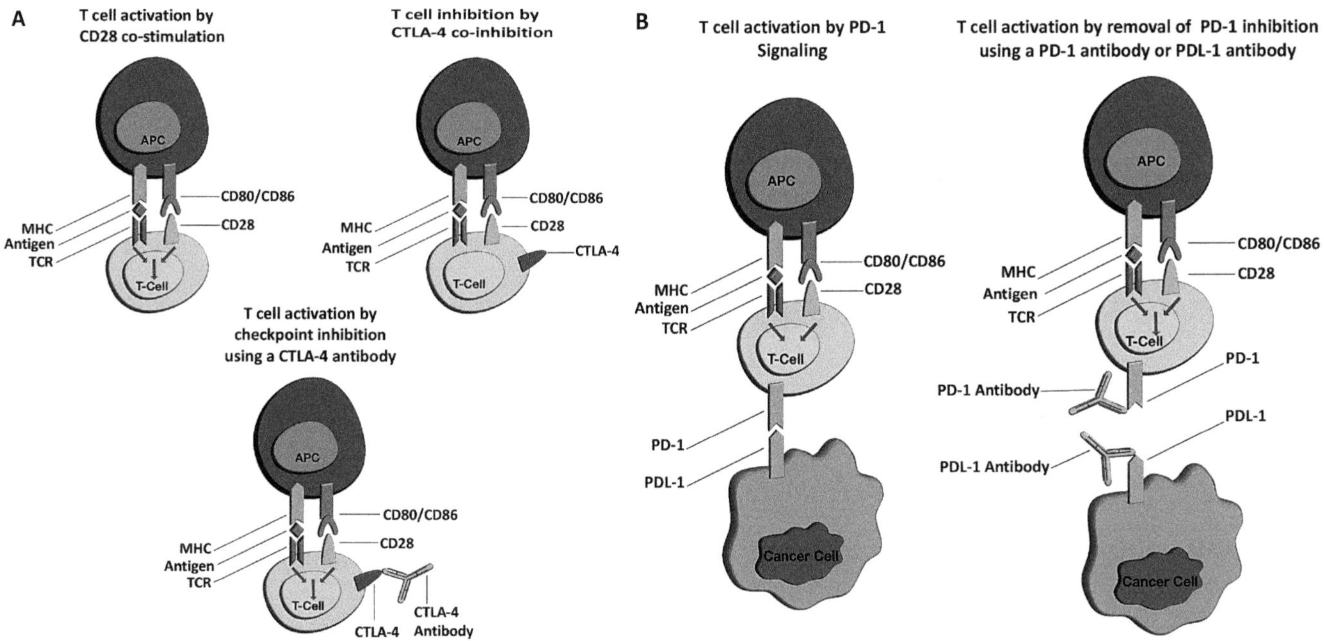

Figure 5.2 (A) The co-stimulation of T cells by CD28, their inhibition by CTLA-4 and their reactivation by CTLA-4 antibodies. (B) Inhibition of T cells by cancer cells using PD-1/PDL-1 signaling and reactivation of T cells by using PD-1 or PDL-1 antibodies.

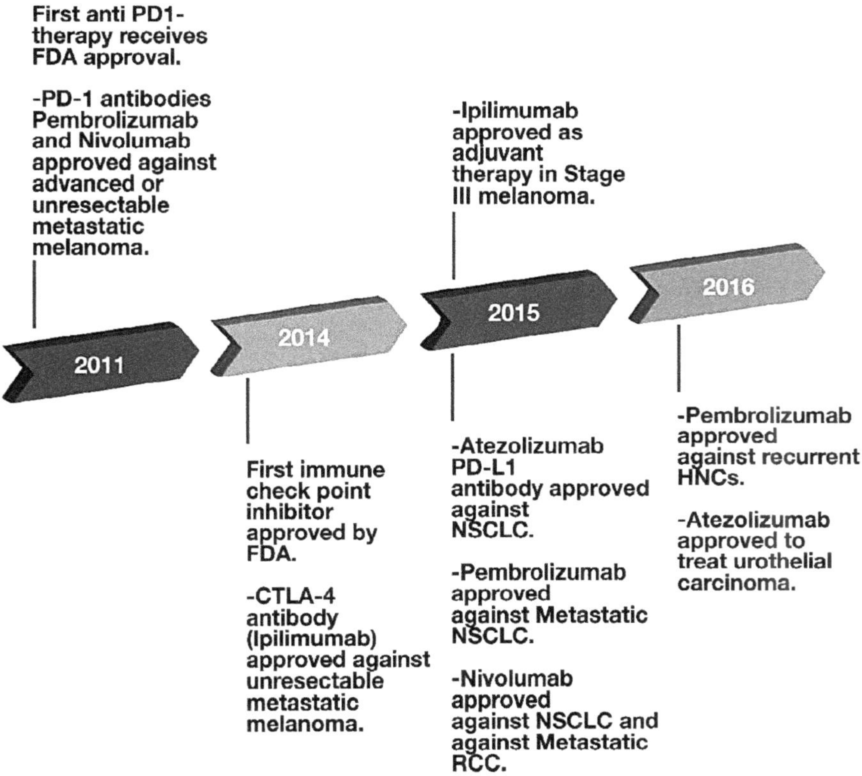

Figure 5.3 Time line showing the discovery and approval of different checkpoint inhibitors. NSCLC, non-small cell lung carcinoma; RCC, renal cell carcinoma; HNC, head and neck carcinoma.

Durvalumab, known as Imfinzi, is a PDL-1 antibody developed by AstraZeneca. These therapies have become the standard of care for melanoma, NSCLC, HNCs, and GUCs since they have demonstrated outcome improvements in numerous clinical trials. However, a major limitation of all of the above-mentioned therapies is their cost-effectiveness. For example, nivolumab therapy costs $140,672 per year [54].

As a result, and as discussed later, research is underway to uncover better delivery strategies of these drugs and this may help better target these drugs and save on the dosage.

While several of the antibodies against checkpoint inhibitors are now approved in clinical use, recent preclinical trials are probing checkpoint inhibition using means other than antibodies such as short interfering RNA (siRNA) against the different checkpoint inhibitor molecules [130]. siRNA checkpoint blockade therapy can be used as in vivo or ex vivo immunotherapy and has the potential to overcome some of the limitations of using antibodies. Aptamers against the different checkpoint inhibitors are also being developed. In recent studies, siRNA against check point molecules have been linked to aptamers [132]. In addition, protein traps of check point inhibitors that are encoded by engineered DNA plasmids to produce checkpoint inhibitor antagonist proteins are also being developed [127].

5.3 Drug Delivery in Medicine

To manage, control, and treat a disease, medicine counts on the use of pharmacologically active agents (drugs). The benefit of a drug is coupled to its way of administration. Success of drug delivery is linked to many factors including, and not restricted to, drug pharmacokinetics, duration of effect, distribution, excretion, absorption, metabolism and toxicity [55, 56]. Once a new drug is discovered, there is a need to improve its mode of delivery, and to evaluate its safety and efficacy.

However, drug delivery is not easily controlled and it is difficult to predict its release rate, and its tissue (or cell) specificity and stability. That is why the progress of medical, synthetic, and conjugate chemistry is highly trending in the medical field to provide more effective ways of drug delivery [56].

5.4 Drug Delivery Routes

Drug delivery routes are generally classified into the following categories [56, 57]: oral, transdermal, nasal, and parenteral. These drug delivery routes are compared in Table 5.1. The table also lists the advantages and disadvantages of using each drug delivery route.

Table 5.1 Advantages and disadvantages of the different drug delivery routes

Routes	Advantages	Disadvantages
Oral	• Easy • Preferred by patients • "Slow-release" preparations are available to extend duration of action • Drugs can be formulated for protection from digestive enzymes and acid	• Unsuitable in un-cooperative patients • Slow absorption of most orally administered drugs • Unpredictable absorption due to degradation by stomach acid and enzymes
Parenteral		
➢ ***Subcutaneous or intramuscular***	• Good absorption, especially for drugs with a low oral bioavailability • Onset is more rapid • Can have very long duration of action, e.g., depot antipsychotics and contraceptives	• Absorption is unpredictable at poorly perfused peripheries • Injections usually cause bruises and frighten children and needle-phobics
➢ ***Intravenous***	• Dependable and reproducible effects • Entire administered dose eaches the systemic circulation directly • Dose can be accurately titrated against response	• Requires a functioning cannula • More expensive and labor intensive than other routes • Cannulation is distressing to some patients, especially children • Cannulae are prone to infection • IV injection of drugs may cause local reactions
Transdermal	• Easy • Non-invasive • High levels of patient satisfaction	• Most drugs have a high molecular weight and are poorly lipid soluble, so are not absorbed via skin or mucous membranes • Very slow absorption
Nasal	• Very rapid absorption due to the huge surface area of the respiratory endothelium • Bronchodilators and inhaled steroids can be targeted to lungs with low levels of systemic absorption	• Bioavailability depends on patient's inhaling technique and the size of delivered drug particles

5.5 Classes of Drug Delivery Systems

Variable drug delivery systems (DDSs) were developed in the past 50 years. However, in recent years, researchers in the pharmaceutical industry are focusing on better targeted, safe, controlled, and efficient delivery systems. Most of the drugs used with recent DDSs consist mainly of protein and/or DNA-based drugs. These drugs are characterized by their ability to release their bioactive ingredients or active pharmaceutical ingredients (APIs) at the right time and site in safe concentrations (to control toxicity), which makes traditional drug delivery methods ineffective. Some of the recent delivery systems also allow for the control of drug release, which decreases drug cytotoxicity and immune reaction [58–61].

Most drug delivery systems are chosen to fit the physicochemical and biological properties of the drug. These properties include the molecular weight of the drug, its solubility in water, its cell permeability and the drug stability and half-life [57]. Therefore, development of any potential drug might face several obstacles including biocompatibility, crossing the blood-brain barrier and its route of delivery. Systemic drugs are diffused randomly in most tissues of the body; however, targeted drugs are routed to a specific organ or tissue, which enhances their efficacy, increases their specificity and decreases the risk of a global immune reaction.

DDS are, generally, classified into the following categories:

(i) Implantable delivery systems: They are introduced into body tissues directly and include polymeric devices of various shapes. These DDS have the drawbacks of irritation by the implants and being potential sites of infection. They have been successfully used in diabetes, cancer, and brain diseases [59, 61].

(ii) Variable release delivery systems: allow for the control of drug release thus avoiding a peak of drug level in the blood right upon administration that is followed by a later decrease in drug level. Maintaining a sustainable and precise supply of the drug in the system is sought after since it decreases drug cytotoxicity and immune reaction [58–61].

(iii) Carrier-based delivery systems: APIs in drugs exhibit low bioavailability and require protection inside the body from acid-catalyzed and enzyme-mediated degradation; hence the increasing importance of developing efficient carrier-based DDSs. In addition, a carrier-based DDS makes an API biocompatible. Biocompatibility is a pharmaceutical parameter described by "The ability of a material to perform with an appropriate host response in a specific situation" [62]. As such, drug biocompatibility is attained when the drug interacts with its target site without inducing unacceptable toxic, immunogenic, thrombogenic, and carcinogenic responses [63]. Also, to avoid physiological and pathological side effects, carrier-based DDS are made using biodegradable polymers such as PLGA [poly (D,L-lactide-co-glycolide)]. Moreover, a carrier-based DDS is usually suited for oral administration, which is the most preferred and convenient route for a DDS. Recent research has focused on enhancing carrier-based DDSs ability of crossing biological barriers such as the blood brain barrier and the gastrointestinal (GI) tract [64, 65].

Overall, carrier-based drug delivery is compatible with the physicochemical characteristics of APIs where the carriers are often designed to increase the load ability of APIs, reduce their side effects, and protect the API against host conditions. Encapsulation of APIs in carrier-based vesicular or particulate systems is often used to preserve the API physicochemical properties. This encapsulation is an innovative and alternative approach to minimize APIs side effects and maintain their efficacy. Furthermore, microencapsulation has been employed in the production of microspheres, manufactured with biocompatible polymers, that can entrap both hydrophilic or hydrophobic APIs [66–68]. Oftentimes, an additional factor for using these carriers is the ability to control the release of the API, and hence maintain effective therapeutic API levels over specified periods of time while minimizing systemic absorption of the APIs [67, 69]. The different classes of drug delivery systems are also summarized in Fig. 5.4.

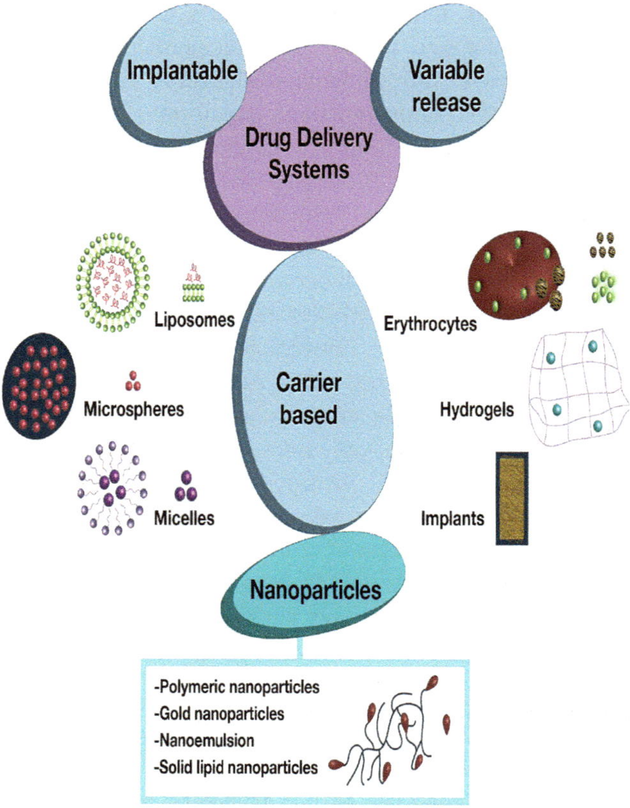

Figure 5.4 Drug delivery systems. Carrier-based drug delivery systems are shown with examples.

5.6 Drug Delivery Carriers

Drug carriers can be defined as any substance that can enhance delivery, effectiveness, specificity, or safety of a drug. In the past few years, innovations in drug delivery systems have allowed easier routes of delivery of carrier-based DDSs. Also, now drug carriers of carrier-based DDSs are used to target drugs to specific tissues, in what is called targeted drug delivery [70–73]. Different drug carriers are currently used in carrier-based DDSs due to their benefits and few limitations (as detailed in the above section about Carrier-based delivery systems). A large variety of drug carriers exists and includes liposomes, niosomes, lipospheres, vesicles, dendrimers, polymeric micelles, liquid crystals, microspheres, hydrogels, implants, nanoparticles, among others. Below we describe some of these drug carriers and discuss nanoparticles in detail.

5.6.1 Liposomes

Liposomes are sphere-shaped particles the size of which varies between two to tens of micrometers [74, 75]. Their bilayer membrane entraps associated compounds. The bilayer membrane phospholipids are derived either from pure artificial lipids or from phospholipids with mixed lipid chains and heads. A drug may be entrapped either at the liposome core or within the phospholipids bilayer. Liposomes are valuable in enhancing the therapeutic outcome of anticancer drugs by increasing drug concentrations at sites of rapidly growing cells and reducing the exposure of normal cells to the drug. Liposomes enhance the solubility, bioavailability, targeting sites and the prolonged control of a drug. They have been used in the delivery of drugs for cancer therapy, infectious diseases and for vaccine preparation. Their disadvantage is lower encapsulation capacity, and leakage with poor controlled release [76, 77].

5.6.2 Microspheres

Microspheres are synthesized from solid chemical matrices. They are small and spherical in shape and their size ranges between 1–300 μm. Many kinds of biodegradable polymers are used for the fabrication of microspheres like dextran, gelatin, albumin, starch, and polypropylene. Microspheres as drug delivery agents are affected by the polymer kind and the matrix size. The microsphere biodegradable polymer is filled with the drug that is dispersed homogenously throughout the core. When the polymer is degraded, the drug is released for the desired time. The precise benefits of microsphere delivery is that they can be injected or ingested, and provide sustained drug release and site-specific delivery. In addition, microspheres can be used for targeted endo-venous and intra-arterial drug delivery systems [70, 78, 79].

5.6.3 Polymeric Micelle Formulations

Polymeric micelle formulations consist of fine molecules that are simply controlled. Polymeric micelles contain an outer hydrophilic ring with an inner hydrophobic core. The inner core is stabilized by hydrophobic chemical polymers that contain hydrophilic chains that point towards the aqueous exterior. The outer ring surface acts as an interface between the inner core and the micelle surroundings. The inner core is capable of holding lipophilic APIs, and the whole micelle is held in the surrounding aqueous environment by the outer ring [74, 75, 80, 81].

5.6.4 Implants

Implants are injected or impregnated with biodegradable polymers and form very viscous liquids or semi-solid structures that can be administered directly within the body fluids. Implants can be used to supply elevated drug concentrations or to control and measure sustained API release at target sites. Implants are usually used when chronic medical aid is needed, for instance, during chemical castration of prostatic adenocarcinoma [82–84].

5.6.5 Hydrogels

Hydrogels exist as a network of water-insoluble polymer chains that sometimes form a colloidal gel when dispersed in a water medium. Hydrogels are photopolymers or copolymers and can be manufactured from natural or synthetic polymeric compounds. Hydrogels can absorb enormous amounts of water or biological fluids to form hydrophilic networks that are similar to biological tissues. Thus, hydrogels have decent biomedical applications; by fine-tuning the physicochemical properties of hydrogels, suitable drug delivery systems can be generated [59, 85].

5.6.6 Drug-Loaded Erythrocytes

This is a powerful drug delivery system. Patient erythrocytes can be isolated and loaded with different biologically active substances, and then returned to the patient's blood [86]. Biocompatibility and biodegradability are among the most important advantages offered by drug-loaded erythrocytes. In addition, they have the ability to carry a wide variation of drugs in large quantities using the small volume of the cells. Drug-loaded erythrocytes have limitations as well. They are eliminated in vivo by the reticuloendothelial system, which shortens their useful time as drug carriers. Also, some encapsulated substances can leak from the loaded erythrocytes [86].

5.7 Types of Nanoparticles

Nanoparticles (NPs) are solid amorphous or crystalline particles with size ranges between 10 and 200 nm. NPs are widely used as drug carriers mainly due to their stability and long-term half-life. They entrap or encapsulate a drug and protect it from degradation and from harsh chemical environments [87–89].

NPs and nanoformulations have already been used as DDSs with great success; nanoparticles as DDSs have been applied or can potentially be applied as antitumor therapy, gene therapy agents, and in radiotherapy. They have been used in the delivery of nucleic acids, proteins, vaccines, antibiotics, and as carriers to bypass the blood-brain barrier [76, 90, 91]. NPs exact mechanism of crossing into the brain is still not well known, however, it is thought that NPs brain crossing efficiency is determined by material composition, particle size, and structure of NPs. Certain NPs mimic molecules that are usually transported to the brain. Polysorbate-coated NPs mimic low-density lipoproteins (LDLs), enabling their transport through the capillary wall into the brain by "hitching a ride on the LDL receptor" [77].

Nanoparticles offer great advantages as DDSs with better drug delivery, release, and targeting. In addition, there is the potential to combine the use of NPs in both diagnosis and therapy, and this property makes NPs a major tool of current nanomedicine.

NPs are highly economical at delivering both hydrophobic and deliquescent medications. Recently, NPs made using biodegradable polymers have gained huge attention as drug carriers. They were successfully used in targeted drug delivery and as a result enhanced the drug efficacy while reducing its side effects. Also, NPs were successfully used to administer proteins, and/or peptides, or DNA through the pre-oral route, offering a cheaper and easier delivery route [87–89].

Overall, NPs mediate the bio-distribution of APIs, improve their stability in tissues, and enhance APIs targeting, loading, release, transport, and interaction with biological barriers such as their ability to cross the blood-brain barrier. However, cytotoxicity of nanoparticles and/or their degradation products remains the major problem of nanoparticles as DDSs. Improving NPs biocompatibility are obviously a main concern of current and future research [77, 92]. NPs are manufactured using a broad number of materials such as polyacetates, poly (alkylcyanoacrylates) (pacas), polysaccharides, and copolymers [88]. Many types NPs have been manufactured as drug carriers based on the above polymers. Some of the NPs used as drug carriers are discussed below.

5.7.1 Polymeric Nanoparticles

Polymer-based NPs are polymeric colloidal particles with submicron-sizes. Within its matrix, a therapeutic agent can be encapsulated or taken up. Also, a therapeutic agent can be adsorbed or conjugated onto the surface of a polymeric nanoparticle [92].

Polymeric NPs can effectively reduce adverse immune responses against an API. Polymers used in polymeric NPs have good biocompatibility, low toxicity and are biodegradable. These properties make polymeric NPs an efficient DDS [93].

Similar to polymeric micelles, the main advantage of polymeric NPs is that they provide encapsulation of APIs and can protect them from hydrolytic or enzymatic degradation. As an example, insulin-embedded polymeric NPs have maintained insulin activity and reduced blood glucose for up to 14 days in diabetic rats following their administration by the oral route [94].

Furthermore, polymeric NPs can be used to boost robust T cell immune responses [95]. Luo et al. reported the use of a nanovaccine, made up of physically mixing an antigen and synthetic polymeric NPs, that generated a more robust CTL response with low systemic cytokine expression than the antigen alone [95].

5.7.2 Gold Nanoparticles

Gold nanoparticles (GNPs) are solid balls made of gold with a diameter that varies from 5 to 100 nm. Depending on their application, GNPs are made in different forms, sizes, and structures, [96]. GNPs have been advantageous in biomedical imaging and have been found to enhance radiation doses [96]. Additionally, GNPs, in numerous cases, were used to improve a tumor vaccine efficacy [97]. To achieve better therapeutic effects, GNPs were combined with peptides. For instance, in vivo, GNP administration of a peptide vaccine induced an antitumor immune reaction in prophylactic and therapeutic tumor models and, in vitro, high-density sub-100 nm peptide-GNP complexes improved vaccine presentation by DCs [97].

The effect of GNP surface modifications was also tested. Fytianos et al. studied the uptake efficiency of GNP modified with either PEG (polyethylene glycol) or PVA (polyvinyl alcohol) or their combination [98]. They found that, in vitro, surface-modified GNP uptake efficiency into human DCs does not correlate with functional changes and cytokine secretion of human DCs [98].

In addition, GNPs have been successfully used, in combination with cancer radiotherapy, to induce localized hyperthermia and destroy tumor cells [96].

5.7.3 Nanoemulsion

Nanoemulsions are nanoscale systems of sizes between 1–100 nm, made by mixing of water, oil, and an amphiphile to make a stable liquid solution. As such nanoemulsions can encapsulate high concentrations of an API and preserve its stability and its curative effects. Compared to the available classical emulsions, nanoemulsions are more efficient, smaller in size, and have high biocompatibility with peptides, proteins and biological tissues as well [99]. Notwithstanding, so far there are few reports on the use of nanoemulsions as tumor peptide vaccines. Nanoemulsions have low encapsulation efficiency and lack stability, limitations that need to be overcome for further development of nanoemulsions into efficient drug carriers.

5.7.4 Solid Lipid Nanoparticles

Solid lipid nanoparticles (SLNs) are nanocolloids of sizes between 50 and 1000 nm. SLNs are made up of a solid lipid core with a shell of phospholipid monolayer. The lipophilic tails of the phospholipids are embedded in the lipid core. SLNs can be administered through different routes. SLNs are very promising oral drug delivery agents and can enhance and control drug delivery. Usually lipophilic APIs can be entrapped within SLNs, but they can be also used to deliver hydrophilic drugs. SLN solid core protects chemically labile APIs and prolongs drug release. SLNs are biocompatible showing low cytotoxicity against mammalian cells. SLNs have been orally administered as drug carriers for the treatment of different cancers, central nervous system diseases, cardiovascular diseases, bacterial/viral infection, and inflammation [100, 101]. SLNs have also been used for drug administration through the ocular route [102].

5.8 Nanoparticle Delivery of Immune Checkpoint Inhibitors

Though immune checkpoint inhibitors therapy has been effective against several cancers, the efficacy of these inhibitors requires improvement. As such therapy with nivolumab, PD-1 antibody, showed a 3-year overall survival rate of around 50% only [103], and around 21% of the patients develop nivolumab-related adverse outcomes [103]. In general, therapy-related adverse outcomes take place during PD-1 inhibition therapy. These adverse side effects could lead to death and include interstitial nephritis, liver and pancreatic toxicity, and inflammatory pneumonitis [104, 105]. Side effects also include type 1 diabetes and heart attacks [106]. The culprit during these cytotoxicities is immune over-activation due to the systemic application of checkpoint antibodies [107, 108]. Systemically applied checkpoint mAbs may disturb the balance of immune tolerance and cause inflammatory/autoimmune-like side -effects in normal tissues and organ systems [109, 110]. These immune-induced toxicities associated with systemic delivery of checkpoint inhibitors remain one of the major obstacles to problem-free checkpoint blockade therapies against cancers. Also, PD-1 antibodies are required in high doses to achieve curative outcomes and this substantially increases the treatment cost burden [103].

Many studies focused on enhancing the efficacy of checkpoint inhibitors, including, as an example, the use of microneedle patch–assisted delivery [111], hydrogels as biomaterial-based delivery systems [112], and nanoparticle delivery [113–115].

NP delivery of immune checkpoint inhibitors, offers an opportunity to overcome the above checkpoint blockade therapy limitations, mainly because nanoparticles can be used as targeted delivery agents. Nanoparticles targeted delivery of checkpoint inhibitors is being tried as a potential solution for side effects, systemic adverse immune responses and high treatment cost associated with systemic use of checkpoint inhibitors. Diverse kinds of nanoparticles are being tested, to determine their potential as delivery carriers and if they can maintain sustained checkpoint inhibitors release at tumor sites [116]. In this respect, several pre-clinical trials determined that checkpoint inhibitors targeted delivery improves their efficacy as well as decreases their toxicity [116]. However, so far there are no reports of the use of targeted nanoparticle delivery of check point inhibitors in clinical trials.

NPs have numerous other advantages, such as enhancing checkpoint inhibitors biodistribution and pharmacokinetics, improving their intracellular delivery capacity and tissue penetration, and better site-specific buildup of systemically delivered therapeutic drugs [117, 118]. One more benefit of using nanoparticles is to reduce the cost of treatment, because of the use of lower doses of checkpoint inhibitors. These advantages justify the clinical investigation of nanoparticle delivery of immune checkpoint inhibitors.

An emerging problem of using checkpoint inhibitors is that using them alone as a monotherapy showed primary as well as adaptive resistances to therapy and had limited durable therapeutic outcomes in a small fraction of the treated patients only [119]. As a result, immune checkpoint inhibitors are being evaluated in combination therapy with other anticancer treatments such as chemotherapy, radiotherapy, and photothermal therapy to attain maximal therapeutic outcomes of checkpoint inhibition [120]. Some of these combination therapies are tried using checkpoint inhibitors conjugated to nanoparticles. Xu et al. used polyethylene glycol-poly(ε-caprolactone) nanoparticles (PEG-PCL NPs) conjugated with PDL-1 mAbs and found that the PDL-1 antibodies conjugated to nanoparticles had higher cellular uptake. Also, when PEG-PCL-PDL-1 mAb nanoparticles were loaded with the chemotherapeutic agent docetaxel, the NPs were more efficient at reducing PDL-1 expression in gastric cancer cells and enhanced docetaxel and PDL-1 mAb delivery level and their overall chemotherapeutic efficacy against the gastric cancer cells [121].

Methoxy poly(-ethylene glycol)$_{3000}$poly(ε-caprolactone)$_{20000}$ [MPEG-PCL] and maleimide-poly (ethylene glycol)$_{3400}$-poly(ε-caprolactone)$_{20000}$ [Mal-PEG-PCL] nanoparticles were conjugated to the CD peptide to form CD NPs, CD-NP. The CD peptide is a PD-1 targeting peptide that consisted of DPPA-1 peptide, a peptide derived from PDL-1, and the peptide CGKRK, a tumor vasculature affinity peptide. The CD-NPs were further conjugated to the paclitaxel (PTX) chemotherapeutic agent to form CD-NP-PTX nanoparticles. The combination therapy nanoparticles, CD-NP-PTX, could improve the tumor targeting, delivery, efficacy, and antitumor effects of checkpoint inhibitors against glioma cells in vivo [122].

A nanoscale kinase-inhibiting supramolecular therapeutic structure, made up of the active parts of the kinase inhibitors selumetinib and PI103, were used in combination with PD-1/PDL-1 inhibitors. The supramolecular nanostructure offered more stability and better drug loading capacity in vitro, when compared to other conventional lipid nanoparticles [123]. The combination improved the antitumor effects of checkpoint inhibitors against melanoma and breast cancer in vivo in animal models [123].

Regulatory T cells (Tregs) have immunosuppressive functions and inhibit CTLs and other immune cells from performing antitumor effector functions against cancer cells. Elimination or suppression of Tregs can be used to induce anticancer immune responses by removing the inhibition of Tregs on CTLs. In one study, nanoparticles were used to deliver a combination of a Tregs inhibitory molecule (tLyp1 peptide) and a checkpoint inhibitor (CTLA-4 antibody). Administration of the combination NPs decreased the number of intra-tumoral Tregs, increased the number of tumor-specific CTLs, exhibited greater tumor suppression, and prolonged survival of mice injected with melanoma cells as compared to NPs carrying the Tregs inhibitor alone or the checkpoint inhibitor alone [3, 124]. This combination NP can be used in cancers that are resistant to checkpoint inhibition therapy, however, a thorough assessment of in vivo side effects should be warranted, as the risk for immune tolerance related adverse responses will increase.

Another strategy that employ nanoparticles for combination therapy using immune checkpoint inhibitors is demonstrated by Chen et al. [120]. They have used PLGA nanoparticles that were co-encapsulated with indocyanine green, a near-infrared (NIR) stimulated dye, and the TLR7 agonist imiquimod [120]. Indocyanine green would enable photodynamic therapy and imiquimod would act as an adjuvant to activate immune responses. Upon administration of the nanoparticles, NIR photothermal stimulation will ablate the primary tumor cells, releasing tumor-associated antigens (TAAs) or neo-antigens that acted with the nanoparticle adjuvant imiquimod to mount a vaccine-like function. A robust immune response was mounted and resulted in the suppression of Tregs, which usually suppress an immune response. Most importantly, when these nanoparticles were used in combination with anti-CTLA-4 antibody they cleared any remaining distant tumor cells in a mouse model of breast cancer. This combination therapy may potentially be used against different tumor types. Furthermore, this strategy has an immunological memory effect, may offer protection against relapse of the initial tumor [120]. A similar strategy uses nanoparticles in photodynamic therapy. The NPs are co-loaded with zinc pyrophosphate (ZnP) and a photosensitizer molecule, pyrolipid, to form ZnP-pyro NPs. Inducing the ZnP-pyro nanoparticles delivered in combination with PDL-1 antibody improved the efficacy of PDL-1 mAbs and made the tumor cells more sensitive to PDL-1 mAbs. The combination therapy with PDL-1 checkpoint blockade not only eliminated the light-irradiated primary tumor cells but also completely prevented the growth of distant untreated tumor cells by generating systemic antitumor CTL responses [125]. In addition, the combination therapy inhibited metastasis to the lung in a mouse model. This result indicates that photodynamic therapy in combination with checkpoint blockade has the ability to potentiate the systemic actions of checkpoint inhibition through the activation of the innate and adaptive immune responses in the cancer microenvironments [125].

In a recent report, delivery of GNPs loaded with anti-PD-1 antibodies offered an opportunity not only for enhancing targeting and delivery of immune checkpoint inhibitors but also for the early identification of potentially responding versus non-responding patients to checkpoint blockade therapy. The delivery of anti-PD-1 GNPs effectively inhibited tumor growth in a colon cancer mouse model, at much lower concentrations than the standard PD 1 antibody dosage used in the clinic. Above all, PD-1 GNPs, in combination with computed tomography imaging, were able to predict a therapeutic response, as early as 48 hours following treatment [126]. Thus, these NPs can be used as a noninvasive tool to make better decisions about the efficacy of a treatment course. In another study, NPs were also used to monitor therapy. Polymeric nanoparticles that can monitor the response of tumor cells to checkpoint inhibition in real time was designed and tried by Kulkarni et al. [123]. They used NPs that were loaded with the chemotherapeutic agent paclitaxel, and also surface modified with PEG which was used to conjugate anti-PD-L1 antibodies. These NPs had an improved antitumor activity and reduced breast and lung tumor volumes in animal models, when compared to conventional NPs. These NPs were also loaded with a reporter molecule (5-FAM-QSY-7–based FRET pair) that can monitor the progress and efficacy of the anticancer therapy; thus this NP had a self-reporting activity as well as an effector anticancer activity [123].

An additional roadblock to the clinical efficacy of checkpoint inhibitors is the lack of response to PD-1/PDL-1 immunotherapy in certain cancers. Among these non-responsive cancers are the mismatch repair (MMR)-capable or microsatellite stable (MSS) solid tumors, such as certain colorectal cancers and pancreatic ductal epithelial adenocarcinoma [127]. Song et al. demonstrated that the chemotherapeutic agent oxaliplatin (OXA) boosts anti-PDL-1 mAb therapy of mouse MSS colorectal cancer, which usually does not respond to PDL-1 mAb treatment. By loading a DNA plasmid encoding a PDL-1 protein trap into lipid-protamine-DNA NPs and applying these NPs following OXA treatment, the PDL-1 trap is expressed, transiently and locally, at the tumor site, and synergizes with OXA to inhibit tumor growth. OXA and PDL-1 trap combination therapy does not induce significant Th17 cells buildup in the spleen, a sign of more tolerance and less tendency to produce immune adverse effects as in the systemic application of PDL-1 mAbs that produces exaggerated systemic adverse immune responses [127]. A similar report used binary synergistic tumor microenvironment-activated nanoparticle prodrug in an effort to trigger immunogenic cell death and stimulate anticancer immunity. The used NPs consisted of an oxaliplatin homodimer in combination with NLG919, an inhibitor of indoleamine-2,3-dioxygenase. These activated OXA NPs stimulated intra-tumoral assembly of CTLs by inducing immunogenic cell death in the cancer cells; the indoleamine-2,3-dioxygenase inhibitor decreased the number of Tregs, which usually dampen the immune response, in the tumor microenvironment. The binary prodrug nanoparticle was more efficient than a combination of IDO-1 inhibitor and OXA at preventing breast or colorectal cancer metastasis [128].

One more strategy to reduce adverse immune reactions associated with systemic delivery of mAbs against checkpoint inhibitors is to use nanoparticles loaded with siRNAs that silence checkpoint molecules. In one application, silencing of PDL-1 and PDL-2 in DCs can enhance immunogenic function of DCs, which induces stronger CTL responses in cells in culture and better antitumor immune responses in several cancer models in mice in vivo [129, 130]. Due to their efficacy, certain checkpoint silencing siRNAs-loaded nanoparticles are in phase I/II trials to evaluate their clinical efficacy such as trial NCT02528682 (www.clinicaltrials.gov).

Ordikhani et al. reasoned that the spleen would be a practical target for immune checkpoint inhibitors delivery. The spleen is populated with resident T cells and APCs that can effectively present antigens to the resident T cells and activate them, the activated T cells can move to distant sites such as tumor sites. The spleen is also a site where immune tolerance and suppression of immune responses could be modulated [131]. Ordikhani et al. delivered PLGA nanoparticles loaded with anti-PD-1 antibody directly into the spleen of a mouse model of melanoma. The spleen-delivered anti-PD-1 NPs resulted in enhanced anticancer immune responses, even though the PD-1 inhibitor was used at low concentrations [131]. This study paves the way for enhancing delivery and efficacy of NP-loaded checkpoint inhibitors by choosing better delivery sites or routes.

5.9 Conclusion

The use of novel drug delivery systems such as carrier-based drug delivery has enhanced drug efficacy and reduced drug side effects. A main advantage of carrier-based drug delivery systems is that they can be used in targeted therapy, which eliminates many of the adverse effects of systemic drug delivery. Nanoparticles have shown better efficiency than counterpart carrier-based drug delivery systems. Nonetheless, proven and tested protocols should be designed to determine short- as well as long-term toxicities associated with certain nanoparticle formulations. Recently, nanoparticles have been used for the delivery of immune checkpoint inhibitors. Checkpoint inhibitor-loaded nanoparticles have shown enhanced efficacy and reduced adverse immune reactions under a multitude of experimental preclinical settings. It remains to be seen whether these effects will prove beneficial in clinical settings. In the near future, we should witness a surge of clinical trials employing nanoparticles for the delivery of check point inhibitors.

References

1. National Cancer Instiitute. Available online: https://www.cancer.gov/about-cancer/treatment/types (accessed on March 7th 2019).
2. Mellman, I., Coukos, G., Dranoff, G. (2011). Cancer immunotherapy comes of age, *Nature*, **480**, pp. 480–489, doi:10.1038/nature10673.
3. Patel, J. D., Krilov, L., Adams, S., Aghajanian, C., Basch, E., Brose, M. S., Carroll, W. L., de Lima, M., Gilbert, M. R., Kris, M. G., et al. (2014). Clinical cancer advances 2013: Annual report on progress against cancer from the American society of clinical oncology, *J. Clin. Oncol.*, **32**, pp. 129–160.
4. Zeng, C., Wen, W., Morgans, A. K., Pao, W., Shu, X. O., Zheng, W. (2015). Disparities by race, age, and sex in the improvement of survival for major cancers: Results from the National Cancer Institute Surveillance, Epidemiology, and end results (SEER) program in the United States, 1990 to 2010, *JAMA Oncol.*, **1**, pp. 88–96.
5. Shih, K., Arkenau, H. T., Infante, J. R. (2014). Clinical impact of checkpoint inhibitors as novel cancer therapies, *Drugs*, **74**, pp. 1993–2013.

6. National Cancer Institute. Available online: https://www.cancer.gov/about-cancer/treatment/types/immunotherapy/bio-therapies-fact-sheet (accessed on March 7th 2019).
7. Coley, W. B. (1891). II. Contribution to the knowledge of sarcoma, *Ann. Surg.*, **14**, pp. 199–220.
8. Busch, W. (1868). Aus der Sitzung der medicinischen Section vom 13 November 1867, *Berl Klin Wochenschr*, **5**, p. 137.
9. Fehleisen, F. (1882). Ueber die Züchtung der Erysipelkokken auf künstlichem Nährboden und ihre Übertragbarkeit auf den Menschen, *Dtsch. Med. Wochenschr.*, **8**, pp. 553–554.
10. Morales, A., Eidinger, D., Bruce, A. W. (2002). Intracavitary Bacillus Calmette-Guerin in the treatment of superficial bladder tumors. 1976, *J. Urol.*, **116**(2), pp. 180–183.
11. Diamond, M. S., Kinder, M., Matsushita, H., Mashayekhi, M., Dunn, G. P., Archambault, J. M., Lee, H., Arthur, C. D., White, J. M., Kalinke, U., et al. (2011). Type I interferon is selectively required by dendritic cells for immune rejection of tumors, *J. Exp. Med.*, **208**, pp. 1989–2003.
12. Fuertes, M. B., Kacha, A. K., Kline, J., Woo, S. R., Kranz, D. M., Murphy, K. M., Gajewski, T. F. (2011). Host type I IFN signals are required for antitumor CD8+ T cell responses through CD8{alpha}+ dendritic cells, *J. Exp. Med.*, **208**, pp. 2005–2016.
13. Tian, T., Olson, S., Whitacre, J. M., Harding, A. (2011). The origins of cancer robustness and evolvability, *Int. Biol. (Camb.)*, **3**, pp. 17–30.
14. Palucka, K., Banchereau, J. (2012). Cancer immunotherapy via dendritic cells, *Nat. Rev. Cancer*, **12**, pp. 265–277.
15. Melero, I., Berman, D. M., Aznar, M. A., Korman, A. J., Perez Gracia, J. L., Haanen, J. (2015). Evolving synergistic combinations of targeted immunotherapies to combat cancer, *Nat. Rev. Cancer*, **15**, pp. 457–472.
16. Linsley, P. S., Clark, E. A., Ledbetter, J. A. (1990). T-cell antigen CD28 mediates adhesion with B cells by interacting with activation antigen B7/BB-1, *Proc. Natl. Acad. Sci. U. S. A.*, **87**, pp. 5031–5035.
17. Martin, P. J., Ledbetter, J. A., Morishita, Y., June, C. H., Beatty, P. G., Hansen, J. A. (1986). A 44 kilodalton cell surface homodimer regulates interleukin 2 production by activated human T lymphocytes, *J. Immunol.*, **136**, pp. 3282–3287.
18. Townsend, S. E., Allison, J. P. (1993). Tumor rejection after direct costimulation of CD8+ T cells by B7-transfected melanoma cells, *Science*, **259**, pp. 368–370.

19. Mullard, A. (2013). New checkpoint inhibitors ride the immunotherapy tsunami, *Nat. Rev. Drug Discov.*, **12**, pp. 489–492.

20. Hanahan, D., Weinberg, R. A. (2011). Hallmarks of cancer: The next generation, *Cell*, **144**, pp. 646–674.

21. Chen, D. S., Mellman, I. (2013). Oncology meets immunology: The cancer-immunity cycle, *Immunity*, **39**, pp. 1–10.

22. Brunet, J. F., Denizot, F., Luciani, M. F., Roux-Dosseto, M., Suzan, M., Mattei, M. G., Golstein, P. (1987). A new member of the immunoglobulin superfamily —CTLA-4, *Nature*, **328**, pp. 267–270.

23. Lindsten, T., Lee, K. P., Harris, E. S., Petryniak, B., Craighead, N., Reynolds, P. J., Lombard, D. B., Freeman, G. J., Nadler, L. M., Gray, G. S., et al. (1993). Characterization of CTLA-4 structure and expression on human T cells, *J. Immunol.*, **151**, pp. 3489–3499.

24. Linsley, P. S., Bradshaw, J., Greene, J., Peach, R., Bennett, K. L., Mittler, R. S. (1996). Intracellular trafficking of CTLA-4 and focal localization towards sites of TCR engagement, *Immunity*, **4**, pp. 535–543.

25. Linsley, P. S., Brady, W., Urnes, M., Grosmaire, L. S., Damle, N. K., Ledbetter, J. A. (1991). CTLA-4 is a second receptor for the B cell activation antigen B7, *J. Exp. Med.*, **174**, pp. 561–569.

26. Krummel, M. F., Allison, J. P. (1995). CD28 and CTLA-4 have opposing effects on the response of T cells to stimulation, *J. Exp. Med.*, **182**, pp. 459–465.

27. Walunas, T. L., Lenschow, D. J., Bakker, C. Y., Linsley, P. S., Freeman, G. J., Green, J. M., Thompson, C. B., Bluestone, J. A. (1994). CTLA-4 can function as a negative regulator of T cell activation, *Immunity*, **1**, pp. 405–413.

28. Tivol, E. A., Borriello, F., Schweitzer, A. N., Lynch, W. P., Bluestone, J. A., Sharpe, A. H. (1995). Loss of CTLA-4 leads to massive lymphoproliferation and fatal multiorgan tissue destruction, revealing a critical negative regulatory role of CTLA-4, *Immunity*, **3**, pp. 541–547.

29. Waterhouse, P., Penninger, J. M., Timms, E., Wakeham, A., Shahinian, A., Lee, K. P., Thompson, C. B., Griesser, H., Mak, T. W. (1995). Lymphoproliferative disorders with early lethality in mice deficient in Ctla-4, *Science*, **270**, pp. 985–988.

30. Ishida, Y., Agata, Y., Shibahara, K., Honjo, T. (1992). Induced expression of PD-1, a novel member of the immunoglobulin gene superfamily, upon programmed cell death, *EMBO J.*, **11**, pp. 3887–3895.

31. Agata, Y., Kawasaki, A., Nishimura, H., Ishida, Y., Tsubata, T., Yagita, H., Honjo, T. (1996). Expression of the PD-1 antigen on the surface of stimulated mouse T and B lymphocytes, *Int. Immunol.*, **8**, pp. 765-772.

32. Nishimura, H., Honjo, T. (2001). PD-1: An inhibitory immunoreceptor involved in peripheral tolerance, *Trends Immunol.*, **22**, pp. 265–268.

33. Nishimura, H., Nose, M., Hiai, H., Minato, N., Honjo, T. (1999). Development of lupus-like autoimmune diseases by disruption of the PD-1 gene encoding an ITIM motif-carrying immunoreceptor, *Immunity*, **11**, pp. 141–151.

34. Nishimura, H., Okazaki, T., Tanaka, Y., Nakatani, K., Hara, M., Matsumori, A., Sasayama, S., Mizoguchi, A., Hiai, H., Minato, N., et al. (2001). Autoimmune dilated cardiomyopathy in PD-1 receptor-deficient mice, *Science*, **291**, pp. 319–322.

35. Dong, H., Zhu, G., Tamada, K., Chen, L. (1999). B7-H1, a third member of the B7 family, co-stimulates T-cell proliferation and interleukin-10 secretion, *Nat. Med.*, **5**, pp. 1365–1369.

36. Freeman, G. J., Long, A. J., Iwai, Y., Bourque, K., Chernova, T., Nishimura, H., Fitz, L. J., Malenkovich, N., Okazaki, T., Byrne, M. C., et al. (2000). Engagement of the PD-1 immunoinhibitory receptor by a novel B7 family member leads to negative regulation of lymphocyte activation, *J. Exp. Med.*, **192**, pp. 1027–1034.

37. Latchman, Y., Wood, C. R., Chernova, T., Chaudhary, D., Borde, M., Chernova, I., Iwai, Y., Long, A. J., Brown, J. A., Nunes, R., et al. (2001). PD-L2 is a second ligand for PD-1 and inhibits T cell activation, *Nat. Immunol.*, **2**, pp. 261–268.

38. Armand, P. (2015). Immune checkpoint blockade in hematologic malignancies, *Blood*, **125**, pp. 3393–3400, doi:10.1182/blood-2015-02-567453.

39. Garon, E. B., Rizvi, N. A., Hui, R., Leighl, N., Balmanoukian, A. S., Eder, J. P., Patnaik, A., Aggarwal, C., Gubens, M., Horn, L., et al. (2015). Pembrolizumab for the treatment of non-small-cell lung cancer, *N. Engl. J. Med.*, **372**, pp. 2018–2028.

40. Scarpace, S. L. (2015). Metastatic squamous cell non-small-cell lung cancer (NSCLC): Disrupting the drug treatment paradigm with immunotherapies, *Drugs Context*, **4**, p. 212289.

41. Barber, D. L., Wherry, E. J., Masopust, D., Zhu, B., Allison, J. P., Sharpe, A. H., Freeman, G. J., Ahmed, R. (2006). Restoring function in exhausted CD8 T cells during chronic viral infection, *Nature*, **439**, pp. 682–687.

42. Maier, H., Isogawa, M., Freeman, G. J., Chisari, F. V. (2007). PD-1: PD-L1 interactions contribute to the functional suppression of virus-specific CD8+ T lymphocytes in the liver, *J. Immunol.*, **178**, pp. 2714–2720.

43. Tzeng, H. T., Tsai, H. F., Liao, H. J., Lin, Y. J., Chen, L., Chen, P. J., Hsu, P. N. (2012). PD-1 blockage reverses immune dysfunction and hepatitis B viral persistence in a mouse animal model, *PLoS One*, **7**, p. e39179.

44. Velu, V., Titanji, K., Zhu, B., Husain, S., Pladevega, A., Lai, L., Vanderford, T. H., Chennareddi, L., Silvestri, G., Freeman, G. J., et al. (2009). Enhancing SIV-specific immunity in vivo by PD-1 blockade, *Nature*, **458**, pp. 206–210.

45. Catakovic, K., Klieser, E., Neureiter, D., Geisberger, R. (2017). T cell exhaustion: From pathophysiological basics to tumor immunotherapy, *Cell Commun. Signal*, **15**, p. 1.

46. Makkouk, A., Weiner, G. J. (2015). Cancer immunotherapy and breaking immune tolerance: New approaches to an old challenge, *Cancer Res.*, **75**, pp. 5–10.

47. Tsai, H. F., Hsu, P. N. (2017). Cancer immunotherapy by targeting immune checkpoints: Mechanism of T cell dysfunction in cancer immunity and new therapeutic targets, *J. Biomed. Sci.*, **24**, p. 35, doi:10.1186/s12929-017-0341-0.

48. Blackburn, S. D., Shin, H., Haining, W. N., Zou, T., Workman, C. J., Polley, A., Betts, M. R., Freeman, G. J., Vignali, D. A., Wherry, E. J. (2009). Coregulation of CD8+ T cell exhaustion by multiple inhibitory receptors during chronic viral infection, *Nat. Immunol.*, **10**, pp. 29–37.

49. Nguyen, L. T., Ohashi, P. S. (2015). Clinical blockade of PD1 and LAG3—potential mechanisms of action, *Nat. Rev. Immunol.*, **15**, pp. 45–56.

50. Pardoll, D. M. (2012). The blockade of immune checkpoints in cancer immunotherapy, *Nat. Rev. Cancer*, **12**, pp. 252–264.

51. Pauken, K. E., Wherry, E. J. (2015). Overcoming T cell exhaustion in infection and cancer, *Trends Immunol.*, **36**, pp. 265–276.

52. Schietinger, A., Greenberg, P. D. (2014). Tolerance and exhaustion: Defining mechanisms of T cell dysfunction, *Trends Immunol.*, **35**, pp. 51–60.

53. Wherry, E. J. (2011). T cell exhaustion, *Nat. Immunol.*, **12**, pp. 492–499.

54. Verma, V., Sprave, T., Haque, W., Simone, C. B., 2nd, Chang, J. Y., Welsh, J. W., Thomas, C. R., Jr. (2018). A systematic review of the cost and

cost-effectiveness studies of immune checkpoint inhibitors, *J. Immunother. Cancer*, **6**, p. 128.

55. Deng, R., Qu, H., Liang, L., Zhang, J., Zhang, B., Huang, D., Xu, S., Liang, C., Xu, W. (2017). Tracing the therapeutic process of targeted aptamer/drug conjugate on cancer cells by surface-enhanced Raman scattering spectroscopy, *Anal. Chem.*, **89**, pp. 2844–2851.
56. Tibbitt, M. W., Dahlman, J. E., Langer, R. (2016). Emerging frontiers in drug delivery, *J. Am. Chem. Soc.*, **138**, pp. 704–717, doi:10.1021/jacs.5b09974.
57. Leucuta, S. E. (2014). Subcellular drug targeting, pharmacokinetics and bioavailability, *J. Drug Target.*, **22**, pp. 95–115.
58. Naydenov, T., Stoimenova, A., Kassarova, M., Kamusheva, M., Dimitrov, P., Petrova, G. (2016). Opinion of Bulgarian pharmacists on drug delivery systems, orodispersible and pediatric dosage forms, *J. Appl. Pharm.*, **8**, p. 1.
59. Patil, J. (2015). Hydrogel system: An approach for drug delivery modulation, *Adv. Pharmacoepidemiol. Drug Saf.*, **4**, p. 1000e135.
60. Patil, J. (2015). Hydrodynamically balanced gastro-retentive site specific drug delivery system: An innovative approach, *J. Pharmacovigil*, **3**, p. e146.
61. Shanmugan, P., Bandameedi, R. (2015). Chronotherapeutic drug delivery systems, *J. Drug Metab. Toxicol.*, **6**, p. 2.
62. Donaruma, L. G. (1988). Definitions in biomaterials, D. F. Williams, ed., Elsevier, Amsterdam, 1987, 72 pp, *Journal of Polymer Science Part C: Polymer Letters*, **26**, pp. 414–414.
63. Naahidi, S., Jafari, M., Edalat, F., Raymond, K., Khademhosseini, A., Chen, P. (2013). Biocompatibility of engineered nanoparticles for drug delivery, *J. Control. Release*, **166**, pp. 182–194.
64. Jassim-Jaboori, A., Oyewumi, M. (2015). 3D printing technology in pharmaceutical drug delivery: Prospects and challenges, *J. Biomol. Res. Ther.*, **4**, pp. 1–3.
65. Saikia, C., Gogoi, P., Maji, T. (2015). Chitosan: A promising biopolymer in drug delivery applications, *J. Mol. Genet. Med. S*, **4**, p. 006.
66. Awan, B., Fatima, N., Riaz, S., Malik, S., Ahmad, W. (2015). Bacterial and liposomal vector guided drug delivery system via tumor markers carrier gene to treat neoplasm, *J. Appl. Pharm.*, **7**, p. 2.
67. Bhasin, S., Patel, R. (2015). Enhanced oral bioavailability of alitretinoin by lipid drug delivery system, *Pharm. Anal. Acta*, **6**, p. 433.

68. Kamel, R. Transdermal Drug Delivery: Benefits and Challenges. In *Proceedings of Journal of Applied Pharmacy*, p. e103. https://www.longdom.org/open-access/transdermal-drug-delivery-benefits-and-challenges-1920-4159-1000e103.pdf.
69. Donthi, M. R., Dudipala, N., Komalla, D. R., Suram, D., Banala, N. (2015). Design and evaluation of floating multi unit mini tablets (MUMTS) muco adhesive drug delivery system of famotidine to treat upper gastro intestinal ulcers, *J. Pharmacovigilance*, **3**, p. 179.
70. Jethara, S. I., Patel, M. R. (2015). Optimizing oral controlled release drug delivery systems using experimental designs, *Intellect. Property Rights: Open Access*, pp. 1–6.
71. Patil, J. (2015). Novel drug delivery strategies: New concepts, *Adv. Pharmacoepidemiol Drug Saf.*, **4**, p. e134.
72. Repanas, A., Wolkers, W. F., Gryshkov, O., Kalozoumis, P., Mueller, M., Zernetsch, H., Korossis, S., Glasmacher, B. (2015). Coaxial electrospinning as a process to engineer biodegradable polymeric scaffolds as drug delivery systems for anti-inflammatory and anti-thrombotic pharmaceutical agents, *Clin. Exp. Pharmacol.*, **5**, p. 1000192.
73. Zanchetta, B., Chaud, M., Santana, M. (2015). Self-emulsifying drug delivery systems (SEDDS) in pharmaceutical development, *J. Adv. Chem. Eng.*, **5**, pp. 1–7.
74. Hu, D., Tang, S., Peng, H., Wang, Q. (2015). The bright future of liposome mediated drug delivery, *Biochem. Physiol.*, **4**, p. e133.
75. Komano, Y., Yagi, N., Onoue, I., Kaneko, K., Miyasaka, N., Nanki, T. (2012). Arthritic joint-targeting small interfering RNA-encapsulated liposome: Implication for treatment strategy for rheumatoid arthritis, *J. Pharmacol. Exp. Ther.*, **340**, pp. 109–113.
76. Sivakumar, P., Mythily, S., Alamelu, S., Panduranga Rao, K. (1994). Liposome-ibuprofen conjugate and the release characteristics of ibuprofen, *Indian Drugs-Bombay*, **31**, pp. 568–568.
77. Swapna, N. J., Nadkarni, G. (1998). Effect of size and charge of liposomes on biodistribution of encapsuled ggmTc-DTPA in rats, *Indian J. Pharmacol.*, **30**, p. 199.
78. Aminabhavi, T. M., Deshmukh, A. S. (2016). Polysaccharide-based hydrogels as biomaterials. In *Polymeric Hydrogels as Smart Biomaterials*, Springer, pp. 45–71.
79. Farooq, U., Malviya, R., Sharma, P. K. (2015). Design and development of multi particulate system for targeted drug delivery using natural polymer, *Pharm. Anal. Acta*, **6**, p. 366.

80. Agrawal, P. (2014). Significance of polymers in drug delivery system, *J. Pharmacovigil*, **3**, p. e127.

81. Saboktakin, M., Maharramov, A., Ramazanov, M. (2015). pH sensitive chitosan-based supramolecular gel for oral drug delivery of insulin, *J. Mol. Genet. Med.*, 9:170. doi:10.4172/1747-0862.1000170.

82. Crawford, E. D., Moul, J. W., Sartor, O., Shore, N. D. (2015). Extended release, 6-month formulations of leuprolide acetate for the treatment of advanced prostate cancer: Achieving testosterone levels below 20 ng/dl, *Expert Opin. Drug Metab. Toxicol.*, **11**, pp. 1465–1474.

83. Hatefi, A., Amsden, B. (2002). Biodegradable injectable in situ forming drug delivery systems, *J. Control. Release*, **80**, pp. 9–28.

84. Ravivarapu, H. B., Moyer, K. L., Dunn, R. L. (2000). Sustained activity and release of leuprolide acetate from an in situ forming polymeric implant, *AAPS PharmSciTech*, **1**, p. E1.

85. Li, J., Mooney, D. J. (2016). Designing hydrogels for controlled drug delivery, *Nat. Rev. Mater.*, **1**, p. 16071.

86. Patel, P. D., Dand, N., Hirlekar, R. S., Kadam, V. J. (2008). Drug loaded erythrocytes: As novel drug delivery system, *Curr. Pharm. Des.*, **14**, pp. 63–70.

87. Malika, V., Kohli, K., Chaudhary, H., Kumar, V. (2014). Nano-carrier for accentuated transdermal drug delivery, *J. Develop. Drugs*, **3**, p. 2.

88. Singh, R., Vyas, S. P. (1996). Topical liposomal system for localized and controlled drug delivery, *J. Dermatol. Sci.*, **13**, pp. 107–111.

89. Wen, H., Li, Y. (2014). Redox sensitive nanoparticles with disulfide bond linked sheddable shell for intracellular drug delivery, *Med. Chem. (Los Angeles)*, **4**, pp. 748–755.

90. Martin, D., Udupa, N. (1994). Nasal drug delivery of gentamycin sulphate, *Indian Drugs*, **31**, pp. 365–369.

91. Narayani, R., Panduranga Rao, K. (1996). Polymer-coated gelatin capsules as oral delivery devices and their gastrointestinal tract behaviour in humans, *J. Biomater. Sci. Polym. Ed.*, **7**, pp. 39–48.

92. Labhasetwar, V., Song, C., Humphrey, W., Shebuski, R., Levy, R. J. (1998). Arterial uptake of biodegradable nanoparticles: Effect of surface modifications, *J. Pharm. Sci.*, **87**, pp. 1229–1234.

93. Conniot, J., Silva, J. M., Fernandes, J. G., Silva, L. C., Gaspar, R., Brocchini, S., Florindo, H. F., Barata, T. S. (2014). Cancer immunotherapy: Nanodelivery approaches for immune cell targeting and tracking, *Front. Chem.*, **2**, p. 105.

94. Damge, C., Maincent, P., Ubrich, N. (2007). Oral delivery of insulin associated to polymeric nanoparticles in diabetic rats, *J. Control. Release*, **117**, pp. 163–170.

95. Luo, M., Wang, H., Wang, Z., Cai, H., Lu, Z., Li, Y., Du, M., Huang, G., Wang, C., Chen, X., et al. (2017). A STING-activating nanovaccine for cancer immunotherapy, *Nat. Nanotechnol.*, **12**, pp. 648–654.

96. Zhang, X. (2015). Gold nanoparticles: Recent advances in the biomedical applications, *Cell Biochem. Biophys.*, **72**, pp. 771–775.

97. Almeida, J. P. M., Lin, A. Y., Figueroa, E. R., Foster, A. E., Drezek, R. A. (2015). In vivo gold nanoparticle delivery of peptide vaccine induces anti-tumor immune response in prophylactic and therapeutic tumor models, *Small*, **11**, pp. 1453–1459.

98. Fytianos, K., Rodriguez-Lorenzo, L., Clift, M. J., Blank, F., Vanhecke, D., Von Garnier, C., Petri-Fink, A., Rothen-Rutishauser, B. (2015). Uptake efficiency of surface modified gold nanoparticles does not correlate with functional changes and cytokine secretion in human dendritic cells in vitro, *Nanomed. Nanotechnol. Biol. Med.*, **11**, pp. 633–644.

99. Ge, W., Li, Y., Li, Z. S., Zhang, S. H., Sun, Y. J., Hu, P. Z., Wang, X. M., Huang, Y., Si, S. Y., Zhang, X. M., et al. (2009). The antitumor immune responses induced by nanoemulsion-encapsulated MAGE1-HSP70/SEA complex protein vaccine following peroral administration route, *Cancer Immunol. Immunother.*, **58**, pp. 201–208.

100. Lin, C. H., Chen, C. H., Lin, Z. C., Fang, J. Y. (2017). Recent advances in oral delivery of drugs and bioactive natural products using solid lipid nanoparticles as the carriers, *J. Food Drug Anal.*, **25**, pp. 219–234, doi:10.1016/j.jfda.2017.02.001.

101. Mishra, V., Bansal, K. K., Verma, A., Yadav, N., Thakur, S., Sudhakar, K., Rosenholm, J. M. (2018). Solid lipid nanoparticles: Emerging colloidal nano drug delivery systems, *Pharmaceutics*, **10**, p. 191.

102. Khurana, L. K., Singh, R., Singh, H., Sharma, M. (2018). Systematic Development and optimization of an in situ gelling system for moxifloxacin ocular nanosuspension using high-pressure homogenization with an improved encapsulation efficiency, *Curr. Pharm. Des.*, **24**, pp. 1434–1445.

103. Wolchok, J. D., Chiarion-Sileni, V., Gonzalez, R., Rutkowski, P., Grob, J. J., Cowey, C. L., Lao, C. D., Wagstaff, J., Schadendorf, D., Ferrucci, P. F., et al. (2017). Overall survival with combined nivolumab and ipilimumab in advanced melanoma, *N. Engl. J. Med.*, **377**, pp. 1345–1356.

104. Cousin, S., Seneschal, J., Italiano, A. (2018). Toxicity profiles of immunotherapy, *Pharmacol. Ther.*, **181**, pp. 91–100.

105. Zou, W., Wolchok, J. D., Chen, L. (2016). PD-L1 (B7-H1) and PD-1 pathway blockade for cancer therapy: Mechanisms, response biomarkers, and combinations, *Sci. Transl. Med.*, **8**, p. 328rv324.

106. Moslehi, J. J., Salem, J. E., Sosman, J. A., Lebrun-Vignes, B., Johnson, D. B. (2018). Increased reporting of fatal immune checkpoint inhibitor-associated myocarditis, *Lancet*, **391**, p. 933.

107. Iwai, Y., Hamanishi, J., Chamoto, K., Honjo, T. (2017). Cancer immunotherapies targeting the PD-1 signaling pathway, *J. Biomed. Sci.*, **24**, p. 26.

108. LaFleur, M. W., Muroyama, Y., Drake, C. G., Sharpe, A. H. (2018). Inhibitors of the PD-1 pathway in tumor therapy, *J. Immunol.*, **200**, pp. 375–383.

109. Robert, C., Schachter, J., Long, G. V., Arance, A., Grob, J. J., Mortier, L., Daud, A., Carlino, M. S., McNeil, C., Lotem, M., et al. (2015). Pembrolizumab versus ipilimumab in advanced melanoma, *N. Engl. J. Med.*, **372**, pp. 2521–2532.

110. Weber, J. S., D'Angelo, S. P., Minor, D., Hodi, F. S., Gutzmer, R., Neyns, B., Hoeller, C., Khushalani, N. I., Miller, W. H., Jr., Lao, C. D., et al. (2015). Nivolumab versus chemotherapy in patients with advanced melanoma who progressed after anti-CTLA-4 treatment (CheckMate 037): A randomised, controlled, open-label, phase 3 trial, *Lancet Oncol.*, **16**, pp. 375–384.

111. Wang, C., Ye, Y., Hochu, G. M., Sadeghifar, H., Gu, Z. (2016). Enhanced cancer immunotherapy by microneedle patch-assisted delivery of anti-PD1 antibody, *Nano Lett*, **16**, pp. 2334–2340.

112. Li, Y., Fang, M., Zhang, J., Wang, J., Song, Y., Shi, J., Li, W., Wu, G., Ren, J., Wang, Z., et al. (2016). Hydrogel dual delivered celecoxib and anti-PD-1 synergistically improve antitumor immunity, *Oncoimmunology*, **5**, p. e1074374.

113. Kapadia, C. H., Perry, J. L., Tian, S., Luft, J. C., DeSimone, J. M. (2015). Nanoparticulate immunotherapy for cancer, *J. Control. Release*, **219**, pp. 167–180.

114. Li, S. Y., Liu, Y., Xu, C. F., Shen, S., Sun, R., Du, X. J., Xia, J. X., Zhu, Y. H., Wang, J. (2016). Restoring anti-tumor functions of T cells via nanoparticle-mediated immune checkpoint modulation, *J. Control. Release*, **231**, pp. 17–28.

115. Luo, L., Yang, J., Zhu, C., Jiang, M., Guo, X., Li, W., Yin, X., Yin, H., Qin, B., Yuan, X., et al. (2018). Sustained release of anti-PD-1 peptide for perdurable immunotherapy together with photothermal ablation against primary and distant tumors, *J. Control. Release*, **278**, pp. 87–99.

116. Lamichhane, P., Amin, N. P., Agarwal, M., Lamichhane, N. (2018). Checkpoint inhibition: Will combination with radiotherapy and nanoparticle-mediated delivery improve efficacy?, *Medicines (Basel)*, **5**, p. 114.

117. Schmid, D., Park, C. G., Hartl, C. A., Subedi, N., Cartwright, A. N., Puerto, R. B., Zheng, Y., Maiarana, J., Freeman, G. J., Wucherpfennig, K. W., et al. (2017). T cell-targeting nanoparticles focus delivery of immunotherapy to improve antitumor immunity, *Nat. Commun.*, **8**, p. 1747.

118. Shao, K., Singha, S., Clemente-Casares, X., Tsai, S., Yang, Y., Santamaria, P. (2015). Nanoparticle-based immunotherapy for cancer, *ACS Nano*, **9**, pp. 16–30.

119. Mahoney, K. M., Rennert, P. D., Freeman, G. J. (2015). Combination cancer immunotherapy and new immunomodulatory targets, *Nat. Rev. Drug Discov.*, **14**, pp. 561–584.

120. Chen, Q., Xu, L., Liang, C., Wang, C., Peng, R., Liu, Z. (2016). Photothermal therapy with immune-adjuvant nanoparticles together with checkpoint blockade for effective cancer immunotherapy, *Nat. Commun.*, **7**, p. 13193.

121. Xu, S., Cui, F., Huang, D., Zhang, D., Zhu, A., Sun, X., Cao, Y., Ding, S., Wang, Y., Gao, E., et al. (2019). PD-L1 monoclonal antibody-conjugated nanoparticles enhance drug delivery level and chemotherapy efficacy in gastric cancer cells, *Int. J. Nanomed.*, **14**, pp. 17–32.

122. Sun, Z., Zhang, Y., Cao, D., Wang, X., Yan, X., Li, H., Huang, L., Qu, X., Kong, C., Qin, H., et al. (2018). PD-1/PD-L1 pathway and angiogenesis dual recognizable nanoparticles for enhancing chemotherapy of malignant cancer, *Drug Deliv.*, **25**, pp. 1746–1755.

123. Kulkarni, A., Natarajan, S. K., Chandrasekar, V., Pandey, P. R., Sengupta, S. (2016). Combining immune checkpoint inhibitors and kinase-inhibiting supramolecular therapeutics for enhanced anticancer efficacy, *ACS Nano*, **10**, pp. 9227–9242.

124. Ou, W., Thapa, R. K., Jiang, L., Soe, Z. C., Gautam, M., Chang, J. H., Jeong, J. H., Ku, S. K., Choi, H. G., Yong, C. S., et al. (2018). Regulatory T cell-targeted hybrid nanoparticles combined with immuno-

checkpoint blockage for cancer immunotherapy, *J. Control. Release*, **281**, pp. 84–96.

125. Duan, X., Chan, C., Guo, N., Han, W., Weichselbaum, R. R., Lin, W. (2016). Photodynamic therapy mediated by nontoxic core-shell nanoparticles synergizes with immune checkpoint blockade to elicit antitumor immunity and antimetastatic effect on breast cancer, *J. Am. Chem. Soc.*, **138**, pp. 16686–16695.

126. Meir, R., Shamalov, K., Sadan, T., Motiei, M., Yaari, G., Cohen, C. J., Popovtzer, R. (2017). Fast image-guided stratification using anti-programmed death ligand 1 gold nanoparticles for cancer immunotherapy, *ACS Nano*, **11**, pp. 11127–11134.

127. Song, W., Shen, L., Wang, Y., Liu, Q., Goodwin, T. J., Li, J., Dorosheva, O., Liu, T., Liu, R., Huang, L. (2018). Synergistic and low adverse effect cancer immunotherapy by immunogenic chemotherapy and locally expressed PD-L1 trap, *Nat. Commun.*, **9**, p. 2237.

128. Feng, B., Zhou, F., Hou, B., Wang, D., Wang, T., Fu, Y., Ma, Y., Yu, H., Li, Y. (2018). Binary cooperative prodrug nanoparticles improve immunotherapy by synergistically modulating immune tumor microenvironment, *Adv. Mater.*, **30**, p. e1803001.

129. Roeven, M. W., Hobo, W., van der Voort, R., Fredrix, H., Norde, W. J., Teijgeler, K., Ruiters, M. H., Schaap, N., Dolstra, H. (2015). Efficient nontoxic delivery of PD-L1 and PD-L2 siRNA into dendritic cell vaccines using the cationic lipid SAINT-18, *J. Immunother.*, **38**, pp. 145–154.

130. Sioud, M. (2019). Releasing the immune system brakes using siRNAs enhances cancer immunotherapy, *Cancers (Basel)*, **11**, p. 176.

131. Ordikhani, F., Uehara, M., Kasinath, V., Dai, L., Eskandari, S. K., Bahmani, B., Yonar, M., Azzi, J. R., Haik, Y., Sage, P. T., et al. (2018). Targeting antigen-presenting cells by anti-PD-1 nanoparticles augments antitumor immunity, *JCI Insight*, **3**, p. e122700.

132. Wang, H., Lam, C. H., Li, X., West, D. L., Yang, X. (2018). Selection of PD1/PD-L1 X-Aptamers, *Biochimie*, **145**, pp. 125–130.

Chapter 6

A Perspective of Ultrasound-Related Micro/Nano Cancer Therapy

Tingting Zheng,[a,b,*] Yun Chen,[a,b,*] Jiao Peng,[a,*] Yu Shi,[a,b] Jun Zhang,[c] Haitao Xiao,[d] Xiangmei Chen,[a] Yongcan Huang,[a] Tao Pei,[a] Zhuxia Zhang,[a] Xue Zhang,[a] Xiaohe Bai,[a] Li Liu,[a] and Jinrui Wang[e]

[a]*Department of Ultrasound, Peking University Shenzhen Hospital, Shenzhen 518036, China*
[b]*Shenzhen Peking University–The Hong Kong University of Science and Technology Medical Center, Shenzhen 518036, China*
[c]*Queensland Micro- and Nanotechnology Centre, Griffith University, Brisbane QLD 4111, Australia*
[d]*Shenzhen University Health Science Center, Shenzhen University, Shenzhen 518055, China*
[e]*Department of Ultrasound, Peking University Third Hospital, Beijing 100191, China*

dr.tingting.zheng@gmail.com

Ultrasound (US) is currently identified not only as a noninvasive diagnosis tool but also as a promising mineral- and/or non-invasive therapeutic tool in hospitals. According to ultrasound adjustability, which includes adjusting frequency, intensity, mode, and waveform, theranostic treatment can be performed in

---

*These authors contributed equally.

Nanoparticle Drug Delivery Systems for Cancer Treatment
Edited by Hala Gali-Muhtasib and Racha Chouaib

ISBN 978-981-4800-54-9 (Hardcover), 978-0-429-34125-0 (eBook)
www.jennystanford.com

different parameter combinations, of which high-intensity focused ultrasound and low-intensity focused ultrasound are two frequently performed treatments, especially for cancer treatment. Most important traits of cancer theranostic focused ultrasound among others are sonothermal effect, ultrasonic cavitation effect, sonoporation effect, sonodynamic effect, and ultrasound targeted drug delivery. To date, high-intensity focused ultrasound (HIFU) ablation has already been applied clinically; however, low-intensity focused ultrasound (LIFU) targeted cancer therapy is still in preclinical study phase. One of the main challenges for its clinical transition ascribes to its low therapeutic efficiency. Scientists therefore induce hybrid micro-/nanoparticles, which play a role as contrast and/or drug delivery carriers in a microscopic scale. In assistance of hybrid micro-/nanobubbles and correlative smart nanoparticles, ultrasound (e.g., LIFU)-related cancer therapy has been verified as a promising choice with minimal side effects or non-trauma. In this chapter, we will discuss the use of HIFU and LIFU approaches to cancer therapy in combination with micro-/nanoparticles to maximize therapeutic effects.

6.1 Introduction

Ultrasound is quite popular as a mineral- and/or non-invasive approach for disease diagnosis and therapy. In the 1980s, ultrasound was first successfully applied in clinical studies for extracorporeal shock wave lithotripsy (ESWL), but not for cancer therapy in the very beginning [1]. However, scientists noticed that focused ultrasound can gather acoustic energy into a target tissue and kill target cells flashily [2]. In the late 20th century, ultrasound was used for cancer therapy in preclinical and clinical studies, basically utilizing its thermal effect, mechanical effect, biologic effect, nonlinear effect, tumor angioembolism, immunological effect and synergistic effect with chemotherapy [3–9]. In general, there are two correlative noninvasive cancer treatments related to ultrasound, one is called "sonothermal therapy," and the other "sonodynamic therapy" [10]. Sonothermal cancer therapy is mainly achieved with high-intensity focused ultrasound (HIFU), while sonodynamic therapy is mainly associated with low-intensity focused ultrasound (LIFU) in combination with a sonosensitizer [2, 11, 12]. One of the most attractive peculiarities of ultrasound cancer therapy in comparison to laser is that it can penetrate into deep tissues where laser has depth limits. Therefore, ultrasound, especially a focused one, is promising to irradiate specific cancer lesions, kill cancer cells and/or deliver drugs locally, and thus minimize the damage to surrounding healthy tissues [13]. Clinically, HIFU is an attractive and frequently used noninvasive treatment for diseases, including cancer therapy [14–16]. HIFU applies higher intensity ultrasound compared with diagnosis ultrasound. Comparatively, HIFU applies ultrasound frequency ranging 0.8–3.2 MHz, focal intensity 100–10,000 mW cm^{-2}, pressure of 10–30 MPa, while diagnosis ultrasound applies intensity ranges from 0.1–100 mW cm^{-2}, and the pressure is 0.001–0.003 MPa [17–19]. Thermal effects and mechanical effects are most important in HIFU cancer treatment, leading to cell and cancer tissue damage. Basically, the mechanism of HIFU-related cancer therapy includes thermal effects, mechanical effects, biologic effects and thermal mechanisms, nonlinear effects, anti-angiogenesis effects, immunological effects, synergistic effects with chemotherapy, relative merits, and clinical applications, among others. Nevertheless, ultrasound cancer therapy is not yet "perfect." High energy input needs radiation dosage and area control; otherwise healthy tissues (e.g., skin) get burned. This issue can be solved via magnetic resonance imaging (MRI) guided focused ultrasound (MRgFUS) cancer ablation surgery. Yet, HIFU is not suitable for all kinds of cancer therapy; for instance, surgery at the alimentary canal is indeed a challenge. Air bubbles in the alimentary canal not only cause cavitation effects but also lead to energy back scattering, yielding significant areas of tissue burned. In other words, high-energy high-intensity ultrasound cancer surgery is not an easy operation approach, and it highly depends on the experience of the surgeon. In order to reduce and even avoid patient discomfort and pain after cancer ablation, LIFU therapy is attracting more and more attention in preclinical studies [20]. According to the ultrasound irritation range, LIFU can be applied in the form of either low frequency (20–100 kHz) or high frequency (0.8–3.2 MHz). Unlike HIFU cancer therapy, which is mainly related to the ultrasound induced thermal effect, LIFU will not induce much ultrasound energy accumulation at the focal zone. The key mechanism of

LIFU cancer therapy is based on ultrasound improving the tumor targeted drug control and release, and ultrasound-related cellular biological effects. Basically, the mechanism of LIFU-related cancer therapy focuses on, for instance, the ultrasound cavitation effect, targeted drug and gene delivery, and sonodynamic therapy, among others.

In ultrasound-related micro-/nano cancer theranostics, ultrasound cavitation induced micro-/nanobubbles are most commonly applied. Basically, bubbles are composed of lipidate layer and gas content. The outer layer is commonly made of phospholipid, but in unique situations outer layers can be biomacromolecules (e.g., albumin, saccharides, nonionic surfactants, gelatin, and so on) [21, 22]. Besides, the gas in the core usually consists of inert gases (e.g., sulfur hexafluoride, perfluocarbon, perfluorinated propane, among others), but in some cases oxygen, sulfuretted hydrogen, and nitrogen are included [23–25]. Commercially, plane microbubbles (e.g., Sonovue) only have less than 15 min half-life in systemic circulation [26]. The number of detectable microbubbles from US imaging decreases rapidly, due to gas leakage from microbubbles as well as non-specific viscera accumulations, especially in lung, liver, and spleen [27–29]. Hence, in the case of accurate and specific ultrasound contrast imaging and/or drug controlled release, the use of a targeting ligand and/or antibody modification is necessary.

HIFU cancer therapy is usually related to sonothermal effects, achieving tumor ablation, while LIFU cancer therapy frequently involves targeted sonodynamic therapy, achieving cellular targeting drug release. Further, cavitation effects are important in both LIFU and HIFU cancer therapy.

6.2 Micro-/Nanobubble: Composition, Preparation and Functionalization

Surfactant composed micro-/nanocavities with gas-filled-in, so called micro-/nanobubbles were first being generated as agents for contrast-enhanced ultrasound (CEUS) sonography [30]. Bubbles are polydispersed under conventional technology. Colloids are always a commixture of micro- and nanobubbles. The amount of microbubbles component in the commixture plays a key role in the quality of ultrasound contrast imaging. In liquid, for instance, blood flow, bubbles are relatively stable until equilibrium between the internal and external pressure is achieved.

Microbubbles are bubbles with sizes ranging from 1 to 100 μm. Larger bubbles (>20 μm) are not stable as smaller ones (1–20 μm); however, the smaller ones shrink to form more effective and stable nanobubbles (<1 μm) (Fig. 6.1) [31]. Suitable size of microbubbles resonate with medical ultrasound transdunacer frequencies in the range of 0.2–15 MHz, yielding in sharp contrast signals in microbubbles perfusion angiography [32]. Once microbubbles shrink into nanobubbles, they do not improve contrast quality of US imaging anymore. Nevertheless, both micro-/nanobubbles are promising candidates for drug delivery; even more, nanobubbles have higher efficiency acting as a drug/gene carrier due to their larger specific surface area compared with microbubbles. One of the mostly applied surfactants for surface coatings of bubbles are phospholipids, which have a hydrophilic head and two hydrophobic tails, which spontaneously form into a monolayer around gas cores (Fig. 6.2A). Abundant modifications can be performed on the amphiphilic lipid surface, including antibody surface decoration (Fig. 6.2B), drug/gene loading (Fig. 6.2C,D) and nanoparticle encapsulation (Fig. 6.2E).

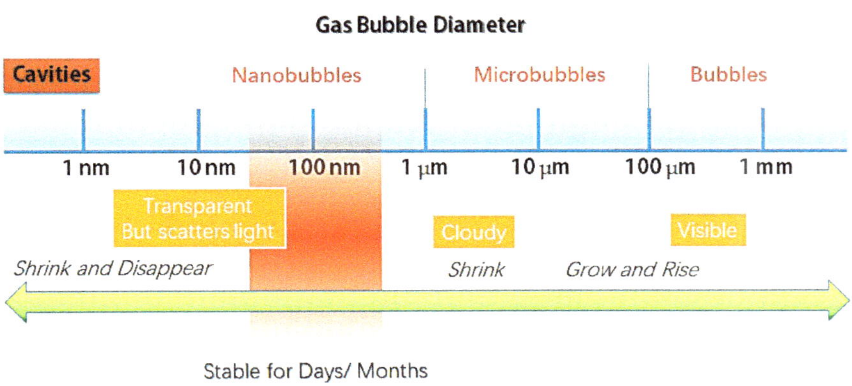

Figure 6.1 Schematic diagram of the physical and chemical properties and stabilities of gas bubbles of different sizes.

To date, many conjugation approaches have been reported for micro-/nanobubbles surface targeted ligand modifications, mainly involving covalent and non-covalent methods. Typical non-covalent conjugation methods is about electrostatic interaction (e.g., ionic, hydrogen bonding, halogen bonding), hydrophobic interaction, Van der Waals forces, and Π-Π stacking, among others [33–36]. On the other hand, targeting ligands could be immobilized on the distal part of hydrophilic monolayer assembled PEG (polyethylene glycol) chain via biotin/avidin coupling, thioether, amide binding and/or disulfide binding (Fig. 6.2) [37–42].

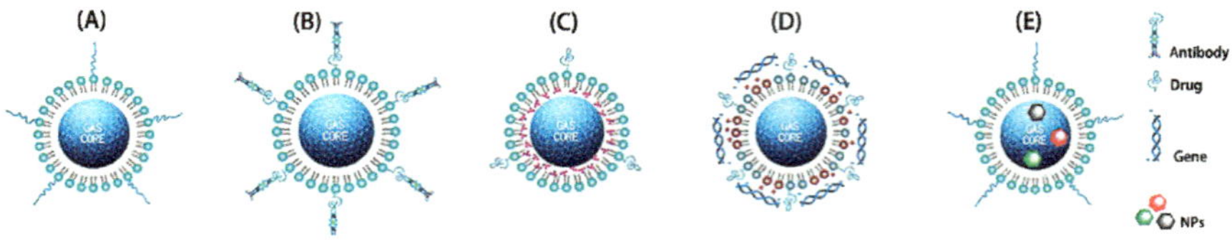

Figure 6.2 Schematic diagram of different types of microbubbles. Gas cores are all encapsulated in a monolayer of phospholipid shell. (A) indicates plane shelled microbubble for ultrasound imaging contrast enhancement; (B) indicates microbubbles with antibody conjugation which is designed for targeting imaging; (C) indicates a microbubble with drug encapsulation design for drug delivery; (D) indicates a microbubble with positive charged modification for gene delivery; (E) indicates a microbubble with nanoparticles encapsulated for synergistic treatment. (Blue content indicates gas core, NPs is short for nanoparticles).

Cancer tissues are rich with abundant blood vessels, thus micro-/nanobubbles can be accumulated briefly via enhanced permeability and retention (EPR) effects [44–46]. For a longer stay in blood vessels, vessel targeting microbubbles are used. Inspired by naturally flowing particles, there are many methods that could be applied for efficient vessel binding [47, 48]. Figure 6.3A shows a binding technique which is achieved by using tethers. Molecules and/or supramolecules (e.g., PEG) are extended as a nanometer scaled spacer on the surface of micro-/nanobubbles, and microbubble localization is achieved via ligand-receptor binding. Micrometer sized surface anchor modification on microbubbles can be easily achieved via 1 min limited compression ~100 KPa with microbubbles gas-core-diffusion [49]. On account of internal gas leakage, excess lipid could be folded on the surface of micro-/nanobubbles. Such folding is used to provide more surface and contact area, thus extending the total length of reachable binding site, and improving targeted tethering of mechanical conditions of micro-/nanobubbles to the vessel wall (Fig. 6.3B). However, in real blood vessels, the blood flushes very fast, leading to the rapid passage of micro-/nanobubbles. For instance, a flow rate of ~1 m/s, ~1 ns dwell time for a 1 nm binding site, is nonsufficient time for the ligand to "knob into holes" of the receptor. Scientists therefore learned from nature, and introduced polymers (e.g., Lewis oligosaccharide derivative, P-selectin glycoprotein ligand-1 (PSGL-1) as grooves to load ligands locally with high density (Fig. 6.3C), in order to achieve a more stable targeted vessel binding.

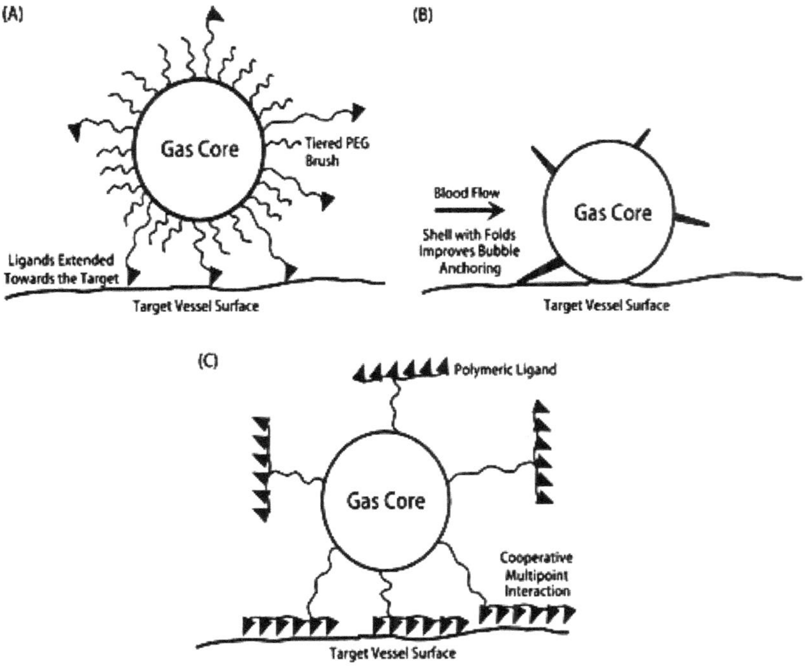

Figure 6.3 Schematic diagram about targeted microbubbles. (A) Tiered PEG is described as coated with microbubbles. (B) Membrane folds that are formed after partial deflation of microbubbles. Under such circumstances, it can act as tethers and then firmly binds with the target substrate. (C) Polymeric version of the ligand on the bubble increases surface density of targeting receptor [48]. Copyright Springer Link.

6.3 Micro-/Nanobubble-Related Cancer Therapy

Detection and propagation of high-frequency (>20 kHz) acoustic wave in tissue is quite a mature technique in clinical diagnosis [50]. Ultrasound impedance could distinguish among different tissues. The echo received could be processed to show 2D and/or 3D tissue images [51]. Besides, damped acoustic energy that accumulates in tissue could cause fluid streaming, tissue motion and heat in situ. For therapeutic adhibition, heating enables ultrasound applications in thermal ablation and/or local physiotherapy; while fluid steaming is related to transdermal sonophoresis [52]. One of the most typical ultrasound therapy effects is cavitation. Once there is a rapid decrease of local pressure, vaporization and/or evolution of dissolved gases show up. However, when there is an increase in local pressure mediated by ultrasound, bubbles form and then collapse violently, yielding significant damage at local and surrounding tissues or structures [53]. Moreover, pre-existing bubbles can be applied in combination with low amplitude ultrasound, leading to prominent mechanical effects.

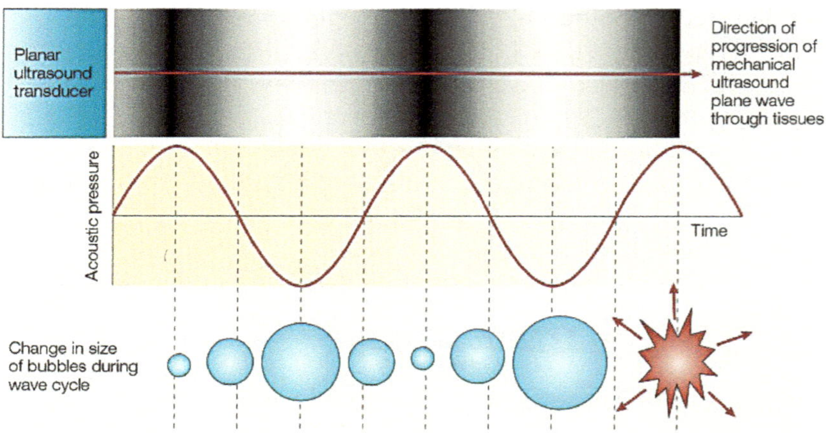

Figure 6.4 The principles of inertial cavitation. Alternating cycles of fluctuating pressure is the result of mechanical ultrasound wave progression, which proceeds through tissues. During the progress, gas is drawn out of the solution during rarefaction, and bubbles are then created as a result [11]. Copyright Nature Publishing Group.

It has been verified that cancer tissue is rich in vascularity which leads to sufficient oxygen, nutrients and energy supply. Thus, compared with healthy tissues, cancers grow much faster. Decreasing oxygen, and nutrient supply and/or increasing local drug concentrations in the tumor microenvironment are promising approaches for inducing cancer apoptosis. HIFU thermal effect (e.g., ablation) can lead to occlusion effects in tumor vessels; this makes the previous two suggestions of tumor starvation therapy possible [54]. Many studies show that a combination of microbubble applications with HIFU results in more effective cancer ablation, owing to the cavitation effect (Fig. 6.4) which is one of the most important properties that microbubbles show with suitable ultrasound frequency [55–59].

6.3.1 Ultrasound Thermal Delivery

The mechanism of Ultrasound Thermal Delivery is as follows: energy is absorbed leading to heating, especially when ultrasound passes through tissues. During this process, heat is used as a main tool for the treatment and cure. Particularly, thermal ablation of abnormal tissues is a special type of treatment for solid tumors [60]. As mentioned before, HIFU devices have come to use tightly focused ultrasonic waves targeted at thermally ablating the tissue [19, 61, 62]. Its combination with MRI has been regarded as a very efficient and instrumental approach to treat prostate cancers [49, 64]. To be more specific, the tightly focused ultrasound allows and even encourages lethal hyperthermia while destruction of healthy tissue is being minimized. On the other hand, the focal zone is way smaller than the target tumor, which could lead to long-term treatment.

As a result, therefore, combining HIFU with chemotherapy to help eliminate remaining cells has resulted in some interest in this field. To be specific, it has been shown that chemotherapy drugs which are currently used in HIFU are very likely to achieve greater therapeutic effect. Research in this area is broad, with many drugs being under investigation. These include doxorubicin, docetaxel, endostatin, paclitaxel and estramustine, adriamycin, mitomycin C, 5-fluorouracil, and cisplatin, and gemcitabine [65–74]. In addition to ablation, mild hyperthermia of tumors could also result from HIFU. This is because increased temperatures result in correspondingly enhanced permeability of blood vessels [75–81]. In addition, when tumor temperatures are increased, drug delivery and efficacy may be improved as a consequence [82, 83]. The immune system is another variable that matters, which has a huge impact on tumor progression [84–86].

Mild hyperthermia is often adopted by HIFU via the release of drugs from loaded nanoparticles. A number of nanoparticles have been designed so that the locally applied heat could be adapted to a tumor location [87–90]. To explain, nanoscale phase change agents encased in block copolymers have already been used to increase damage in a tumor during HIFU [91–94]. Vaporization of the particles that follow the last step would result in potential harm and destruction to the blood supply. Furthermore, it has been shown that both iron and gold nanoparticles loaded onto polymer-based agents would only incur more damage to HIFU ablation [95–97].

6.3.2 Ultrasound Mechanical Delivery

One of the most important and commonly applied ultrasound mechanical effects is the cavitation effect (Fig. 6.5). This may induce cytoskeleton destruction effects, mediate apoptosis, oxidative damage, drug transport enhancement, drug resistance reduction and immunomodulatory effects. Ultrasound disrupts the cytoskeleton and decreases cellular activity when it is applied to cells both in vitro and in vivo. Unstable bubbles ranging from 1–10 micrometer in diameter are generated when ultrasound penetrates into liquid medium, which is so called "Cavitation Effect." Cavitation can be manipulated via changing the tuning intensity and frequency of the ultrasound (Fig. 6.6). Under certain conditions, ultrasonic cavitation bubbles start vibrating, which leads to around 10 min reversible openness of cytomembranes with 150 nm ultrasonic aperture. This phenomenon is well known as "ultrasonic perforation," which can be performed either for cancer cell killing, or for increasing microvascular permeability to help deliver drugs more effectively.

Different from ultrasound thermal effect, which is about heat, ultrasound is believed to be able to transfer momentum to the tumor, and this is due to the existence of the pressure gradient [101–104]. As a result, this process would then lead to acoustic streaming and result in additional stoichiology effect for further tumor theranostics (Fig. 6.7) [105–110]. More so, the sonophoresis effect of acoustic radiation force (ARF) is well adopted in transdermal drug delivery applications [111, 112]. It has been suggested that ARF is able to enhance the tumoral uptake and effect of small-drug molecules [113–117]. On the other hand, it is best achieved with high intensities since ARF is determined by energy absorption.

As has been suggested in the previous sections, cavitation nucleating agents are another alternative adopted in drug delivery applications. They are used to reduce the intensity required for cavitation. Whilst a lot of molecules can function as ultrasound cavitation nuclei, the best studied types are the microbubbles [118–124]. Since microbubbles are highly echogenic, concurrent imaging can be performed as well, which could lead to better target site recognition [125–130].

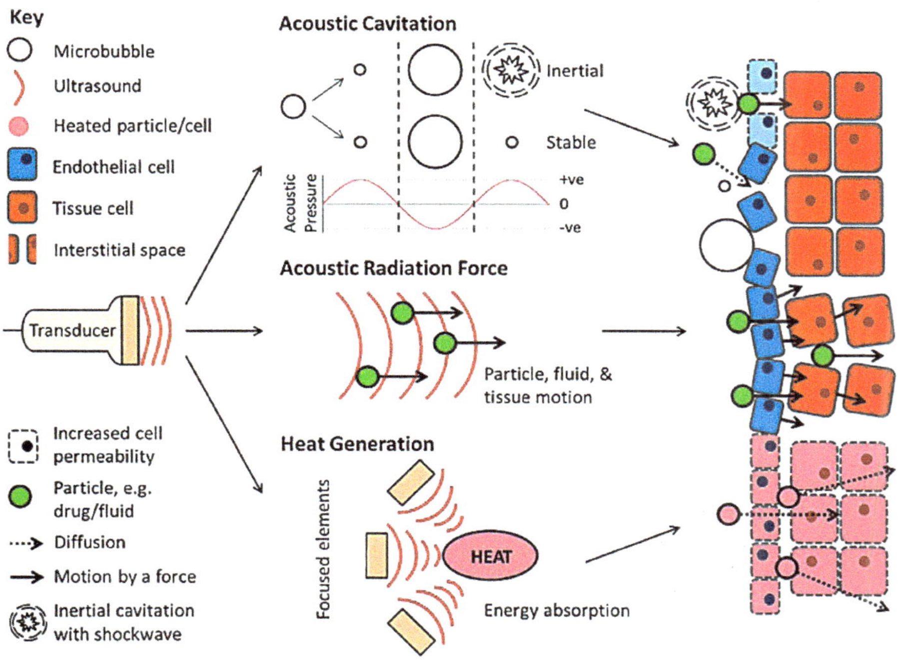

Figure 6.5 Ultrasound-mediated delivery is shown in the demonstration. These are categorized into the mechanical effects of acoustic cavitation and acoustic radiation force, in addition to thermal effects [98].

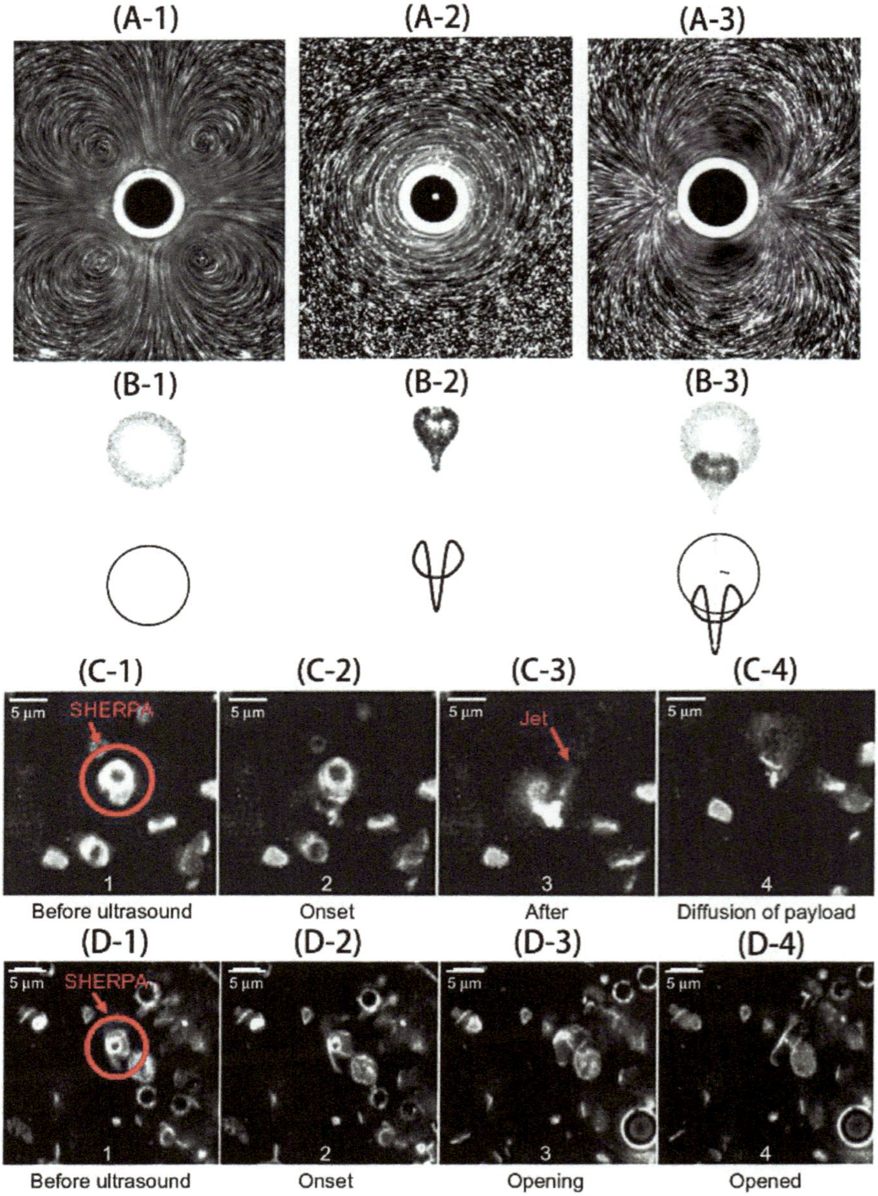

Figure 6.6 (A) Different microstreaming fluid flow patterns around microbubbles oscillating in diameter under the influence of ultrasound.

The microbubbles are the dark circles with white outlines in the center of each frame. The little white dots are particles that mark the streamlines of fluid flow patterns in the surrounding water. The microbubbles are all roughly the same size in each frame, and the different flow patterns shown in frames (A1-3) arise from different driving ultrasound frequencies. Copyright Elsevier. (B) Still images from a high-speed video of a microbubble undergoing asymmetric collapse during inertial cavitation creating a fluid jet. (B-1 up) the original microbubble before ultrasound exposure. (B-1 down) A schematic of the microbubble outline before ultrasound exposure. (B-2 up) the microbubble during ultrasound exposure, undergoing an asymmetric collapse where the left side actually collapsed through the airspace of the microbubble and penetrated through the right side. (B-2 down) schematic showing the microbubble outline during collapse. (B-3 up) Overlaid images of the microbubble before ultrasound exposure and during collapse. (B-3 down) Schematic of the before and during ultrasound exposure outlines of the microbubble [99]. Copyright Medical Physics. (C) and (D) are rupture and payload release patterns of drug delivery vehicles containing a microbubble. (C) The SHRPA drug delivery particle contained a microbubble within a fluorescently labeled liposome and is indicated by the red circle. Upon ultrasound exposure, the microbubble underwent inertial cavitation and shattered the surrounding lipid membrane, creating a debris field of lipid particles and a jet. (D) at lower intensity, the ultrasound pulse caused the microbubble to undergo microstreaming instead of inertial cavitation, causing a single rupture in the surrounding lipid membrane with subsequent slow opening as a single piece [100]. Copyright Elsevier.

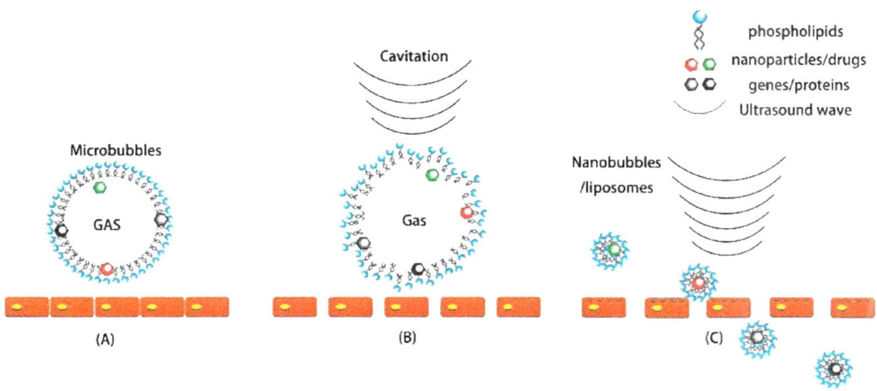

Figure 6.7 Representation of ultrasound induced targeted drug/gene delivery.

Usually, microbubbles are made up of a high-molecular-weight gas core on the order of 1 to 10 μm in diameter, encapsulated in a lipid, protein or, less commonly, a polymer shell [131]. Their high compressibility would be very much likely to incur significant linear and nonlinear scattering of incident acoustic waves [132–138]. To date, it has been found that ultrasound-responsive nanodroplets of liquefied perfluorocarbons (PFCs) for drug delivery has an ability to extravasate in leaky tumors [139–142]. More so, solid nanoparticles with entrained gas pockets are adopted for similar purposes as well [143, 144].

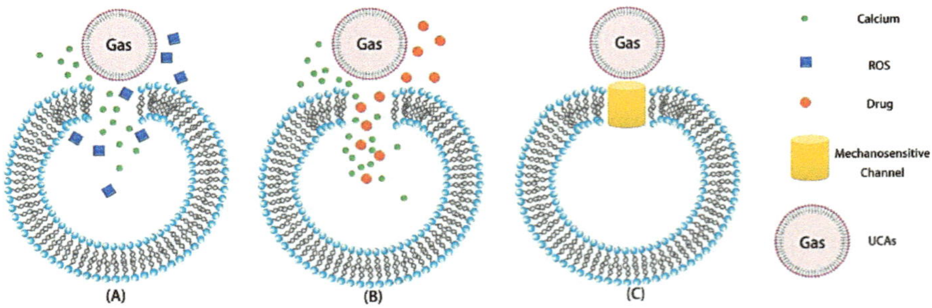

Figure 6.8 Biological effects of stable and inertial cavitating ultrasound contrast agents (UCAs). (A) Reactive oxygen species (ROS) are produced during cavitation. (B) The mechanical effects of stable and inertial cavitating UCAs: reversible pores are formed and allows the uptake of drugs and calcium. (C) The microstreamings generated by stably cavitating UCAs and the corresponding shear stresses can deform the cell membrane. This leads to cytoskeletal rearrangements and differences in membrane tension.

Cavitation agents can be driven to undergo stable oscillation or inertial collapse, depending upon the acoustic parameters and agent in use (Fig. 6.8) [145–147]. In general, driving agents to inertial collapse is accepted as promoting the greatest drug delivery, leading to temporarily increased permeability and delivery of any co-delivered therapeutic agent [21, 148–155]. However, the potential severity of blood vessel disruptions could be undesirable in certain locations in the body, for instance, the brain or kidneys [156–160]. In these areas, stable oscillation generates sufficient shear stress to disrupt cell membranes and increase vascular permeability; particularly in the highly selective blood brain barrier, recently reviewed in [154, 161–166].

Endocytosis pathways are also known to be activated by the acoustically driven shear stresses, and these are also implicated in delivery [167–171].

Apart from co-administration, cavitation agents can be changed and adapted to load the therapeutic drug of interest [172–177]. Generally speaking, the delivery, effect, and release would be promoted by colocalizing the drug or nanocarrier to the site of microbubble activity [178–183]. Last but not least, appropriate characterization of the cavitation agent and drug loading is of overriding significance in this field [184–186].

6.3.3 Ultrasound-Related Drug and Gene Delivery

As we showed above, micro-/nanobubbles are good carriers for drug and/or gene constructs delivery in vivo. In assisted ultrasound irradiation at fixed points, for instance, a tumor in situ, micro-/nanobubbles would result in the targeted release of the drug/gene.

It is no doubt that ultrasound can penetrate into the body in a deeper way as compared with laser therapy. To be specific, it can be concluded from Fig. 6.9 that vehicles which pass through the ultrasound focal volume would go through a remarkably higher intensity than in other conditions (Fig. 6.9).

In general, there are two main ways for drug/gene micro-/nanovehicle traits design. It can be observed from the figure that the first set of properties consist of the opposing drug containment requirements. However, the other set is that the vehicle should have long-circulation property and good cell penetration effect [22]. The vehicle must be able to encapsulate the drug/gene without significant leakage during its body circulations. Therefore, it could be demonstrated that an activation of microbubble-based drug delivery vehicles was achieved only inside the focal region of ultrasound (Fig. 6.9).

Under such circumstances, it can be spotted that a channel was molded. Fluorescently labeled drug delivery vehicles that contained a microbubble were designed with biotinylated lipids on the inside. What should be noticed here is that the vehicles fragmented could only be found in the focal zone. After that, these fluorescent biotinylated fragments are bound to the avidin, by which the location of activation would be marked.

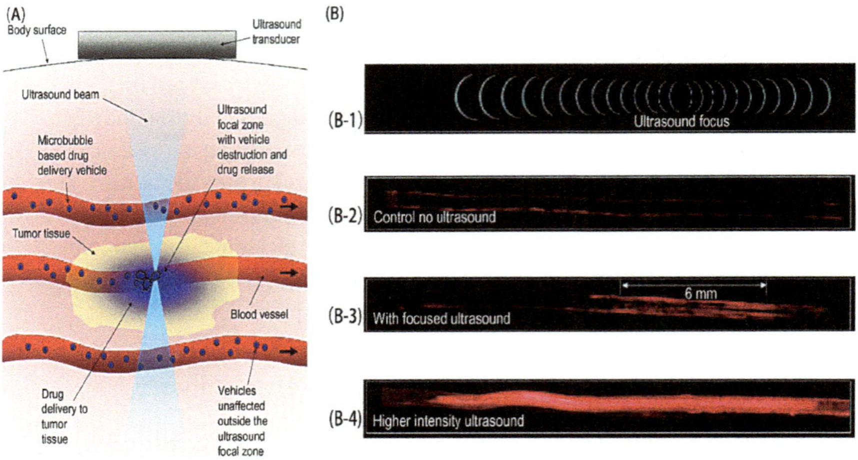

Figure 6.9 (A) Schematic diagram on ultrasound use with targeted drug delivery [22]. Copyright Dovepress. (B) In vitro demonstration of the localized activation of microbubble-based drug delivery vehicles. An agar tissue phantom with a molded channel down the center was used to model a blood vessel. Avidin was coated on the surface of the entire channel to simulate a targeting molecule. Fluorescently labeled drug delivery vehicles containing a microbubble were introduced into the entire length of the channel. The inside of the vehicles was biotinylated to cause binding to the avidin-coated channels when exposed by ultrasound-induced vehicle rupture. (B-1) the focal zone of the ultrasound beam. (B-2) the fluorescent vehicles were introduced to the channel then washed out with water with no ultrasound exposure to demonstrate the background level of nonspecific binding. (B-3) When low intensity ultrasound was applied, only vehicles in the focal zone were activated were left behind after washing. (B-4) Higher intensity ultrasound caused a larger number of vehicle activations and expanded the effective focal zone over which activation would occur [100]. Copyright Elsevier.

Microbubble cavitation drug delivery has been verified with mechanical targeting properties. Patients suffering from chemotherapy side effects would benefit from that.

6.3.4 Sonodynamic Therapy

Sonodynamic therapy (SDT) is one of the most important ultrasound-related cancer treatments [187–189]. Compared with photodynamic therapy (PDT), acoustic waves can penetrate deeper into tissues than laser; therefore SDT has attracted more attention in preclinical studies. There are three key points to trigger SDT. These include ultrasound, sonosensitizers, and oxygen molecules [13, 187, 190, 191].

The cancer microenvironment has relatively high oxygen concentrations than healthy tissues, once we induce sonosensitizer in ultrasound focal area, oxygen molecules chemically generate reactive oxygen species (ROS) which destroy local cellular membranes and their relevant biomolecules (e.g., proteins, genes and so on) [192]. ROS is cytotoxic when it reacts with healthy cells, and damages them as well. Therefore, focused ultrasound not only gives sonosensitizer energy support, but also enables targeted ROS release (Fig. 6.10). In order to help sonosensitizer molecules accumulate in cancers via EPR effects, micro-/nano-sized particles are more effective. Either the sonosensitizer is self-assembled into micro-/nano-sized particles, or micro-/nanoparticles are applied as micro-/nanocapsules. Owing to the excellent tissue penetration of ultrasound, PDT is a promising cancer treatment for wider array of deeper and/or less accessible tumors [193].

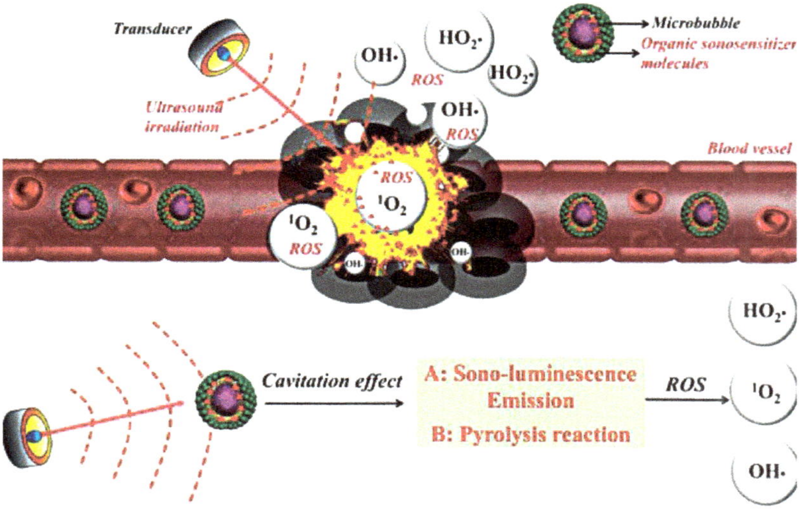

Figure 6.10 Schematic diagram about SDT mechanism. Microbubbles loaded nano-sonosensitizers accumulate in tumor blood vessels via EPR effect, and then SDT is triggered by ultrasound. Microbubbles release sonosensitizers by cavitation effect and ROS is generated by receiving energy from ultrasound [194].

Typical sonosensitizers include porphyrins, xanthene, and their derivatives, organic and inorganic micro-/nanoparticles augmented sonosensitizers, chemotherapeutics sonosensitizers, cytoskeletal sonosensitizers, among others [187, 194].

However, traditional sonosensitizers are less water-soluble and have faster elimination from circulation, yielding low concentrations at focal zones (Fig. 6.11). Micro-/nano particles (e.g., microbubbles, Au, SiO_2, among others) play important roles to assist sonosensitizers and provide smart control and release functions [190, 195–197]. In addition, some inorganic nanoparticles themselves are intrinsic sonosensitizers (e.g., TiO_2, black phosphorus, etc.) [197–199].

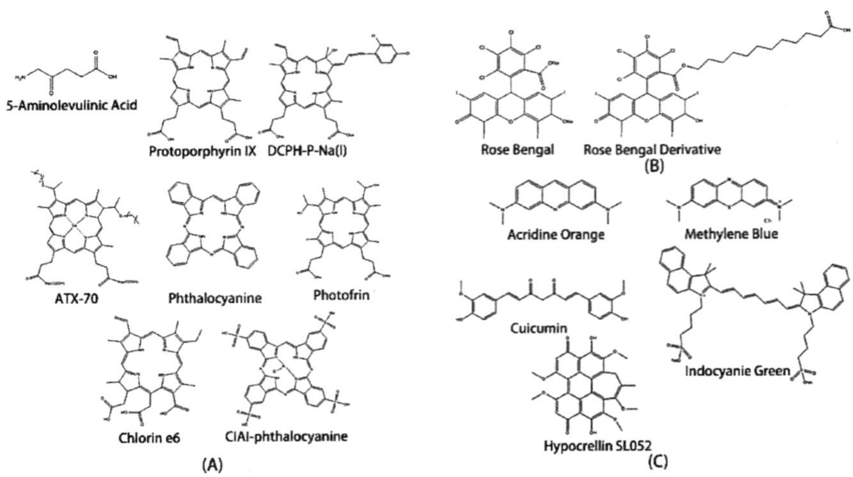

Figure 6.11 (A) Examples of sonosensitizers based on porphyrin and its related structures. (B) Examples of xanthene based sonosensitizers. (C) Other sonosensitizers [200].

FDA-approved nanomedicines are usually good candidates for sonosensitizer nanocarriers. For instance, Chen's group suggested using targeted liposome to encapsulate hydrophobic hematoporphyrin monomethyl ether (HMME). In this way, HMME can have a good colloidal stability in aqueous media, and its release could be triggered with ultrasound irritation on liposome membranes via lipid oxidation [201]. Canaparo's group showed an attractive positively charged polymer nanoparticles which is self-assembled from poly methyl methacrylate core shell structures. They adsorbed sonosensitizer meso-tetrakis (4-sulfonatophenyl) porphyrin (TPPS) on its surface, yielding TPPS-NPs. Cellular uptake of these positively charged nanoparticles significantly increased, leading to the intracellular sonosensitizer accumulation in the target cancer. The systemic input of a shock wave, focusing the ultrasound on the tumor, resulted in 15 fold higher reactive oxygen species release from local cancer tissue as compared with free TPPS sonodynamic effect [202].

Nanoparticles assisted sonosensitizers are commonly used in cancer SDT solutions (Fig. 6.12). Two most typical types of nanoparticles which have been frequently applied in preclinical studies are involved with organic nanoparticles and inorganic nanoparticles. In order to improve either solubility or targeting capabilities of sonosensitizers, organic nanoparticles are introduced in some cases due to their facile preparation, low biotoxicity and relatively acceptable biodegradation speed [203].

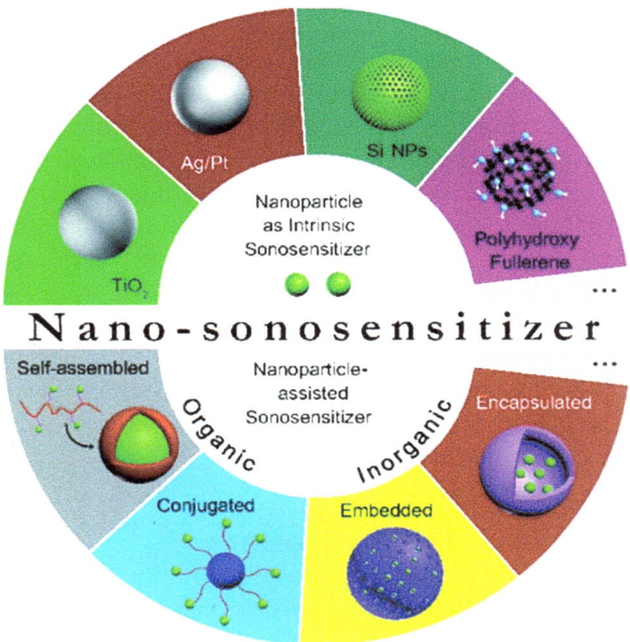

Figure 6.12 Schematic diagram of sonosensitizers used in sonodynamic therapy [189].

In comparison with organic nanocarriers, inorganic nanoparticles are more stable chemically and physiologically, meanwhile, they always have larger specific surface area as well as multifunctional active groups [203]. One of the most extensively used nanodevices are mesoporous silica nanoparticles (MSNs) [204]. Chen's group manipulated graphene oxide nanosheets (nrGO) coated MSNs iron-oxide conjugated rose Bengal nanoparticles (MSN-ION-PEG-RB), yielding nrGO@MSN-ION-PEG-RB, as sonosensitizer, and activated ROS via 2 W cm^{-2}, 1.5 MHz, 1 min focused ultrasound irradiation with starting temperature of 20.7°C [205]. Beside, metal nanoparticles, Au has been reported as a promising sonosensitizer nanocarrier for cancer SDT treatment [190]. Shanei's group modified protoporphyrin IX (PpIX) on the surface of gold nanoparticles for an in vivo investigation based on a colon tumor model. They showed that Au NPs can play as cavitation nuclei, causing the collapse of much more significant cavities [206].

Sonodynamic therapy (SDT) shows great potential in a significant penetration of deep cancer tissue in a non-invasive operation. Aside from this advantage, SDT is considered as a high precision method, having advanced patient compliance, and with less side effects reported. In general, it is a burgeoning noninvasive preclinical tumor treatment. It combines ultrasound with a sonosensitizer for synergistic therapeutic effects. The mechanism of cancer sonodynamic therapy includes ultrasonic cavitation effects, cytoskeleton destruction, cell apoptosis abduction, oxidative damage, drug target delivery, drug resistance reduction, and immune regulation. The sonosensitizer plays an important key in cancer SDT process, deriving from porphyrin to xanthene serial compounds. To date, SDT has been verified to have significantly high cancericidal effects in vitro, ex vivo and in vivo in abundant preclinical studies.

6.4 Summary and Conclusions

Ultrasound-related micro-/nanoparticle cancer therapy is now broadly being verified in preclinical studies. Most attempted treatments are relying on a combination of ultrasound driven NPs targeted mechanical delivery and sonosensitizer based sonothermal effect and/or sonodynamic therapy. Intravenous injection and intratumoral injection are frequently used administrations.

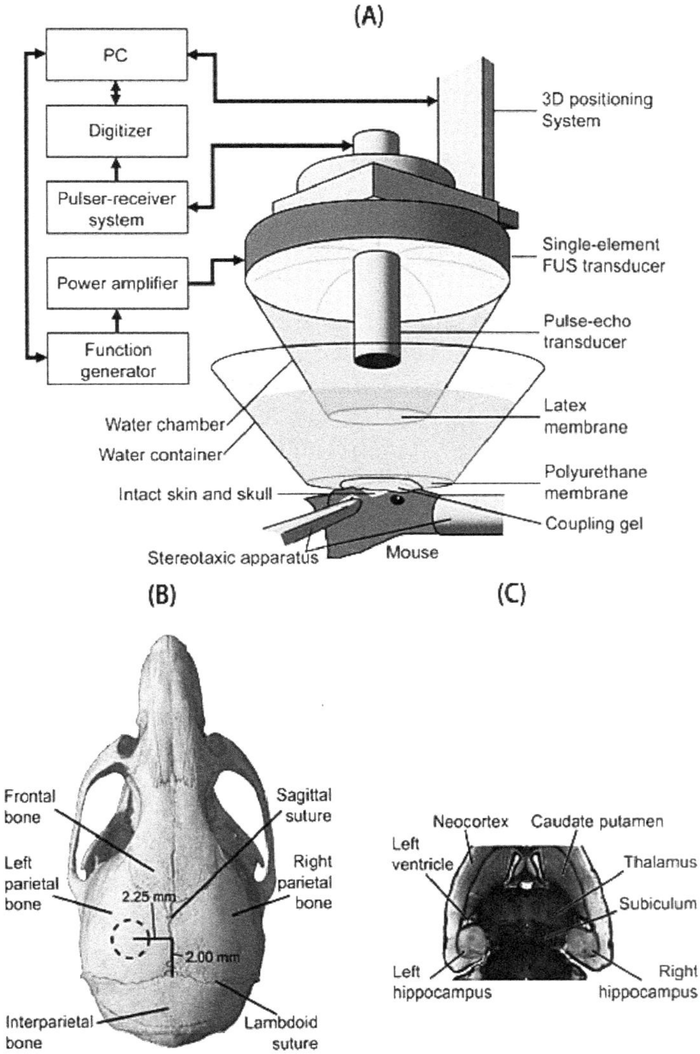

Figure 6.13 (A) Schematic diagram of focused ultrasound induced blood brain barrier opening experimental setup. (B) Acoustic waves propagated through left parietal bone and converged to a 1.32 mm diameter region (dotted circle). (C) Left hippocampus was targeted while the right region acted as a control [207].

Particularly, sonosensitive nanoparticles application in cancer treatment of the central nervous system needs to conquer the blood brain barrier. For instance, focused ultrasound (FUS) need to be conducted to temporary open the blood brain barrier after nano-sonosensitizer application in circulation (Fig. 6.13). According to all the promising applications of ultrasound-related micro-/nanoparticles cancer therapy, we expect further research to facilitate its translation from preclinical studies into the clinical fields.

Acknowledgment

We acknowledge the support of the National Natural Science Foundation of China (81871358); China postdoctoral fund (2018M640807); Ministry of Science and Technology of China (2016YFC0104707); Natural Science Foundation of Guangdong Province, China (2018A0303130228); the Science and Technology Planning Project of Shenzhen Municipality, China (JCYJ20170412171856582) and Joint Project of Peking University-Griffith University No. 036 Research Internal, Health and Family Planning Commission of Shenzhen Municipality (SZSM201512026).

References

1. Martin, X., Mestas, J. L., Cathignol, D., Margonari, J., Dubernard, J. M. (1986). Ultrasound stone localization for extracorporeal shock-wave lithotripsy, *British Journal of Urology*, **58**, pp. 349–352.
2. Kremkau, F. W. (1979). Cancer-therapy with ultrasound–historical review, *Journal of Clinical Ultrasound*, **7**, pp. 287–300.
3. Ali, A., Hamed, A., Hixon, R., Mankbadi, R., Mobarek, A., Rizk, M. (2013). Effect of inflow treatment on acoustic radiation from large-scale structure in a round jet, in: Aerospace Sciences Meeting & Exhibit.
4. Dason, S., Allard, C. B., Hoogenes, J., Orovan, W., Shayegan, B. (2018). High-intensity focused ultrasound (HIFU) as salvage therapy for radio-recurrent prostate cancer: Predictors of disease response, *International Brazilian Journal of Urology*, **44**, pp. 248–257.
5. Ladiges, W. C., Zeidner, N. S. (1980). An overview of feline cancer therapy, *Feline Practice*, **10**, pp. 38–43.
6. Strunk, H. M., Lützow, C., Henseler, J., Mücke, M., Rauch, M., Marx, C., Schild, H. H., Marinova, M. (2018). Mesenteric vessel patency

following HIFU therapy in patients with locally invasive pancreatic cancer, *Ultraschall in Der Medizin*, pp. s-0043-125391.

7. Sun, S. J., Xu, Y. X., Fu, P., Chen, M., Sun, S. H., Zhao, R. R., Wang, J. R., Liang, X. L., Wang, S. M. (2018). Ultrasound-targeted photodynamic and gene dual therapy for effectively inhibiting triple negative breast cancer by cationic porphyrin lipid microbubbles loaded with HIF1 alpha-siRNA, *Nanoscale*, **10**, pp. 19945–19956.
8. Yan, K., Wachsberger, P., Yu, Y. (2009). Investigation of acoustic radiation force for radio-protecting normal tissues during radiation therapy, *Conf Proc IEEE Eng Med Biol Soc*, **2009**, pp. 2304–2307.
9. Yildirim, A., Shi, D., Roy, S., Blum, N. T., Chattaraj, R., Cha, J. N., Goodwin, A. P. (2018). Nanoparticle-mediated acoustic cavitation enables high intensity focused ultrasound ablation without tissue heating, *ACS Applied Materials & Interfaces*, **10**, pp. 36786–36795.
10. Wood, A. K. W., Sehgal, C. M. (2015). A review of low-intensity ultrasound for cancer therapy, *Ultrasound in Medicine & Biology*, **41**, pp. 905–928.
11. Kennedy, J. E. (2005). High-intensity focused ultrasound in the treatment of solid tumors. *Nature Reviews Cancer*, **5**, pp. 321–327.
12. Tempany, C. M. C., Mcdannold, N. J., Hynynen, K., Jolesz, F. A. (2011). Focused ultrasound surgery in oncology: Overview and principles, *Radiology*, **259**, pp. 39–56.
13. Kuroki, M., Hachimine, K., Abe, H., Shibaguchi, H., Kuroki, M., Maekawa, S., Yanagisawa, J., Kinugasa, T., Tanaka, T., Yamashita, Y. (2007). Sonodynamic therapy of cancer using novel sonosensitizers, *Anticancer Research*, **27**, p. 3673.
14. Jingqi, W., Lu, Z., Jun, Z., Yuhong, M., Wei, Y., Lifeng, R., Chengbing, J., Dobromir, D, D., Hui, Z., Kun, Z. (2018). Clinical Usefulness of the Microbubble Contrast Agent SonoVue in Enhancing the Effects of High-Intensity Focused Ultrasound for the Treatment of Adenomyosis, *J Ultrasound Med.*, **37**(12), pp. 2811–2819.
15. Linares-Espinós, E., Carneiro, A., Martínez-Salamanca, J. I., Bianco, F., Castro-Alfaro, A., Cathelineau, X., Valerio, M., Sanchez-Salas, R. (2018). New technologies and techniques for prostate cancer focal therapy: A review of the current literature, *Minerva Urologica E Nefrologica*, **70**.
16. Shah, R., Ostapoff, K. T., Kuvshinoff, B., Hochwald, S. N. (2018). Ablative therapies for locally advanced pancreatic cancer, *Pancreas*, **47**, p. 6.

17. Lang, H. H., Wu, A. L. H. (2017). High intensity focused ultrasound (HIFU) ablation of benign thyroid nodules—a systematic review, *Journal of Therapeutic Ultrasound*, **5**, p. 11.

18. Mahmoud, M. Z., Alkhorayef, M., Alzimami, K. S., Aljuhani, M. S., Sulieman, A. (2014). High-intensity focused ultrasound (HIFU) in uterine fibroid treatment: Review study, *Polish Journal of Radiology*, **79**, pp. 384–390.

19. Wu, Feng. (2018). High-intensity focused ultrasound in the treatment of breast tumours, *Ecancermedicalscience*, **12**, pp.

20. Ahmadi, F., Mcloughlin, I. V., Chauhan, S., Ter-Haar, G. (2012). Bio-effects and safety of low-intensity, low-frequency ultrasonic exposure, *Progress in Biophysics & Molecular Biology*, **108**, pp. 119–138.

21. Ferrara, K., Pollard, R., Borden, M. (2007). Ultrasound microbubble contrast agents: Fundamentals and application to gene and drug delivery, *Annual Review of Biomedical Engineering*, **9**, pp. 415–447.

22. Ibsen, S., Schutt, C. E., Esener, S. (2013). Microbubble-mediated ultrasound therapy: A review of its potential in cancer treatment, *Drug Design Development & Therapy*, **7**, pp. 375–388.

23. Carroll, B. A., Turner, R. J., Tickner, E. G., Boyle, D. B., Young, S. W. (1980). Gelatin encapsulated nitrogen microbubbles as ultrasonic contrast agents, *Investigative Radiology*, **15**, pp. 260–266.

24. Senior, R., Moreo, A., Gaibazzi, N., Agati, L., Tiemann, K., Shivalkar, B., Von, B. S., Galiuto, L., Lardoux, H., Trocino, G. (2013). Comparison of sulfur hexafluoride microbubble (SonoVue)-enhanced myocardial contrast echocardiography with gated single-photon emission computed tomography for detection of significant coronary artery disease: A large European multicenter study, *Journal of the American College of Cardiology*, **62**, pp. 1353–1361.

25. Tang, C., Fang, K., Guo, Y., Li, R., Fan, X., Chen, P., Chen, Z., Liu, Q., Zou, Y. (2017). Safety of sulfur hexafluoride microbubbles in sonography of abdominal and superficial organs: Retrospective analysis of 30,222 cases, *Journal of Ultrasound in Medicine*, **36**, p. 531.

26. Schneider, M. (1999). Characteristics of SonoVuetrade mark, *Echocardiography*, **16**, p. 743.

27. Perkins, A. C., Frier, M., Hindle, A. J., Blackshaw, P. E., Bailey, S. E., Hebden, J. M., Middleton, S. M., Wastie, M. L. (1997). Human biodistribution of an ultrasound contrast agent (Quantison) by radiolabelling and gamma scintigraphy, *British Journal of Radiology*, **70**, p. 603.

28. Tartis, M. S., Kruse, D. E., Zheng, H., Zhang, H., Kheirolomoom, A., Marik, J., Ferrara, K. W. (2008). Dynamic microPET imaging of ultrasound contrast agents and lipid delivery, *Journal of Controlled Release Official Journal of the Controlled Release Society*, **131**, pp. 160–166.

29. Walday, P., Tolleshaug, H., Gjøen, T., Kindberg, G. M., Berg, T., Skotland, T., Holtz, E. (1994). Biodistributions of air-filled albumin microspheres in rats and pigs, *Biochemical Journal*, **299** (Pt 2), pp. 437–443.

30. Stride, E., Edirisinghe, M. (2008). Novel microbubble preparation technologies, *Soft Matter*, **4**, pp. 2350–2359.

31. Temesgen, T., Bui, T. T., Han, M., Kim, T. I., Park, H. (2017). Micro and nanobubble technologies as a new horizon for water-treatment techniques: A review, *Advances in Colloid and Interface Science*, **246**, pp.

32. Deelman, L. E., Anne-Emilie, D., Rychak, J. J., Kumar, S. (2010). Targeted renal therapies through microbubbles and ultrasound, *Advanced Drug Delivery Reviews*, **62**, pp. 1369–1377.

33. Johnson, D., Hilal, N., Waters, K., Hadler, K., Cilliers, J. (2009). Measurements of interactions between particles and charged microbubbles using a combined micro-and macroscopic strategy, *Langmuir the ACS Journal of Surfaces & Colloids*, **25**, p. 4880.

34. Matsuura, K., Uchida, T., Ogawa, S., Guan, C. (2015). Surface interaction of microbubbles and applications of hydrogen-bubble method for cleaning and separation, in: *International Symposium on Micro-Nanomechatronics & Human Science.*

35. Cavalieri, F., Micheli, L., Zhou, M., Tortora, M., Palleschi, G., Ashokkumar, M. (2013). Electrochemical investigation of the interaction between; lysozyme-shelled microbubbles and vitamin C, *Analytical & Bioanalytical Chemistry*, **405**, pp. 5531–5538.

36. Xing, Z. W., Ke, H. T., Liu, S. Q., Dai, Z. F., Wang, J. R., Liu, J. B. (2008). Preparation of polyelectrolyte multilayer film-coated microbubble ultrasound contrast agent, *Acta Academiae Medicinae Sinicae*, **30**, pp. 10–14.

37. Della, M. A., Allémann, E., Bettinger, T., Bussat, P., Lassus, A., Pochon, S., Schneider, M. (2008). Grafting of abciximab to a microbubble-based ultrasound contrast agent for targeting to platelets expressing GP IIb/IIIa–characterization and in vitro testing, *European Journal of Pharmaceutics & Biopharmaceutics*, **68**, pp. 555–564.

38. Kheirolomoom, A., Dayton, P. A., Lum, A. F. H., Little, E., Paoli, E. E., Zheng, H., Ferrara, K. W. (2007). Acoustically-active microbubbles conjugated to liposomes: Characterization of a proposed drug delivery vehicle, *Journal of Controlled Release Official Journal of the Controlled Release Society*, **118**, pp. 275–284.

39. B. B. Goldberg, J. B. Liu and F. Forsberg. (1994). *Ultrasound in Medicine & Biology*, **20**, pp. 319–333.

40. Pantano, P., Kuhr, W. G. (1993). Dehydrogenase-modified carbon-fiber microelectrodes for the measurement of neurotransmitter dynamics. 2. Covalent modification utilizing avidin-biotin technology, *Analytical Chemistry*, **65**, pp. 623–630.

41. Takalkar, A. M., Klibanov, A. L., Rychak, J. J., Lindner, J. R., Ley, K. (2004). Binding and detachment dynamics of microbubbles targeted to P-selectin under controlled shear flow, *Journal of Controlled Release*, **96**, pp. 473–482.

42. Willmann, J. K., Paulmurugan, R., Chen, K., Gheysens, O., Rodriguez-Porcel, M., Lutz, A. M., Chen, I. Y., Chen, X., Gambhir, S. S. (2008). US imaging of tumor angiogenesis with microbubbles targeted to vascular endothelial growth factor receptor type 2 in mice, *Radiology*, **246**, pp. 508–518.

43. Yan, F., Li, L., Deng, Z., Jin, Q., Chen, J., Yang, W., Yeh, C. K., Wu, J., Shandas, R., Liu, X. (2013). Paclitaxel-liposome-microbubble complexes as ultrasound-triggered therapeutic drug delivery carriers, *Journal of Controlled Release*, **166**, pp. 246–255.

44. Jin, W. T., Hiep, T. T., Tung, P. T., Poudel, B. K., Woo, K. B., Dong, S. K., Choi, H. G., Chun, I. K., Jeong, J. H. and Yong, C. S.. (2015). *Asian Journal of Pharmaceutical Sciences*, **10**, pp. 5249–5262.

45. Kim, H. J., Yi, Y., Kim, A., Miyata, K. (2018). Small delivery vehicles of siRNA for enhanced cancer targeting, *Biomacromolecules*, **19**, pp.

46. Nicolson, F., Jamieson, L. E., Mabbott, S., Plakas, K., Shand, N. C., Detty, M. R., Graham, D., Faulds, K. (2018). Through tissue imaging of a live breast cancer tumour model using handheld surface enhanced spatially offset resonance Raman spectroscopy (SESORRS), *Chemical Science*, **9**, p. 3788.

47. Fritz, J., Katopodis, A. G., Kolbinger, F., Anselmetti, D. (1998). Force-mediated kinetics of single P-selectin/ligand complexes observed by atomic force microscopy, *Proceedings of the National Academy of Sciences of the United States of America*, **95**, pp. 12283–12288.

48. Klibanov, A. L. (2009), Preparation of targeted microbubbles: Ultrasound contrast agents for molecular imaging, *Medical & Biological Engineering & Computing*, **47**, pp. 875–882.

49. Rychak, J. J., Lindner, J. R., Ley, K., Klibanov, A. L. (2006). Deformable gas-filled microbubbles targeted to P-selectin, *Journal of Controlled Release Official Journal of the Controlled Release Society*, **114**, pp. 288–299.

50. Culjat, M. O., Goldenberg, D., Tewari, P., Singh, R. S. (2010). A review of tissue substitutes for ultrasound imaging, *Ultrasound in Medicine & Biology*, **36**, pp. 861–873.

51. Souquet, J., Bercoff, J. (2004). Ultrafast ultrasound imaging, *Ultrasound in Medicine & Biology*, **37**, pp. S17–S17.

52. Ryu, Y. C., Dong, I. K., Kim, S. H., Wang, H. M. D., Hwang, B. H. (2018). Synergistic transdermal delivery of biomacromolecules using sonophoresis after microneedle treatment, *Biotechnology & Bioprocess Engineering*, **23**, pp. 286–292.

53. Sauerbruch, T., Delius, M., Paumgartner, G., Holl, J., Wess, O., Weber, W., Hepp, W., Brendel, W. (1987). Fragmentation of gallstones by extracorporeal shock waves, *Seminars in Liver Disease*, **7**, pp. 317–321.

54. Jiang, C. P., Investigation on occlusion effect of tumor vessel in HIFU Surgery, in: *International Conference on Biomedical Engineering & Informatics*, 2009.

55. Chung, D. J., Cho, S. H., Lee, J. M., Hahn, S. T. (2012). Effect of microbubble contrast agent during high intensity focused ultrasound ablation on rabbit liver in vivo, *European Journal of Radiology*, **81**, pp. e519–e523.

56. Hu, J., Wu, F., Sui, Y., Xu, T. (2013). A microbubble contrast agent improves prediction of ablated areas during radiofrequency ablation: A rabbit liver study, *Journal of Ultrasound in Medicine Official Journal of the American Institute of Ultrasound in Medicine*, **32**, pp. 787–793.

57. Luo, W., Zhou, X., Ren, X., Zheng, M., Zhang, J., He, G. (2007). Enhancing effects of SonoVue, a microbubble sonographic contrast agent, on high-intensity focused ultrasound ablation in rabbit livers in vivo, *Journal of Ultrasound in Medicine*, **26**, pp. 469–476.

58. Nouso, K., Shiraga, K., Uematsu, S., Okamoto, R., Harada, R., Takayama, S., Kawai, W., Kimura, S., Ueki, T., Okano, N. (2010). Prediction of the ablated area by the spread of microbubbles

during radiofrequency ablation of hepatocellular carcinoma, *Liver International*, **25**, pp. 967–972.

59. Takegami, K., Kaneko, Y., Watanabe, T., Maruyama, T., Matsumoto, Y., Nagawa, H. (2005). Erythrocytes, as well as microbubble contrast agents, are important factors in improving thermal and therapeutic effects of high-intensity focused ultrasound, *Ultrasound in Medicine & Biology*, **31**, pp. 385–390.
60. Chu, K. F., Dupuy, D. E. (2014). Thermal ablation of tumours: Biological mechanisms and advances in therapy, *Nature Reviews Cancer*, **14**, pp. 199–208.
61. Hill, C. R., Ter Haar, G. R. (1995). Review article: High intensity focused ultrasound–potential for cancer treatment, *British Journal of Radiology*, **68**, p. 1296.
62. Jones, T. A., Chin, J., Mcleod, D., Barkin, J., Pantuck, A., Marks, L. S. (2018). High-intensity focused ultrasound for radio-recurrent prostate cancer: A North American Clinical Trial, *J Urol*, **199**(1), pp. 133–139.
63. Endo, Y., Nwawka, O. K., Smith, S., Burket, J. C. (2018). Tarsometatarsal joint communication during fluoroscopy-guided therapeutic joint injections and relationship with patient age and degree of osteoarthritis, *Skeletal Radiology*, **47**, pp. 271–277.
64. Hynynen, K. (2011). MRIgHIFU: A tool for image-guided therapeutics, *Journal of Magnetic Resonance Imaging Jmri*, **34**, pp. 482–493.
65. Chen, X., Cvetkovic, D., Ma, C. M., Chen, L. (2012). Quantitative study of focused ultrasound enhanced doxorubicin delivery to prostate tumor in vivo with MRI guidance, *Medical Physics*, **39**, pp. 2780.
66. Cline, H. E., Schenck, J. F., Hynynen, K., Watkins, R. D., Souza, S. P., Jolesz, F. A. (1992). MR-guided focused ultrasound surgery, *Journal of Computer Assisted Tomography*, **16**, pp. 956–965.
67. Fang, T., Zhuang, Y., Cao, J. (2005). Rosiglitazone enhanced growth inhibition of cisplatin in human lung adenocarcinoma cell, *Chinese Pharmacological Bulletin*, **21**, pp.
68. Lv, W., Yan, T., Wang, G., Zhao, W., Zhang, T., Zhou, D. (2016). High-intensity focused ultrasound therapy in combination with gemcitabine for unresectable pancreatic carcinoma, *Therapeutics & Clinical Risk Management*, **12**, pp. 687–691.
69. Paparel, P., Chapelon, J. Y., Bissery, A., Chesnais, S., Curiel, L., Gelet, A. (2008). Influence of the docetaxel administration period (neoadjuvant or concomitant) in relation to HIFU treatment on the growth of

Dunning tumors: Results of a preliminary study, *Prostate Cancer & Prostatic Diseases*, **11**, p. 181.

70. Paparel, P., Curiel, L., Chesnais, S., Ecochard, R., Chapelon, J. Y., Gelet, A. (2015). Synergistic inhibitory effect of high-intensity focused ultrasound combined with chemotherapy on Dunning adenocarcinoma, *Bju International*, **95**, pp. 881–885.

71. Staruch, R. M., Ganguly, M., Tannock, I. F., Hynynen, K., Chopra, R. (2012). Enhanced drug delivery in rabbit VX2 tumours using thermosensitive liposomes and MRI-controlled focused ultrasound hyperthermia, *International Journal of Hyperthermia the Official Journal of European Society for Hyperthermic Oncology North American Hyperthermia Group*, **28**, pp. 776–787.

72. Wang, R. S., Liu, L. X., Gu, Y. H., Lin, Q. F., Guo, R. H., Shu, Y. Q. (2010), The effect of endostatin and gemcitabine combined with HIFU on the animal xenograft model of human pancreatic cancer, *Biomedicine & Pharmacotherapy*, **64**, pp. 309–312.

73. Yu, T., Zhang, Y., He, H., Zhou, S., Liu, Y., Huang, P. (2011). Anticancer potency of cytotoxic drugs after exposure to high-intensity focused ultrasound in the presence of microbubbles and hematoporphyrin, *Molecular Pharmaceutics*, **8**, pp. 1408–1415.

74. Yu, X., Zhang, L., Chen, J. (2018). Effectiveness of treatment with endostatin in combination with emcitabine, carboplatin, and gemcitabine in patients with advanced non-small cell lung cancer: A retrospective study, *Open Medicine*, **13**, pp. 142–147.

75. Bae, I., Fan, S., Bhatia, K., Kohn, K. W., Fornace, A. J., O'connor, P. M. (1995). Relationships between G1 arrest and stability of the p53 and p21Cip1/Waf1 proteins following γ-irradiation of human lymphoma cells, *Cancer Research*, **55**, pp. 2387–2393.

76. Bischof, J. C., Padanilam, J., Holmes, W. H. (1995). Dynamics of cell membrane permeability changes at supraphysiological temperatures, *Biophysical Journal*, **68**, pp. 2608–2614.

77. Vajkoczy, P., Menger, M. D., Vollmar, B., Schilling, L., Schmiedek, P., Hirth, K. P., Ullrich, A. and Fong, T. A. T. (1999). *Neoplasia*, **1**, 31–41.

78. Dudar, T. E., Jain, R. K. (1984). Differential response of normal and tumor microcirculation to hyperthermia, *Cancer Research*, **44**, pp. 605–612.

79. Feizi, S., Naderan, M., Ownagh, V., Sadeghpour, F. (2018). Distribution of the anterior, posterior, and total corneal astigmatism in healthy eyes, *International Ophthalmology*, **38**, p. 481.

80. Los, G., Vugt, M. J., Van, Pinedo, H. M. (1994). Response of peritoneal solid tumours after intraperitoneal chemohyperthermia treatment with cisplatin or carboplatin, *British Journal of Cancer,* **69**, pp. 235–241.

81. Marmor, J. B. (1979). Interactions of hyperthermia and chemotherapy in animals, *Cancer Research*, **39**, p. 2269.

82. Sen, A., Capitano, M. L., Spernyak, J. A., Schueckler, J. T., Thomas, S., Singh, A. K., Evans, S. S., Hylander, B. L., Repasky, E. A. (2011). Mild elevation of body temperature reduces tumor interstitial fluid pressure and hypoxia, and enhances efficacy of radiotherapy in murine tumor models, *Cancer Research*, **71**, pp. 3872–3880.

83. Vaupel, P., Horsman, M. R. (2010). Tumour perfusion and associated physiology: Characterization and significance for hyperthermia, *International Journal of Hyperthermia the Official Journal of European Society for Hyperthermic Oncology North American Hyperthermia Group*, **26**, pp. 209–210.

84. Bull, J. M. C. (2017). A review of immune therapy in cancer and a question: Can thermal therapy increase tumor response?, *International Journal of Hyperthermia the Official Journal of European Society for Hyperthermic Oncology North American Hyperthermia Group*, p. 1.

85. E. l. Majzoub I., Qdaisat, A., Thein, K. Z., Win, M. A., Han, M. M., Jacobson, K., Chaftari, P. S., Prejean, M., Reyes-Gibby, C., Yeung, S. J. (2018). Adverse Effects of Immune Checkpoint Therapy in Cancer Patients Visiting the Emergency Department of a Comprehensive Cancer Center, *Ann Emerg Med*, **73**(1), pp. 79–87.

86. Werthmöller, N., Frey, B., Rückert, M., Lotter, M., Fietkau, R., Gaipl, U. S. (2016). Combination of ionising radiation with hyperthermia increases the immunogenic potential of B16-F10 melanoma cells and, *International Journal of Hyperthermia the Official Journal of European Society for Hyperthermic Oncology North American Hyperthermia Group*, **32**, p. 1.

87. Browning, R. J., Pjt, R., Parhizkar, M., Pedley, R. B., Edirisinghe, M., Knowles, J. C., Stride, E. (2017). Drug delivery strategies for platinum based chemotherapy, *ACS Nano*, **11**, p. 8560.

88. Chen, Y. (2000). Combinational immunotherapy for established tumors with engineered tumor vaccines and adenovirus-mediated gene transfer, *Cancer Gene Therapy*, **7**, pp. 1023–1033.

89. Grüll, H., Langereis, S. (2012). Hyperthermia-triggered drug delivery from temperature-sensitive liposomes using MRI-guided

high intensity focused ultrasound, *Journal of Controlled Release*, **161**, pp. 317–327.

90. Ouyang, J., Liu, R. Y., Chen, W., Liu, Z., Xu, Q., Zeng, K., Deng, L., Shen, L., Liu, Y. N. (2018). A black phosphorus based synergistic antibacterial platform against drug resistant bacteria, *Journal of Materials Chemistry B*, p. 6(39).
91. Jin, H., Wang, X., Chen, Z., Li, Y., Liu, C., Xu, J. (2018). Fabrication of β-conglycinin-stabilized nanoemulsions via ultrasound process and influence of SDS and PEG 10000 co-emulsifiers on the physicochemical properties of nanoemulsions, *Food Research International*, **106**.
92. Kopechek, J. A., Park, E., Mei, C. S., Mcdannold, N. J., Porter, T. M. (2015). Accumulation of phase-shift nanoemulsions to enhance MR-guided ultrasound-mediated tumor ablation in vivo, *Journal of Healthcare Engineering*, **4**, pp. 109–125.
93. Shao, Y., Wu, C., Wu, T., Li, Y., Chen, S., Yuan, C., Hu, Y. (2018). Eugenol-chitosan nanoemulsions by ultrasound-mediated emulsification: Formulation, characterization and antimicrobial activity, *Carbohydrate Polymers*, **193**, p. S0144861718303655.
94. You, Y., Wang, Z., Ran, H., Zheng, Y., Wang, D., Xu, J., Wang, Z., Chen, Y., Li, P. (2016). Nanoparticle-enhanced synergistic HIFU ablation and transarterial chemoembolization for efficient cancer therapy, *Nanoscale*, **8**, p. 4324.
95. Lu, M., Cheng, X., Jiang, J., Li, T., Zhang, Z., Tsauo, C., Liu, Y., Wang, Z. (2018). Dual-modal photoacoustic and magnetic resonance tracking of tendon stem cells with PLGA/iron oxide microparticles in vitro, *Plos One*, **13**, p. e0193362.
96. Sun, Y., Zheng, Y., Ran, H., Zhou, Y., Shen, H., Chen, Y., Chen, H., Krupka, T. M., Li, A., Li, P. (2012). Superparamagnetic PLGA-iron oxide microcapsules for dual-modality US/MR imaging and high intensity focused US breast cancer ablation, *Biomaterials*, **33**, pp. 5854–5864.
97. Zhou, H., Huang, X., Guo, Y., Shang, T., Wang, Z., Ultrasound, D. O. (2016). Preparation of streptavidin-coated and paclitaxel-loaded phase-shifting PLGA nanoparticles and ultrasound imaging in vitro, *Chinese Journal of Interventional Imaging & Therapy*, 2016, p, 012.
98. Browning, R. J., Rajkumar, V., Pedley, R. B., Eckersley, R. J., Blower, P. J. (2014). Prospects for enhancement of targeted radionuclide therapy of cancer using ultrasound, *Journal of Labelled Compounds & Radiopharmaceuticals*, **57**, pp. 279–284.

99. Postema, M., Van Wamel, A., Ten Cate, F. J., De Jong, N. (2005). High-speed photography during ultrasound illustrates potential therapeutic applications of microbubbles, *Medical Physics*, **32**, pp. 3707–3711.

100. Ibsen, S., Benchimol, M., Simberg, D., Schutt, C., Steiner, J., Esener, S. (2011). A novel nested liposome drug delivery vehicle capable of ultrasound triggered release of its payload, *Journal of Controlled Release*, **155**, pp. 358–366.

101. Leabanks, H., Teo, B., Stride, E., Coussios, C. C. (2016). The effect of particle density on ultrasound-mediated transport of nanoparticles, *Physics in Medicine & Biology*, **61**, pp. 7906–7918.

102. Monteiro, S., Silva, E. K., Alvarenga, V. O., Moraes, J., Freitas, M. Q., Silva, M. C., Raices, R., Sant'ana, A. S., Meireles, M., Cruz, A. G. (2018). Effects of ultrasound energy density on the non-thermal pasteurization of chocolate milk beverage, *Ultrasonics Sonochemistry*, **42**.

103. Raghavan, R. (2018). Theory for acoustic streaming in soft porous matter and its applications to ultrasound-enhanced convective delivery, *Journal of Therapeutic Ultrasound*, **6**, p. 6.

104. Thiriet, M., Solovchuk, M., Sheu, W. H. (2014). HIFU treatment of liver Cancer—reciprocal effect of blood flow and US Studied from a Patient-Specific Configuration, Computation Modeling of Objects Presented in Images. *Fundamentals, Methods, and Applications*, 2014, pp. 1–11.

105. Mesiwala, A. H., Farrell, L., Wenzel, H. J., Silbergeld, D. L., Crum, L. A., Winn, H. R., Mourad, P. D. (2002). High-intensity focused ultrasound selectively disrupts the blood-brain barrier in vivo, *Ultrasound in Medicine & Biology*, **28**, pp. 389–400.

106. Alonso, A., Reinz, E., Jenne, J. W., Fatar, M., Schmidtglenewinkel, H., Hennerici, M. G., Meairs, S. (2010). Reorganization of gap junctions after focused ultrasound blood–brain barrier opening in the rat brain, *Journal of Cerebral Blood Flow & Metabolism Official Journal of the International Society of Cerebral Blood Flow & Metabolism*, **30**, pp. 1394–1402.

107. Hancock, H. A., Smith, L. H., Cuesta, J., Durrani, A. K., Angstadt, M., Palmeri, M. L., Kimmel, E., Frenkel, V. (2009). Investigations into pulsed high-intensity focused ultrasound-enhanced delivery: Preliminary evidence for a novel mechanism, *Ultrasound in Medicine & Biology*, **35**, pp. 1722–1736.

108. O'neill, B. E., Vo, H., Angstadt, M., Li, K. P. C., Quinn, T., Frenkel, V. (2009). Pulsed high intensity focused ultrasound mediated nanoparticle

delivery: Mechanisms and efficacy in murine muscle, *Ultrasound in Medicine & Biology*, **35**, pp. 416–424.

109. Seidl, M., Steinbach, P., Worle, K., Hofstadter, F. (1994). Induction of stress fibres and intercellular gaps in human vascular endothelium by shock-waves, *Ultrasonics*, **32**, pp. 397–400.

110. Zhao, Y., Xin, Y., He, Z., Hu, W. (2018). Function of connexins in the interaction between glial and vascular cells in the central nervous system and related neurological diseases, *Neural Plasticity*, 2018, p. 6323901.

111. Ita, K. (2016). Percutaneous penetration of anticancer agents: Past, present and future, *Biomedicine & Pharmacotherapy*, **84**, p. 1428.

112. Prausnitz, M. R., Langer, R. (2008). Transdermal drug delivery, *Nature Biotechnology*, **26**, pp. 1261–1268.

113. Chauvet, D., Marsac, L., Pernot, M., Boch, A. L., Guillevin, R., Salameh, N., Souris, L., Darrasse, L., Fink, M., Tanter, M. (2013). Targeting accuracy of transcranial magnetic resonance-guided high-intensity focused ultrasound brain therapy: A fresh cadaver model, *Journal of Neurosurgery*, **118**, pp. 1046–1052.

114. Gnanaskandan, A., Hsiao, C. T., Chahine, G. (2018). Modeling of Microbubble Enhanced High Intensity Focused Ultrasound. arXiv.org.

115. Lü, F., Huang, W., Benditt, D. G. (2018). A feasibility study of non-invasive ablation of ventricular tachycardia using high-intensity focused ultrasound, *Journal of Cardiovascular Electrophysiology*, **29**(5), pp. 788–794.

116. Larkin, J. O., Casey, G. D., Tangney, M., Cashman, J., Collins, C. G., Soden, D. M., O'sullivan, G. C. (2008). Effective tumor treatment using optimized ultrasound-mediated delivery of bleomycin, *Ultrasound in Medicine & Biology*, **34**, pp. 406–413.

117. Frenkel V., Etherington A., Greene M., Quijano J., Xie J., Hunter F., Dromi S., Li K. C. (2006). Delivery of liposomal doxorubicin (Doxil) in a breast cancer tumor model: Investigation of potential enhancement by pulsed-high intensity focused ultrasound exposure, *Academic Radiology*, **13**, pp. 469–479.

118. Cheri, X. D., Fred, S., Hua, P., Jianmin, C. (2004). Ultrasound-induced cell membrane porosity, *Ultrasound in Medicine & Biology*, **30**, pp. 519–526.

119. Deng, J., Zhang, Y., Feng, J., Wu, F. (2010). Dendritic cells loaded with ultrasound-ablated tumour induce in vivo specific antitumour immune responses, *Ultrasound in Medicine & Biology*, **36**, pp. 441–448.

120. Gummadi, S., Eisenbrey, J. R., Lyshchik, A. (2018). A narrative review on contrast-enhanced ultrasound in aortic endograft endoleak surveillance, *ultrasound quarterly*, **46**, p. 255.

121. Jong, N. D. (1996). Improvements in ultrasound contrast agents, *IEEE Engineering in Medicine & Biology Magazine*, **15**, pp. 72–82.

122. Schneider, M., Arditi, M., Barrau, M. B., Brochot, J., Broillet, A., Ventrone, R., Yan, F. (1995). BR1: A new ultrasonographic contrast agent based on sulfur hexafluoride-filled microbubbles, *Investigative Radiology*, **30**, pp. 451–457.

123. Sponheim, N., Hoff, L., Waaler, A., Muan, B., Morris, H., Holm, S., Myrum, M., Jong, N. D., Skotland, T. (1993). Albunex-a new ultrasound contrast agent, in: *International Conference on Acoustic Sensing and Imaging*, pp. 103–108.

124. Stanziola, A., Leow, C. H., Bazigou, E., Weinberg, P. D., Tang, M. X. (2018). ASAP: Super-contrast vasculature imaging using coherence analysis and high frame-rate contrast enhanced ultrasound, *IEEE Transactions on Medical Imaging*, **37**(8), pp. 1847–1856.

125. Alter, J., Sennoga, C. A., Lopes, D. M., Eckersley, R. J., Wells, D. J. (2009). Microbubble stability is a major determinant of the efficiency of ultrasound and microbubble mediated in vivo gene transfer, *Ultrasound in Medicine & Biology*, **35**, pp. 976–984.

126. Browning, R. J., Mulvana, H., Tang, M., Hajnal, J. V., Wells, D. J., Eckersley, R. J. (2011). Influence of needle gauge on in vivo ultrasound and microbubble-mediated gene transfection, *Ultrasound in Medicine & Biology*, **37**, pp. 1531–1537.

127. Browning, R. J., Mulvana, H., Tang, M. X. (2012). Effect of albumin and dextrose concentration on ultrasound and microbubble mediated gene transfection in vivo, *Ultrasound in Medicine & Biology*, **38**, pp. 1067–1077.

128. Chin, A., Crooke, B., Heywood, L., Brijball, R., Pelecanos, A. M., Abeypala, W. (2018). A randomised controlled trial comparing needle movements during combined spinal-epidura anaesthesia with and without ultrasound assistance, *Anaesthesia*, **73**(4), pp. 466–473

129. Fokong, S., Theek, B., Wu, Z., Koczera, P., Appold, L., Jorge, S., Resch-Genger, U., Van, Z. M., Storm, G., Kiessling, F. (2012). Image-guided, targeted and triggered drug delivery to tumors using polymer-based microbubbles, *Journal of Controlled Release*, **163**, pp. 75–81.

130. Schoenherr, J. R., Waechter, J., Millington, S. J. (2018). Subjective awareness of ultrasound expertise development: Individual experience

as a determinant of overconfidence, *Advances in Health Sciences Education*, pp. 1–17.

131. Sirsi, S. R., Borden, M. A. (2014). State-of-the-art materials for ultrasound-triggered drug delivery, *Advanced Drug Delivery Reviews*, **72**, pp. 3–14.

132. Christensenjeffries, K., Browning, R. J., Tang, M. X., Dunsby, C., Eckersley, R. J. (2017). In vivo acoustic super-resolution and super-resolved velocity mapping using microbubbles, *IEEE Transactions on Ultrasonics Ferroelectrics & Frequency Control*, **64**, pp. 1478–1486.

133. Couture, O., Hingot, V., Heiles, B., Muleki-Seya, P., Tanter, M. (2018). Ultrasound localization microscopy and super-resolution: A state of the art, *IEEE Transactions on Ultrasonics Ferroelectrics & Frequency Control*, **65**, pp. 1304–1320.

134. Efthymiou, K., Pelekasis, N., Butler, M. B., Thomas, D. H., Sboros, V. (2018). The effect of resonance on transient microbubble acoustic response: Experimental observations and numerical simulations, *Journal of the Acoustical Society of America*, **143**, pp. 1392–1406.

135. Mehta, K. S., Lee, J. J., Taha, A. A., Avgerinos, E., Chaer, R. A. (2017). Vascular applications of contrast-enhanced ultrasound imaging, *Journal of Vascular Surgery*, **66**.

136. Morgan, K. E., Allen, J. S., Dayton, P. A., Chomas, J. E. (2000). Experimental and theoretical evaluation of microbubble behavior: Effect of transmitted phase and bubble size, *IEEE Transactions on Ultrasonics, Ferroelectrics, and Frequency Control*, **47**, pp. 1494–1509.

137. Quarles, C. C., Bell, L. C., Stokes, A. M. (2018). Imaging vascular and hemodynamic features of the brain using dynamic susceptibility contrast and dynamic contrast enhanced MRI, *Neuroimage*, **187**, pp. 32–55.

138. Song, W., Cong, W., Wei, Y., Zhang, X., Wei, W. (2018). Experimental study of microbubble drag reduction on an axisymmetric body, *Modern Physics Letters B*, **32**, p. 1850035.

139. Cao, Y., Chen, Y., Yu, T., Guo, Y., Liu, F., Yao, Y., Li, P., Wang, D., Wang, Z., Chen, Y. (2018). Drug release from phase-changeable nanodroplets triggered by low-intensity focused ultrasound, *Theranostics*, **8**, pp. 1327–1339.

140. Du, L., Jin, Y., Zhou, W., Zhao, J. (2011). Ultrasound-triggered drug release and enhanced anticancer effect of doxorubicin-loaded poly(D,L-lactide-co-glycolide)-methoxy-poly(ethylene glycol) nano-droplets, *Ultrasound in Medicine & Biology*, **37**, pp. 1252–1258.

141. Rapoport, N., Nam, K. H., Gupta, R., Gao, Z., Mohan, P., Payne, A., Todd, N., Liu, X., Kim, T., Shea, J. (2011). Ultrasound-mediated tumor imaging and nanotherapy using drug loaded, block copolymer stabilized perfluorocarbon nanoemulsions, *Journal of Controlled Release*, **153**, pp. 4–15.

142. Rapoport, N. Y., Kennedy, A. M., Shea, J. E., Scaife, C. L., Nam, K. H. (2009). Controlled and targeted tumor chemotherapy by ultrasound-activated nanoemulsions/microbubbles, *Journal of Controlled Release*, **138**, pp. 268–276.

143. Bhatnagar, S., Kwan, J. J., Shah, A. R., Coussios, C. C., Carlisle, R. C. (2016). Exploitation of sub-micron cavitation nuclei to enhance ultrasound-mediated transdermal transport and penetration of vaccines, *Journal of Controlled Release*, **238**, pp. 22–30.

144. Kwan, J. J., Myers, R., Coviello, C. M., Graham, S. M., Shah, A. R., Stride, E., Carlisle, R. C., Coussios, C. C. (2015). Ultrasound-propelled nanocups for drug delivery, *Small*, **11**, pp. 5305–5314.

145. Hoff, L., Sontum, P. C., Hovem, J. M. (2000). Oscillations of polymeric microbubbles: Effect of the encapsulating shell, *Journal of the Acoustical Society of America*, **107**, pp. 2272–2280.

146. Loughran, J., Eckersley, R. J., Tang, M. X. (2012). Modeling non-spherical oscillations and stability of acoustically driven shelled microbubbles, *Journal of the Acoustical Society of America*, **131**, p. 4349.

147. Sboros, V. (2008). Response of contrast agents to ultrasound, *Advanced Drug Delivery Reviews*, **60**, pp. 1117–1136.

148. Farhadi, A., Ho, G., Kunth, M., Ling, B., Lakshmanan, A., Lu, G. J., Bourdeau, R. W., Schröder, L., Shapiro, M. G. (2018). Recombinantly expressed gas vesicles as nanoscale contrast agents for ultrasound and hyperpolarized MRI, *Aiche Journal*, pp.

149. Guo, J. C., Yang, Y. J., Guo, M., Jiang, X. L. (2018). Ultrasound-guided intertumoral injection of contrast agents combined with human p53 gene for the treatment of breast cancer, *Kaohsiung Journal of Medical Sciences*, **34**(8), pp. 438–446.

150. Hernot, S., Klibanov, A. L. (2008). Microbubbles in ultrasound-triggered drug and gene delivery, *Advanced Drug Delivery Reviews*, **60**, pp. 1153–1166.

151. Tachibana, K., Tachibana, S. (2001). The use of ultrasound for drug delivery, *Echocardiography*, **18**, pp. 323–328.

152. Li, Y., Xia, Z., Luo, W., Wang, D., Li, Y., Wang, J., Li, Z., Zhang, S., Luo, S., Ying, W. (2018). Dual-functionalized nanoparticles loaded microbubbles for enhancement of drug uptake, *Ultrasonics*, **87**, p. S0041624X17307849.

153. Matsuo, M., Yamaguchi, K., Jr, L. B. F., Endo, H., Ogawa, K., Tachibana, K., Nakayama, J. (2011). Synergistic inhibition of malignant melanoma proliferation by melphalan combined with ultrasound and microbubbles, *Ultrasonics Sonochemistry*, **18**, pp. 1218–1224.

154. Van, W. A., Kooiman, K., Harteveld, M., Emmer, M., Ten Cate, F. J., Versluis, M., De, J. N. (2006). Vibrating microbubbles poking individual cells: Drug transfer into cells via sonoporation, *Journal of Controlled Release*, **112**, pp. 149–155.

155. Xi, X. P., Zong, Y. J., Ji, Y. H., Wang, B., Liu, H. S. (2018). Experiment research of focused ultrasound combined with drug and microbubble for treatment of central nervous system leukemia, *Oncotarget*, **9**, pp. 5424–5434.

156. Dell'oro, S., Verderio, M., Incerti, M., Mastrolia, S. A., Cozzolino, S., Vergani, P. (2018). 2D versus 3D real time ultrasound with live xPlane imaging to visualize aortic and ductal arches: Comparison between methods, *Peerj*, **6**, p. e4561.

157. Jr, W. J., Galen, K. P., Wojdyla, J. K., Hughes, M. S., Klibanov, A. L., Brandenburger, G. H. (2002). Microbubbles induce renal hemorrhage when exposed to diagnostic ultrasound in anesthetized rats, *Ultrasound in Medicine & Biology*, **28**, pp. 1535–1546.

158. Mcdannold, N., Vykhodtseva, N., Hynynen, K. (2006). Targeted disruption of the blood-brain barrier with focused ultrasound: Association with cavitation activity, *Physics in Medicine & Biology*, **51**, p. 793.

159. Sanches, P. G., Rossin, R., Böhmer, M., Tiemann, K., Grüll, H. (2013). Real-time imaging and kinetics measurements of focused ultrasound-induced extravasation in skeletal muscle using SPECT/CT, *Journal of Controlled Release*, **168**, pp. 262–270.

160. Tung, Y. S., Vlachos, F., Choi, J. J., Deffieux, T., Selert, K., Konofagou, E. E. (2010). In vivo transcranial cavitation threshold detection during ultrasound-induced blood–brain barrier opening in mice, *Physics in Medicine & Biology*, **55**, pp. 6141–6155.

161. Alli, S., Figueiredo, C. A., Golbourn, B., Sabha, N., Wu, M. Y., Bondoc, A., Luck, A., Coluccia, D., Maslink, C., Smith, C. (2018). Brainstem blood brain barrier disruption using focused ultrasound: A demonstration

of feasibility and enhanced doxorubicin delivery, *Journal of Controlled Release*, p. S0168365918302621.

162. Cottier, K. E., Galloway, E. A., Calabrese, E. C., Tome, M. E., Liktorbusa, E., Kim, J., Davis, T. P., Vanderah, T. W., Largentmilnes, T. M. (2018). Loss of blood-brain barrier integrity in a KCl-induced model of episodic headache enhances CNS drug delivery, *Eneuro*, **5**.

163. Jones, R. M., Deng, L., Leung, K., Mcmahon, D., O'reilly, M. A., Hynynen, K. (2018). Three-dimensional transcranial microbubble imaging for guiding volumetric ultrasound-mediated blood-brain barrier opening, *Theranostics*, **8**, p. 2909.

164. Mcmahon, D., Mah, E., Hynynen, K. (2018). Angiogenic response of rat hippocampal vasculature to focused ultrasound-mediated increases in blood-brain barrier permeability, *Scientific Reports*, **8**, pp.

165. Ting, C. Y., Fan, C. H., Liu, H. L., Huang, C. Y., Hsieh, H. Y., Yen, T. C., Wei, K. C., Yeh, C. K. (2012). Concurrent blood-brain barrier opening and local drug delivery using drug-carrying microbubbles and focused ultrasound for brain glioma treatment, *Biomaterials*, **33**, pp. 704–712.

166. Wu, S. Y., Aurup, C., Sanchez, C. S., Grondin, J., Zheng, W., Kamimura, H., Ferrera, V. P., Konofagou, E. E. (2018). Efficient blood-brain barrier opening in primates with neuronavigation-guided ultrasound and real-time acoustic mapping, *Scientific Reports*, **8**, pp. 47–54.

167. Lentacker, I., Wang, N., Vandenbroucke, R. E., Demeester, J., De Smedt, S. C., Sanders, N. N. (2009). Ultrasound exposure of lipoplex loaded microbubbles facilitates direct cytoplasmic entry of the lipoplexes, *Molecular Pharmaceutics*, **6**, p. 457.

168. Meijering, B. D., Juffermans, L. J., Van, W. A., Henning, R. H., Zuhorn, I. S., Emmer, M., Versteilen, A. M., Paulus, W. J., Van Gilst, W. H., Kooiman, K. (2009). Ultrasound and microbubble-targeted delivery of macromolecules is regulated by induction of endocytosis and pore formation, *Circulation Research*, **104**, pp. 679–687.

169. Schlicher, R. K., Radhakrishna, H., Tolentino, T. P., Apkarian, R. P., Zarnitsyn, V., Prausnitz, M. R. (2006). Mechanism of intracellular delivery by acoustic cavitation, *Ultrasound in Medicine & Biology*, **32**, pp. 915–924.

170. Walton, C. B., Shohet, R. V. (2009). Tiny bubbles and endocytosis?, *Circulation Research*, **104**, pp. 563–565.

171. Zhang, N., Yan, F., Liang, X., Wu, M., Shen, Y., Chen, M., Xu, Y., Zou, G., Jiang, P., Tang, C. (2018). Localized delivery of curcumin into brain with polysorbate 80-modified cerasomes by ultrasound-targeted

microbubble destruction for improved Parkinson's disease therapy, *Theranostics*, **8**, pp. 2264–2277.

172. Kumar, K., Vulugundam, G., Jaiswal, P. K., Shyamlal, B. R. K., Chaudhary, S. (2017). Efficacious cellular codelivery of doxorubicin and EGFP siRNA mediated by the composition of PLGA and PEI protected gold nanoparticles, *Bioorganic & Medicinal Chemistry Letters*, **27**, pp. 4288–4293.

173. Mullin, L. B., Phillips, L. C., Dayton, P. A. (2013). Nanoparticle delivery enhancement with acoustically activated microbubbles, *IEEE Transactions on Ultrasonics Ferroelectrics & Frequency Control*, **60**, pp. 65–77.

174. Nesbitt, H., Sheng, Y., Kamila, S., Logan, K., Thomas, K., Callan, B., Taylor, M. A., Love, M., O'rourke, D., Kelly, P. (2018). Gemcitabine loaded microbubbles for targeted chemo-sonodynamic therapy of pancreatic cancer, *Journal of Controlled Release*, **279**, pp. 8–16.

175. Ngo, T. H., Barnard, R. J., Cohen, P., Freedland, S., Tran, C., Degregorio, F., Elshimali, Y. I., Heber, D., Aronson, W. J. (2003). Effect of isocaloric low-fat diet on human LAPC-4 prostate cancer xenografts in severe combined immunodeficient mice and the insulin-like growth factor axis, *Clinical Cancer Research*, **9**, pp. 2734–2743.

176. Xu, Jeff, S., Huang, J., Qin, R., Hinkle, George, H., Povoski, Stephen, P., Martin, Edward, W. (2010). Synthesizing and binding dual-mode poly (lactic-co-glycolic acid) (PLGA) nanobubbles for cancer targeting and imaging, *Biomaterials*, **31**, pp. 1716–1722.

177. Zhang, X., Zheng, Y., Wang, Z., Huang, S., Chen, Y., Jiang, W., Zhang, H., Ding, M., Li, Q., Xiao, X. (2014). Methotrexate-loaded PLGA nanobubbles for ultrasound imaging and Synergistic Targeted therapy of residual tumor during HIFU ablation, *Biomaterials*, **35**, pp. 5148–5161.

178. Burke, C. W., Hsiang, Y. H., Th, A. E., Kilbanov, A. L., Price, R. J. (2011). Covalently linking poly(lactic-co-glycolic acid) nanoparticles to microbubbles before intravenous injection improves their ultrasound-targeted delivery to skeletal muscle, *Small*, **7**, pp. 1227–1235.

179. Burke, C. W., Timbie, K., Kilbanov, A. L., Price, R. J. (2014). Ultrasound-activated agents comprised of 5FU-bearing nanoparticles bonded to microbubbles inhibit solid tumor growth and improve survival, *Molecular Therapy*, **22**, pp. 321–328.

180. Kamaly, N., Yameen, B., Wu, J., Farokhzad, O. C. (2016). Degradable controlled-release polymers and polymeric nanoparticles: Mechanisms of controlling drug release, *Chemical Reviews*, **116**, p. 2602.

181. Liu, J., Wang, S., Cai, X., Zhou, S., Liu, B. (2018), Hydrogen peroxide degradable conjugated polymer nanoparticles for fluorescence and photoacoustic bimodal imaging, *Chemical Communications*, **54**, pp. 10.1039.C1037CC09856A.

182. Wu, P., Jia, Y., Qu, F., Sun, Y., Wang, P., Zhang, K., Xu, C., Liu, Q., Wang, X. (2017). Ultrasound-responsive polymeric micelles for sonoporation-assisted site-specific therapeutic action, *ACS Applied Materials & Interfaces*, **9**, pp. 25706–25716.

183. Yang, Y., Hou, W., Liu, S., Sun, K., Li, M., Wu, C. (2018). Biodegradable polymer nanoparticles for photodynamic therapy by bioluminescence resonance energy transfer, *Biomacromolecules*, **19**, p. 201.

184. Izadifar, Z., Babyn, P., Chapman, D. (2018). Ultrasound cavitation/microbubble detection and medical applications, *Journal of Medical & Biological Engineering*, pp. 1–18.

185. Khalili, S., Mahdi, M. (2018). Numerical analyses of nonlinear behavior of microbubble contrast agents in ultrasound field and effective parameters, *Journal of the Acoustical Society of America*, **143**, pp. 2111–2118.

186. Mulvana, H., Browning, R. J., Luan, Y., Jong, N. D., Tang, M. X., Eckersley, R. J., Stride, E. (2017). Characterization of contrast agent microbubbles for ultrasound imaging and therapy research, *IEEE Transactions on Ultrasonics Ferroelectrics & Frequency Control*, **64**, pp. 232–251.

187. Canavese, G., Ancona, A., Racca, L., Canta, M., Dumontel, B., Barbaresco, F., Limongi, T., Cauda, V. (2018). Nanoparticle-assisted ultrasound: A special focus on sonodynamic therapy against cancer, *Chemical Engineering Journal*, pp. 155–172.

188. Pan, X., Bai, L., Wang, H., Wu, Q., Wang, H., Liu, S., Xu, B., Shi, X., Liu, H. (2018). Metal-organic-framework-derived carbon nanostructure augmented sonodynamic cancer therapy, *Advanced Materials*, p. 1800180.

189. Pan, X., Wang, H., Wang, S., Sun, X., Wang, L., Wang, W., Shen, H., Liu, H. (2018). Sonodynamic therapy (SDT): A novel strategy for cancer nanotheranostics, *Science China Life Sciences*, **61**, p. 415.

190. Rosenthal, I., Sostaric, J. Z., Riesz, P. (2004). Sonodynamic therapy–a review of the synergistic effects of drugs and ultrasound, *Ultrasonics Sonochemistry*, **11**, pp. 349–363.

191. Tachibana, K., Jr, L. B. F., Ikeda-Dantsuji, Y. (2008). Sonodynamic therapy, *Ultrasonics*, **48**, pp. 253–259.

192. Dewaele, M., Maes, H., Agostinis, P. (2010). ROS-mediated mechanisms of autophagy stimulation and their relevance in cancer therapy, *Autophagy*, **6**, pp. 838–854.

193. Pl, D. S. J., Dad, C. M., Porcacchia, A. S., Pmm, F., Jorge, S. D., Araldi, R. P., Ferreira, A. K. (2017). The roles of ROS in cancer heterogeneity and therapy, *Oxidative Medicine & Cellular Longevity*, **2017**, p. 2467940.

194. Canavese, G., Ancona, A., Racca, L., Canta, M., Dumontel, B., Barbaresco, F., Limongi, T., Cauda, V. (2018). Nanoparticle-assisted ultrasound: A special focus on sonodynamic therapy against cancer, *Chemical Engineering Journal*, **340**, pp. 155–172.

195. Ameneh, S., Ahmad, S., Tayyebi, M. N., Hossein, E., Nassirli, H. P. D. (2011). A novel nanosonosensitizer for sonodynamic therapy, *Journal of Ultrasound in Medicine*, **30**, pp. 1321–1329.

196. Nomikou, N., Fowley, C., Byrne, N. M., Mccaughan, B., Mchale, A. P., Callan, J. F. (2012). Microbubble-sonosensitiser conjugates as therapeutics in sonodynamic therapy, *Chemical Communications*, **48**, pp. 8332–8334.

197. Xi, W., Wang, W., Yu, L., Yang, T., Cao, J., Yu, C. (2017). Site-specific sonocatalytic tumor suppression by chemically engineered single-crystalline mesoporous titanium dioxide sonosensitizers, *Journal of Materials Chemistry B*, **5**, pp.

198. Ouyang, J., Deng, L., Chen, W., Sheng, J., Liu, Z., Wang, L., Liu, Y. N. (2018). Two dimensional semiconductors for ultrasound-mediated cancer therapy: The case of black phosphorus nanosheets, *Chemical Communications*, **54**, p. 2874.

199. Shen, S., Ding, B., Wu, L., Qi, X. Y., Ge, Y. R. (2015). Fe_3O_4@TiO_2 Nanocomposites for magnetic targeting sonodynamic therapy of cancer, *Key Engineering Materials*, **636**, pp. 129–132.

200. Costley, D., Mc, E. C., Fowley, C., Mchale, A. P., Atchison, J., Nomikou, N., Callan, J. F. (2015). Treating cancer with sonodynamic therapy: A review, *International Journal of Hyperthermia*, **31**, pp. 107–117.

201. Chen, M. J., Xu, A., He, W., Ma, W., Song, S. (2017). Ultrasound triggered drug delivery for mitochondria targeted sonodynamic therapy, *Journal of Drug Delivery Science & Technology*, **39**, pp. 501–507.

202. Canaparo, R., Varchi, G., Ballestri, M., Foglietta, F., Sotgiu, G., Guerrini, A., Francovich, A., Civera, P., Frairia, R., Serpe, L. (2013). Polymeric nanoparticles enhance the sonodynamic activity of meso-tetrakis (4-sulfonatophenyl) porphyrin in an in vitro neuroblastoma model, *International Journal of Nanomedicine*, **8**, pp. 4247–4263.

203. Qian, X., Zheng, Y., Chen, Y. (2016). Micro/nanoparticle-augmented sonodynamic therapy (SDT): Breaking the depth shallow of photoactivation, *Advanced Materials*, **28**, pp. 8097–8129.

204. Osminkina, L. A., Nikolaev, A. L., Sviridov, A. P., Andronova, N. V., Tamarov, K. P., Gongalsky, M. B., Kudryavtsev, A. A., Treshalina, H. M., Timoshenko, V. Y. (2015). Porous silicon nanoparticles as efficient sensitizers for sonodynamic therapy of cancer, *Microporous & Mesoporous Materials*, **210**, pp. 169–175.

205. Chen, Y. W., Liu, T. Y., Chang, P. H., Hsu, P. H., Liu, H. L., Lin, H. C., Chen, S. Y. (2016). A theranostic nrGO@MSN-ION nanocarrier developed to enhance the combination effect of sonodynamic therapy and ultrasound hyperthermia for treating tumor, *Nanoscale*, **8**, pp. 12648–12657.

206. Shanei, A., Sazgarnia, A., Meibodi, N. T., Eshghi, H., Hassanzadeh-Khayyat, M., Esmaily, H., Kakhki, N. A. (2012). Sonodynamic therapy using protoporphyrin IX conjugated to gold nanoparticles: An in vivo study on a colon tumor model, *Iranian Journal of Basic Medical Sciences*, **15**, pp. 759–767.

207. Choi, J. J., Feshitan, J. A., Baseri, B., Wang, S., Tung, Y. S., Borden, M. A., Konofagou, E. E. (2010). Microbubble-size dependence of focused ultrasound-induced blood-brain barrier opening in mice in vivo, *IEEE Transactions on Biomedical Engineering*, **57**, pp. 145–154.

Chapter 7

Applications of Nanoparticles in the Treatment of Gliomas

Gerardo Caruso, Elena Fazzari, Salvator M. Cardali, and Maria Caffo

Department of Biomedical and Dental Sciences and Morphofunctional Imaging, Unit of Neurosurgery, University of Messina, Italy
A.O.U. Policlinico "G. Martino," via Consolare Valeria 1, 98125 Messina, Italy

gcaruso@unime.it

Malignant gliomas are the most common primary brain tumors in adults; they include 78% of all primary malignant brain tumors. The efficacy of current multimodal therapeutic strategies in gliomas is limited by the lack of specific therapies against the malignant cells. The prognosis in patients affected by primary brain tumors is still very unfavorable. Late diagnosis and the limitation of conventional therapies, which may result from inefficient delivery of the therapeutic or contrast agent, are major challenges of glioma treatment. The difficulty in overcoming the blood–

Nanoparticle Drug Delivery Systems for Cancer Treatment
Edited by Hala Gali-Muhtasib and Racha Chouaib

ISBN 978-981-4800-54-9 (Hardcover), 978-0-429-34125-0 (eBook)
www.jennystanford.com

brain barrier is one of the biggest obstacles for the delivery of chemotherapeutics. Nanotechnology-based drugs that can cross the blood–brain barrier and carry drugs or genes targeted against brain tumors could revolutionize the therapy of malignant gliomas. Nanotechnology has quickly found its application in clinical methodologies including imaging, diagnostic, and therapeutics. Nanoparticle systems are versatile and provide prolonged drug delivery directly to the tumor following direct intracerebral injection or by functionalizing the material surface with peptides and ligands allowing the drug-loaded material to be specifically targeted to the tumor. In this chapter, we describe the principal applications of nanomedicine in malignant gliomas and discuss various preclinical and/or clinical studies used to treat these aggressive brain tumors.

7.1 Introduction

Malignant brain tumors are an important cause of morbidity and mortality. Brain tumors have an incidence of 28.57 per 100,000 population for adults age 20+ years and 5.57 per 100,000 population if we consider children and adolescents (range 0–19 years); among the age 85+ years the incidence rate is the highest (83.14 per 100,000 population) [1]. The most frequent malignant tumors of the central nervous system (CNS) are the gliomas [2]. Gliomas account for 27% of all tumors and 80% of malignant tumors [1]. According to WHO, the three major classes of gliomas are oligodendrogliomas, astrocytomas, and oligoastrocytomas [3]. Astrocytomas are the most common CNS tumors. In fact, they account for ~75% of all gliomas. Among astrocytomas, grade III anaplastic astrocytomas and grade IV glioblastomas (GBM) represent the nosological entities most notorious and are associated with very poor prognosis [4]. In particular, GBM is the third most frequently reported histology and it represents the 15.1% of all brain tumors and the 46.1% of malignant brain tumors; its incidence increases with age and it is about 3.20 per 100,000 population [1]. GBM histopathology includes cellular pleomorphisms and high mitotic activity [5] and it is characterized by a wide infiltration of the healthy parenchyma [6].

GBM's aggressiveness and its rate of recurrence are mainly due to the infiltrative nature resulting in incomplete resection, the chemoresistance and the poor delivery of therapeutics across the blood–brain barrier (BBB) [7]. GBM can be divided into three types: (1) GBM isocitrate dehydrogenase-wild-type (IDH-mutant) (about 90% of cases): corresponds with the primary GBM and it prevails in patients over 55 years of age; (2) GBM IDH-mutant (about 10% of cases): corresponds to secondary GBM (a tumor with a history of prior lower grade diffuse glioma) and prevails in younger patients; (3) GBM NOS (not otherwise specified): tumors for which it is not possible to perform a full IDH evaluation [8]. Survival estimates for GBM are low; in fact 5 years survival is only 5.1% (with a higher survival for patients who are diagnosed under age of 20 years) [1].

The choice of treatment depends on several factors, such as the histology, the size of the lesion, patient's age and neurological status [2]. The standard treatment of gliomas is multimodal and embraces surgical treatment, radiotherapy, and chemotherapy [9, 10] (Fig. 7.1).

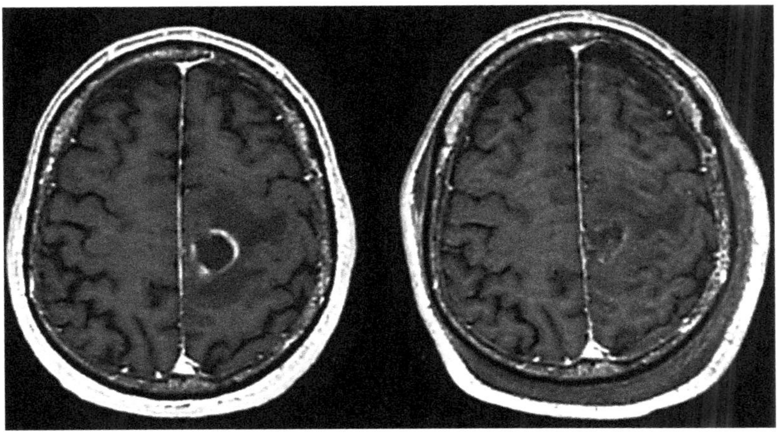

Figure 7.1 T1 MRI (Magnetic Resonance Imaging) weighted images showing GBM pre and post-operatively.

The extent of resection is a prognostic factor, but a radical surgery is not always achievable because of the need to preserve vital nervous and vascular structures to ensure a good quality of life and due to diffuse malignant infiltration of the white matter [11]. Although aggressive treatment increases survival, it could be associated with a significant impairment in the quality of life [12]. Several techniques have been designed to improve tumor detection and to increase the chance of total tumor resection, such as neuronavigation, the use of 5-aminolevulinic acid (5-ALA), intra-operative MRI (iMRI) [13]. There is evidence showing that the combined use of these techniques improves the rates of successful complete resection of tumors to 96.2% [14]. In particular, Coburger et al. showed an important volumetric increase of extent of resection using both 5-ALA and iMRI if compared to iMRI alone [15]. Although radiotherapy and chemotherapy are less invasive options, they are subject to important side effects, such as post-radiation leukoencephalopathy [16] and nerve damage, hair loss, vomiting, infertility insomnia and skin rash, respectively. Radiotherapy and chemotherapy can be used on their own or in combination as neoadjuvant therapy or as adjuvant therapy. Radiation therapy can cause an important side effect [17]. The state-of-the art treatment of GBM is the STUPP protocol which consists of radiotherapy plus continuous daily temozolomide, followed by six cycles of adjuvant temozolomide [16]. The effectiveness of chemotherapy is unfortunately limited by a number of factors, such as toxic effects on healthy cells, tumor cell chemoresistance and poor selectivity of anticancer drugs. Recent research demonstrates a role for immunotherapy in the treatment of gliomas. The therapeutic strategies used include the use of inhibitors of immune check points, cytokines, dendritic cell, and lymphocytes. Unfortunately, it is not easy to translate such research into clinical trials, as it is difficult to determine the maximum tolerable dose in humans and to assess every aspect of the immune response [18, 19]. Finally, hyperthermia is an additional treatment strategy; it allows the destruction of cancer cells through the diffusion of heat (41–46°C) in specific target sites. In fact, it determines physical changes in cells that ultimately trigger apoptotic mechanisms. The methods used are different and include microwaves, infrared rays, and ultrasound. Unfortunately, these strategies have some side effects, such as excessive warming of the healthy tissues, dispersion of the heat in the blood and difficult targeting of the diseased sites. Despite these strategies patient median survival times is still low and, in GBM for example, it ranges between 3 and 16 months. The effectiveness of existing treatments, especially regarding gliomas, is burdened by the presence of the BBB that severely restricts the passage of drugs and their potential to reach the target.

Nanomedicine is the application of nanotechnology in healthcare. It is a revolutionary discovery in disease diagnosis and therapy with potential application in cancer research and treatment. In particular, nanoparticles (NPs) allow entry of drugs within the BBB and, thus reduce doses administered, side effects and resistance [10, 20]. Their nanometric size, electrostatic charge, and lipophilic characteristics allow them to freely penetrate into the brain tissue. This review highlights the recent studies on the development of nanocarrier drug delivery systems for the treatment of gliomas.

7.2 Biology of Gliomas

Tumor development is essentially related to two categories of genes: tumor suppressor genes and proto-oncogenes. They are respectively down-regulated and up-regulated during tumorigenesis. In particular, gliomas show mostly one pattern of invasion and show no metastasis outside the brain [21]. Gliomagenesis consists of a complex process that involves different events starting with the activation of the growth factor receptors pathways, namely the activation of the epidermal growth factor receptor (EGFR), platelet-derived growth factor receptor (PDGFR), c-Kit and vascular endothelial growth factor receptor (VEGFR); all of which are over-expressed in gliomas [22–24]. These tumors are characterized by high invasion and angiogenesis. The invasion consists of an active passage of glioma cells through the host cellular and extracellular matrix (ECM) which depends on three crucial steps: gliomatose cells adhesion to ECM, ECM degradation by protease secreted by glioma cells, and migration through the ECM. The degradation of the ECM, a fundamental step of invasiveness, involves the secretion of proteases, such as metalloproteases (MMPs) that are produced and secreted by tumor and stromal cells. Among these, the enzyme most involved in gliomagenesis is MMP1; it is capable of degrading the interstitial collagen (types 1, 2, and 3) and is located at the endothelial level, but only in tumor vessels. An in vitro study demonstrated an inhibition of glioma growth after decreased secretion of MMPs [25]. MMP-1 is also involved in the cleavage of the $\alpha\upsilon\beta3$ integrin, increasing its ligand-binding activity. Integrins are critical in allowing cell adhesion and, consequently, survival, proliferation, migration and differentiation of gliomas. The $\alpha\upsilon\beta3$ integrin is overexpressed in endothelial cells of tumor vessels and is directly involved in the pathway that regulates endothelial cell migration. A recent study also demonstrated that the overexpression of CCN-1, a ligand for integrins, results in promoting angiogenesis and is a poor prognostic factor for patient survival [26]. Angiogenesis is a complex mechanism of formation of new blood vessels from pre-existing vasculature and is a histologic marker of tumor malignancy. Glioma vessels are structurally and functionally abnormal and this is the basis of presence of vasogenic edema and increased interstitial pressure. The transition from a non-vascularized tumor to a tumor mass characterized by the presence of new vessels is defined as the angiogenic switch and is determined by a precarious balance between molecules that regulate angiogenesis, an altered balance of proangiogenic and antiangiogenic factors. These factors are produced by glioma cells, fibroblasts, leukocytes and microglial cells. Promoters of angiogenesis are vascular endothelial growth factor (VEGF), hepatocyte growth factor (HGF), basic fibroblast growth factor (bFGF), transforming growth factor α and β (TGF), platelet-derived growth factor (PDGF), angiopoietins, tumor necrosis factor α and β (TNF), prostaglandin E1 (PGE1) and prostaglandin E2 (PGE2), integrins, selectins and cadherins. VEGF is a mitogenic factor responsible for increased survival of endothelial cells. In fact, it increases vascular permeability and subsequent extravasation of plasma proteins and reduces the number of pericytes of the vessels. VEGF is overexpressed in gliomas due to tumor hypoxia and presence of proto-oncogenes. It is secreted by cancer, stromal and inflammatory cells. The link between ECM and VEGF is mediated by the presence of proteoglycans or glycosaminoglycans. The interaction of VEGF and its receptor activates the signaling pathway involving the phosphatidylinositol-3-kinase (PI3K) and protein kinase B (PKB/Akt). PI3K/Akt pathway increases expression of receptors involved in cell adhesion. Therapeutic strategies include targeting VEGF and VEGFR, thus reducing the link between the receptor and VEGF leading to decreased vessel density.

7.3 Blood–Brain Barrier

The effectiveness of chemotherapy is limited by the presence of the BBB that limits the passage of a large number of anticancer drugs. The CNS is the protected organ of the whole human body. In fact, the exchange between the blood and the cerebrospinal fluid (CSF) is strictly regulated by the BBB [27, 28]. It is known that, for neural function and synaptic signaling, it is necessary to maintain a stable environment and an optimal ionic composition [2, 10]. BBB is a high selective diffusion barrier and consists of the choroid plexus epithelium, the arachnoid epithelium and the blood–brain interstitial fluid [7]. It can be divided into two portions: an inner endothelial cell layer that is the wall of the capillaries (it contains tight junctions) and a basal membrane upon which astrocytic and pericytes feet processes lie [29, 30]. The astrocytes provide support to the endothelial cells, to regulate blood flow, to provide nutrients to the CNS and to maintain extracellular balance. Pericytes have mainly a structural role as they ensure the integrity of the vessels, but are also involved in angiogenesis, differentiation of the vessels and in the formation of tight junctions [2]. The microvessel endothelial cells of the BBB, unlike those in the rest of the body, are not equipped with fenestrations and possess more TJs and adherent junctions (AJs). Tight junctions consist of a complex set of transmembrane and intracellular proteins and are located on the apical portion of the endothelial cells [29]; these junctions limit the paracellular flux of hydrophilic molecules. The tightness of the BBB is also guaranteed by adhesion junctions. The latter are localized in the basal portion of the lateral plasma membrane of endothelial cells and are formed by cadherins and transmembrane glycoproteins anchored to the cytoskeleton via cytoplasmic proteins [20].

The portion of the BBB located between the tumor mass and the cerebral capillary network is called the blood-brain tumor barrier (BBTB). Its characteristics vary during the course of malignant tumor growth and traverse three phases. During the first phase, the capillaries have normal morphology and provide for the energetic needs of the tumor. The second phase is characterized by an important tumor growth that involves the invasion of healthy tissues. The capillaries therefore become insufficient to provide adequate energy intake and the consequent hypoxia stimulates the neoangiogenesis process. The capillaries newly formed have fenestrations and allow the passage of substances that are smaller than 12 nm. In the third and final phase, the morphology of the capillaries is totally overturned with increased vascular permeability [31]. Vascular permeability in the tumor site is increased because the BBB is poorly developed and is also impaired. In fact, recent studies have shown that tight junctions are opened in glioma cells because the junction protein claudin-1 is lost and claudin-5 is down-regulated. Increased vascular permeability underlies cerebral edema. The increased vascular permeability, caused by structural alterations of the BBB, allows passive diffusion of drugs; however, alterations of the BBB are not present evenly across the lesion, so some areas are difficult to reach by passive targeting [32]. Thanks to the complex organization of the BBB, the majority of drugs cannot reach target sites and carry out their therapeutic action. There are different mechanisms by which molecules can cross the BBB which are outlined below.

(i) Passive diffusion: This kind of mechanism allows molecules to move between cells or across cells. Paracellular diffusion is a non-saturable and non-competitive movement of molecules between cells. Transcytosis, instead, allows movement across the cells. Both types of passive diffusion described depends on the electrochemical gradient and the molecular weight of compounds and do not require the use of energy (ATP). Only water-soluble and lipid soluble molecules with a molecular weight below 400 Daltons can cross the BBB, but the majority of drugs have a molecular weight higher than 400 Daltons [2]. Facilitated diffusion is a type of transport down a concentration gradient in which the passage through the membrane is allowed by molecular binding to a protein carrier and the subsequent conformational change of the protein.

(ii) Active diffusion: Almost all substances need an active transport; ATP provides the energy required to transport molecules against a gradient. The protein carrier-mediated transport (CMT) is a process which requires specific proteins for the influx of solutes within the membrane. This means of transport can be used for brain drug-delivery, if the drugs are modified in such a way as to mimic the endogenous ligand. Endocytosis is a transport mechanism that allows entry of molecules through formation of vesicles that are destined to reach the endosomes or lysosomes. There are two types of endocytosis: bulk-phase endocytosis and mediated endocytosis (receptor or absorptive). Bulk-phase endocytosis or pinocytosis is a non-competitive and non-saturable transport specific to the extracellular fluid. Receptor-mediated endocytosis is a highly specific transport, that is clathrin-dependent, and that allows eukaryotic cells to uptake only selected molecules. Transcytosis refers to the transcellular movement of molecules. Adsorptive endocytosis/transcytosis differs from the others because adsorptive processes depend upon electrostatic interactions between the substrate and the cell membrane. A new transport mechanism, cell-mediated transcytosis, through the BBB has been recently discovered. It allows the immune system cells, such as macrophages or monocytes, to cross the BBB.

7.3.1 Brain Drug Delivery and Strategies to Cross the Blood–Brain Barrier

As already mentioned, the effectiveness of antitumor therapies is directly proportional to their ability to reach the target sites. However, this is largely hampered by the presence of the BBB; in addition, the bioavailability of drugs within the brain is conditioned by the plasma half-life, plasma protein binding and rapid clearance in the extracellular fluid [21]. In order to increase the concentration of antitumor drugs in the brain, there are invasive and non-invasive strategies. An invasive and effective way to allow the entry of medications within the brain is the transient destruction of the BBB. The methods used are different and include chemical, ultrasound, and magnetic destruction.

7.3.1.1 Intranasal delivery

Intranasal delivery is an important way to bypass the BBB. In fact, the respiratory region of the nasal cavities is considered to be the site of best pharmacological absorption of systemic circulation. The types of transport exploited are the classical transcellular, paracellular, transcytosis, and carrier-mediated transport. This method has the advantages of rapid absorption, avoiding first-passage metabolism and the lack for the need of alterations in the BBB. Medications are released in the cerebrospinal fluid or in the brain through the olfactory mucosa, exploiting the connective tissue surrounding the olfactory nerve or the axons that form it. A recent study investigated the efficacy of the use of nanocapsules for the release of olanzapine through intranasal administration. The nanocapsules come into contact with the olfactory mucosa through continuous washes with solutions containing these NPs. However, intranasal delivery shows important limitations; the olfactory region of the nasal epithelium corresponds to about 5% in humans and, therefore, in many cerebral regions, the release of medication does not permit the achievement of therapeutic doses [33]. Intranasal mucosa is damaged by repeated use of this route of administration and, moreover, a discrete dose of medication is often eliminated by the mucociliary clearance system [33].

7.3.1.2 Chemical methods

An invasive and effective way to allow the entry of medications within the brain is the destruction of the BBB. The methods used are different and include osmotic destruction, ultrasound destruction and magnetic destruction. Mannitol is the most widely used substance among hypertonic solutions and allows a temporary opening of the BBB. Factors that influence the destruction of the BBB are the mannitol concentration and the rate of administration. Mannitol causes a dehydration of endothelial cells, thus reducing their volume with subsequent destruction of the tight junctions [2], to allow the entry of medications by paracellular passive absorption. However, this method is not free from side effects. In fact, the opening of the BBB determines the non-selective passage of toxic substances and the hypertonic solution itself can damage the central nervous system if its concentration is too high [32]. In addition to mannitol, other substances such as leukotrienes, bradykinin, VEGF and histamines are able to create a transient and reversible damage to the BBB. These substances have a greater effect on cerebral gliomas than healthy tissues, due to increased vascular permeability of tumor tissues. An important strategy is to exploit the chemical properties of the BBB. For this purpose, antitumor drugs can be chemically modified to bypass the barrier. For example, hydrophobic groups can be added to the medications, so as to make them lipid-soluble to facilitate transport through the BBB via passive transport. This process is called lipidization. The obstacles to overcome are, essentially, two: First, the modified medications, must undergo different enzymatic transformations before they can carry out their function. Second, lipidization makes these molecules able to traverse every lipid membrane and thus causes their spread in tissues. Prodrugs are inactive molecules, which are the result of modifications of biologically active molecules. Generally, the alterations of the medication are aimed at obtaining a better passage through the membranes. At the target site, prodrugs are converted into the biologically active medication. One drawback of the increased lipophilicity of prodrugs consists of their sequestration in other tissues and thereby in a consequent cytotoxicity in peripheral tissues and in a low bioavailability in the target sites.

7.3.1.3 Energy-based physical methods

It has been recently demonstrated that even the use of ultrasound can temporarily open the BBB. The method allows for the direct concentration of ultrasonic waves in target sites, resulting in minimal effect on the surrounding healthy tissues and consists of the systemic injection of acoustically activated microbubbles, which have the ability to permeabilize the vascular endothelium and the cellular membranes [34]. Finally, the introduction of magnetic NPs that, if exposed to different magnetic fields, generate heat and increase the permeability of the BBB. However, this method does not allow the attainment of sufficient doses within the CNS in vivo [34].

7.3.1.4 Intracerebral direct delivery

Among the invasive methods, the most commonly used is the intravascular or intraventricular administration of the medication by using catheters. Unfortunately, these strategies are burdened by important side effects, such as infections or catheter obstruction. Chemotherapeutic agents can also be administered locally through the implanting of microparticles of polymeric, biodegradable and non-degradable, materials, encapsulating drugs inside them; for example, Gliadel wafer is a polymeric material containing carmustine that, after being positioned in the tumor resection cavity, releases the medication for 5 days. The results of this treatment, in terms of median survival time, were very encouraging; in fact, there was an increase of the median survival time by about two months, compared to untreated patients [35]. However, this method is also subject to different side effects, such as edema caused by the high concentration of carmustine and the obstructive hydrocephalus determined by the dislocation of the wafer [35].

7.3.1.5 Non-invasive methods

Among non-invasive strategies, the most widely used strategies foresee chemical or biochemical modifications of drugs (conjugation with specific antibodies) or, more recently, the use of NPs, which have the ability to cross the BBB after intravenous administration without its disruption [35]. Cell-penetrating peptides (CPPs) are peptides formed by a few basic amino acids that are able to transport within cells small and large molecules, such as plasmids, nucleic acids, liposomes and NPs [34]. In addition to enhancing the internalization of the NPs, they have the ability to allow the passage of many drugs through the BBB. The mechanism used is not certain, but several studies have demonstrated that the most widespread one is endocytosis.

Target drug delivery is a therapeutic system that promotes the accumulation of therapeutic agents in specific sites, thus reducing doses administered and peripheral toxicity. Targeting can be of two types: active or passive. Passive targeting exploits the typical modifications of the tissues affected by the pathology, for example the enhanced permeability and retention (EPR) of brain tumors [34]. An important achievement in tumor therapy with NPs is to increase the EPR effect; in fact, particles smaller than

100 nm extravasate from tumor vasculature, thanks to increased vascular permeability, but those NPs that are greater than 20 nm are retained in intra tumoral tissues. EPR effect can increase the effectiveness of the drug that has a higher exposure time [3]. The EPR effect is exploited for passive targeting in the treatment of many tumors that are characterized by a hypervasculature and an incomplete lymphatic drainage [34]. Active targeting makes use of specific physiological systems, such as antigen-antibody binding, ligand–receptor binding. Transportation systems currently used for this purpose are glucose transporters and insulin, transferrin and glutathione receptors, and antibodies that specifically bind to these receptors. In particular, these systems serve to facilitate transcytosis and receptor-mediated endocytosis of NPs tied on receptors present on endothelial cells [36–38]. Strategies used by NPs to allow the transport of drugs are different: (1) The NPs open tight junction sites between endothelial cells, (2) the NPs conjugated to drugs are transported into cells by transcytosis or endocytosis, and (3) coating agents surrounding the NPs inhibit the efflux from the cell [39]. The use of NPs to facilitate the passage of anticancer drugs across the BBB is the best strategy. In fact, the NPs allow the transport of substances without altering the structural and functional integrity of the BBB. The factors affecting the transport of NPs include the type of polymer or surfactant used, the size of the NPs and the drug conjugated to the NPs and the surface charge. Nanoparticles of sizes less than 200 nm are generally captured via clathrin-mediated endocytosis, while the uptake of NPs of sizes of 500 nm is caveolae-mediated. As for the surface charge, the NPs with a higher rate of uptake are the positively charged ones, as the plasma membrane is charged negatively [35].

7.4 Nanoparticles in Treatment of Gliomas

Nanotechnology is the scientific branch that deals with particles of sizes of the nanometer order. The NPs are substances that have at least one of their dimensions smaller than a micrometer. The characteristics that make these systems revolutionary are their small size, their high solubility, the possibility to change the NP surface and the ability to exploit the characteristics of the tumor tissue for maximizing benefit [40]. The fields of application of NPs are varied: diagnosis, imaging, treatment, and drug delivery (Table 7.1). For such purposes, a series of small molecules can be tied to the NPs, for example by means of molecular contrast enhancement or medications. The methods used for the bond are different: covalent linkage, adsorption, and encapsulation. Drug delivery is both an immense challenge and a great resource for treating cerebral tumors and in particular gliomas. In recent years, research has focused on formulating NPs and modifying their characteristics in order to overcome the BBB and ensure the passage of drugs and their release in the target sites. A nanocarrier system should be biodegradable, biocompatible, should not trigger an immune response and should avoid clearance by the reticuloendothelial system (RES) [31]. The RES consists of monocytes and macrophages that swallow the NPs and destroy them. In particular, the characteristic that mainly affects the clearance of this system is the size of the NPs; in fact, the most eliminated particles are larger than 200 nm. Another key feature for the pharmacokinetics of NPs is the surface charge. In fact, the non-charging NPs are less vulnerable to RES, but they tend to hardly interact with each other and to aggregate, but they do not interact with the cells, threatening to release drugs outside of cells. The charged NPs, however, are more attackable by the RES, but they have an increased number of interactions with the cells. With regard to the brain drug delivery, the positive charge is the one that allows a greater passage through the BBB by adsorption-mediated transcytosis. The plasma half-life of NPs significantly affects the reaching of target sites and the degree of drug release. A method used to increase the plasma half-life is the surface hydration. It consists in binding some compounds to the surface of the NPs, such as polyethylene glycol (PEG) or monosialoganglioside (GM1). In fact, these compounds create an additional outer layer that makes it difficult for the opsonization and uptake by monocytes and macrophages. The most widely used molecule is PEG [31].

7.4.1 Polymer Nanoparticles

Polymer NPs have dimensions ranging from 10 to 1000 nm in diameter. They consist of natural or synthetic polymers; among the first being cellulose, alginate, gliadin; and among the second polylactide (PLA), and poly-(lactide-co-glycolide) (PLGA). In general, synthetic polymers are preferred since their chemical-physical properties, such as solubility or permeability, can be easily quantified. Polymeric NPs can convey a large amount of therapeutic agents, both hydrophilic and lipophilic, both high and low molecular weight (e.g., DNA or antisense oligonucleotides) [2]. The techniques of synthesis of polymeric NPs are different and include adsorption, entrapment, encapsulation, and dissolution. Based on the technique used, we distinguish nanospheres and nanocapsules. In nanospheres, the drugs are entrapped or adsorbed within the polymers, while in nanocapsules the medications are inside the liquid core surrounded by the polymeric membrane. Polymer particles can also be used to coat other NPs; for example, the hydrophilic PEG prevents the recognition of NPs by the endothelial systems and its phagocytosis; hydrogel and dextran guarantee an increase in the plasma half-life of drugs [2]. The disadvantages of polymeric NPs relate to the complexity and the cost of preparation and the concrete possibility of triggering immune responses and allergic reactions [41].

7.4.2 Liposomes

Liposomes are composed of a double lipid layer and a fluid core. They have spherical shape and dimensions including nm and mm. The double lipid layer is formed by phospholipids and cholesterol, and by varying the respective percentages, one can adjust the properties of the NPs and the release of the medications. The most important characteristic of liposomes is its ability to carry both hydrophobic agents (inside the double lipid layer), and water-soluble agents (inside the aqueous core). The main drawback is that liposomes are an easy target of the RES; degradation can be minimized or avoided due to the bond of liposomes with polymer particles such as PEG. Peg coating (PEGylation) prevents the interaction of nanoparticles with the plasma components and reduces renal filtration. Liposomes can also be linked to monoclonal antibodies, which allow the optimal attainment of the target sites. For example, a great candidate for tumor targeting is EGFR and its main variant EGFRvIII (characterized by the deletion of 267 amino acids), overexpressed in more than 40% of the glioblastomas. Several studies have tested the efficacy and safety of doxorubicin encapsulate in PEGylated liposomes [42]. Recently, the use of small-interfering RNAs (fragments of RNA that suppresses gene expression) was also studied. These agents are degraded and eliminated rapidly after injection; their transport within the aqueous core of the liposomes allows for their therapeutic use [2].

7.4.3 Metallic Nanoparticles

Small metal NPs are toxic to cultured cancer cells. Recently, the effectiveness of metallic NPs has been studied also in vivo. Magnetic NPs have been investigated as MRI contrast agents and, in the last two decades, also for their action as targeting agents for cancer therapy. A nontoxic and biocompatible surface coating that can also provide targeted delivery is required for effective therapy by metallic NP [43]. Gold NPs can be used both in the diagnostic field (gold nanoshells as contrast agents) and in the therapeutic field (gold-silica for photothermal ablation of tumor cells). These NPs concentrate on target sites by receptor-mediated endocytosis mechanisms or by passive diffusion. To avoid the uptake and degradation by the reticuloendothelial system, the gold NPs can be PEGylated. The thermal ablation mechanism is expressed due to the peculiar capacitance of the gold NPs to absorb the light between the visible and the near-infrared spectra and to create heat, which destroys the target cells. Nanoshells are activated by a laser tuned to the nanoshell resonance, which determines an increase in temperatures and therefore cellular damage. Recent studies in vivo, conducted on mouse models, have shown a good efficacy of photothermal ablation. The main advantage is not to cause harm to neighboring cells. Titanium dioxide (TiO_2) is a catalyst in the photo-oxidation of the organic substrates. NPs constituted by this material induce cytotoxicity in many tumors [44]. The cytotoxic effect is tied to the activation of the NPs by UV radiation and has been studied by creating a three-dimensional model of glioma cells. In order to assess the effect of the NPs, markers of necrosis and apoptosis were studied, and it was discovered that they primarily induce apoptosis in the first 6 hours. Vodyanoy et al. [44] compared the effectiveness of Zinc (Zn) and copper (Cu) NPs with those of natural metal NPs isolated from animal blood. They found that the changes observed in cells treated with Zn and Cu nanoparticles are similar to those caused by apoptotic agents. It was seen that these NPs act against cultured glioma cancer cells, but not against healthy astrocyte cells.

7.4.4 Carbon Nanotubes

Carbon nanotubes (CNs) are molecules formed by carbon atoms and consist of one or more layers of graphene (single-walled or multi-walled carbon nanotubes) [45]. Among the single-walled nanotubes are classified the nanohorns, which can be produced at lower cost and higher material purity. Drugs are linked to the outer surface of CNs (with covalent or non-covalent bonds, such as hydrophobic and electrostatic interactions) or are included internally within the CNS. The input mechanisms within the cell depends on the size of the CNs; those of a size less than 400 nm are internalized by passive diffusion, while the others by endocytosis. A means of enabling effective targeting is to bind folic acid to the outer surface of the CNs, which is recognized by the cells of tumors that overexpress folate. An important advantage of CNs is its ability to carry siRNAs; in vitro studies have demonstrated that the release of siRNA-conjugated CNs determines inhibition of cellular proliferation. Unfortunately, these NPs cause significant side effects, such as increased oxidative stress and, if inhaled, acute lung injury, inflammation, and fibrosis [2].

7.4.5 Graphene Nanoparticles

The graphene has a two-dimensional hexagonal structure. The biologically most significant and used compound is its derivative, the oxide graphene oxide (GO). GO is applied in numerous biomedical fields including drug delivery, photothermal therapy and imaging [46–48]. In particular, the corresponding nanoparticle (nano-GO/NGO) is used. Several studies have tested the effectiveness of the use of the NGO as a carrier. It was conjugated with the chemotherapeutic drug 1,3-bis(2-chloroethyl)-1-nitrosurea (BCNU) and with the polymeric compound polyacrylic acid to improve the solubility and cellular penetration, as the drug, if not conveyed by NP, had demonstrated poor stability at different temperatures and a short half-life. The use of BCNU conjugated to

NGO has proved promising in terms of efficacy in vitro. Recently, the photothermal properties of the graphene, particularly in combination with chemotherapy agents, have been studied. Doxorubicin was conjugated to GSPI, a targeted-peptide (IP)-modified mesoporous silica coated graphene nanosheet. The results were significantly better than the single use of chemotherapy or photothermal therapy alone.

7.4.6 Dendrimers

The dendrimers are macromolecules with a three-dimensional structure and consist of a core and branches. The set of core, branches and functional groups determines the chemical-physical characteristics of these nanoparticles, such as size, shape and reactivity. Due to these properties, dendrimers can convey encapsulated drugs or agents related to functional groups with electrostatic or covalent bonds. Among the dendrimers, avidimers are particularly important in antitumor therapy; they are compounds that exploit the characteristics of the tumor vascularization thanks to methotrexatepolyamidoamine (PAMAM) bioconjugated platform functionalized with small targeting ligands. Recently experiments have been undertaken with dendriworms which are specific dendrimers developed for the delivery of siRNA. They are of the multimodal dendrimer-conjugated magneto-fluorescent nanoworms and, in a GBM animal model, they caused the suppression of the expression of EGFR [49].

Table 7.1 Summary of the types of NPs for treatment of gliomas and related delivery methods

Nanoparticles	Delivery methods
Polymer nanoparticles	Intranasal delivery, chemical methods, energy based physical methods
Liposomes	Chemical methods, non-invasive methods, energy based physical methods
Metallic nanoparticles	Chemical methods, energy based physical methods
Carbon nanotubes	Chemical methods, energy based physical methods
Graphene nanoparticles	Chemical methods, energy based physical methods
Dendrimers	Chemical methods, energy based physical methods

7.5 Nanomedicine Applications in Cancer Stem Cells

A crucial role in determining chemoresistance is manifested by cancer stem cells (CSCs). In fact, the CSC hypothesis suggests that tumors originate from cells that have stem cell–like properties, e.g., cells that express markers specific for stem cells and self-renewal. These CSCs offer resistance to cytotoxic therapies [50] and to current strategies of treatment [51, 52]. Recent studies have shown that CSCs are more resistant to radiotherapy and chemotherapy and can repair DNA damage better than non-CSC [53, 54]. It is therefore clear that the mechanisms of recurrence involve the survival of CSCs that can ultimately reinitiate tumor formation [55]. Nanomedicine, only recently, considered the use of NPs that can target the CSCs. A recent study has demonstrated the efficacy of cationic polyurethane-short branch polyethylenimine (PU-PEI) NPs combined with a plasmid encoding a microRNA145, a miRNA with suppressive action on the tumor [56]. This therapy showed chemosensitization and radiosensitization of brain tumors and an increase of animal survival. Another example is the use of NPs tied to a specific antibody specific to CD133 [57]. Photothermal therapy using carbon nanotubes bound to the antibody has determined a selective lysis of the CD133-positive GBM CSCs. EGFR is overexpressed in GBM and in GBM CSCs [58]. A clinical study on animals showed the effectiveness of anti-EGF nanomedicine in GBM [59]. The therapeutic agent used was cetuximab (that binds both EGFR and EGFRvIII deletion mutant) conjugated with iron-oxide NPs (IONPs). The result was an increase in survival. Another target is the ligand of TfR (diferric transferrin, Tf) or antibodies anti-TfR. In fact, GBM cells have demonstrated an overexpression of TfR. The binding of the ligand or antibody determines the internalization of receptor–ligand or antibody–receptor complex via receptor-mediated endocytosis [50]. An example is the Tf-conjugated lipopolyplex Nps (Tf-Nps). Using these therapeutic products causes an inhibition of migration and expression of EGFR [60]. These NPs are able to act either on CSCs or on differentiated cancer cells. Finally, the use of p53 is currently in multiple clinical trials. For example, a nanocomplex called SGT-53 is able to delivering the exogenous wild-type tumor suppressor gene p53 across the BBB. The result of this therapy is the sensitization of highly Temozolomide (TMZ)-resistant GBM to TMZ and an extended survival [37]. In addition, a combination of SGT-53 and TMZ causes a significant apoptosis of CD133 positive CSCs [37]. It is important to note that the only action on CSCs may not be effective in eradicating the disease, since the remaining differentiated cells might be able, alone, to sustain tumor growth or even to dedifferentiate into GBM CSCs in response to stress such as hypoxia microenvironment or radiation [60]. It is therefore desirable to develop a more effective therapeutic approach directed towards both CSCs and non-CSCs using a common target (TfR or EGFR for example).

7.6 Nanoparticle Drug Delivery

Drug delivery can be defined as the process of releasing a compound at a specific rate and at a specific site. Novel drugs require effective delivery technologies that will minimize side effects and lead to better patient compliance. Conventional anticancer agents have low molecular weights and high pharmacokinetic volumes of distribution that cause their cytotoxicity. Efficacious doses are obtained with high concentrations and, when administered alone, these drugs lack specificity and cause significant damage to non-cancerous tissues. Over the last two decades, nanotechnology has established new diagnosis and therapeutic strategies. Particular attention was paid to the development of the drug delivery systems. NPs are used as nanocarriers when conjugated with specific drugs for the purpose of improving the efficacy of these therapies. NPs-based drug delivery systems provide better penetration of therapeutic and diagnostic agents into the desired body compartments, enabling enhanced efficacy with reduced doses and systemic concentrations of the drug with minor risks caused by other treatments. NPs-based drug delivery shows the potential to enhance drug bioavailability, improve the timed release of drug molecules, and enable precision drug targeting without altering the structural and functional integrity of the BBB [61]. Latest developments in NPs drug delivery for gliomas involve the use of targeted-NPs using antibodies and peptides as targeted molecules. In this way, specific NPs can carry drugs and release these molecules specifically at cancerous tissues. NPs may be delivered to specific sites by size-dependent passive targeting or by active targeting. Passive targeting consists of a chemical modification of the NPs to increase permeability or stability. PEG is able to increase half-life of nanocarrier drug delivery systems by decreasing their uptake by macrophages due to steric repulsion effects, and inhibition of plasma-protein adsorption [62]. PEGylation has been successfully applied in the majority of drug delivery systems, including lipid, polymeric and inorganic NPs. The term active targeting refers to the use of peripherally conjugated targeting moieties for enhanced delivery of NPs systems. This method has been performed to obtain a high degree of selectivity to specific tissues and to enhance the uptake of NPs into cancer cells and angiogenic microcapillaries. In the active targeting methods, these compounds include the anticancer agent, a targeting moiety-penetration enhancer, such as receptors, receptor ligands, enzymes, antibodies, and surface modifications [63]. However, because it is a non-specific process, the adsorptive process also occurs in the blood vessels and in other organs. This poses a challenge for both achieving therapeutic concentration in the brain and limiting the drug distribution in non-target organ. Cell-penetrating peptides and cationic proteins are investigated to enhance brain drug delivery via adsorptive-mediated transcytosis. A large variety of cargo molecules/materials have been effectively delivered into cells via CPPs, including small molecules, proteins, peptides, fragments of DNA, liposomes and NPs. The transcription factor TAT, involved in the replication cycle of human immunodeficiency virus (HIV), was demonstrated to penetrate into cells [64]. Transporters for the nutritive substances for the brain are usually overexpressed on the BBB and can be used for brain-targeted delivery [65]. The glutathione transporter is highly expressed on the BBB, and thus researchers conjugated it onto liposomes to deliver various drugs to brain. Systemic administration of glycosyl cholesterol derivative liposomes containing coumarin-6 displayed 3.3-fold higher $C_{\max}$ with less cytotoxicity than conventional liposomes to brain capillary endothelial cells [66]. Receptor-mediated transcytosis across the BBB has been more explored because of its high specificity. Therapeutic compounds are able to cross the BBB after association/conjugation to these specific ligands. Receptor-mediated transcytosis has been demonstrated for transferrin, insulin, insulin-like growth factors (IGF-1 and IGF-2), the low-density lipoprotein receptor-related protein (LRP) [34]. Transferrin receptor (Tf-R) is a transmembrane glycoprotein overexpressed in GBM cells. Drug targeting to the Tf-R can be achieved by using the endogenous ligand transferrin, or by using antibodies directed against the Tf-R. Doxorubicin was loaded into Tf-R-NPs and showed anti-tumor effects, with median survival time of 70% longer than that of the doxorubicin solution-treated brain tumor-bearing rats [67]. The endogenous ligands may bind with the receptors hindering the binding efficiency of ligand-modified NPs. To avoid this problem, antibodies against these receptors were developed. The binding site of antibodies to the receptors was different from that of ligands with receptors, and thus ligand competition could be avoided. Ulbrich et al. developed human serum albumin (HSA) NPs coupled to transferrin or TR-mAbs (OX26) for delivery of loperamide and showed efficiency in transporting the drug to the brain using mice in which OX26 was conjugated to HSA NPs. OX26 mAb can avoid competition with endogenous transferrin in the circulation system because it binds to an extracellular domain of TR [68]. LRP-1 and LRP-2 can interact with a diverse range of molecules and mediators including ApoE, tissue plasminogen activator (tPA), plasminogen activator inhibitor 1 (PAI-1), lactoferrin, heparin cofactor II, heat shock protein 96 (HSP-96) and engineered angiopeps [69]. Several drugs that do not cross the BBB, including tubocurarine, loperamide, dalargin, and doxorubicin, show higher concentrations in the brain when associated with polysorbate 80-coated NPs. Polysorbate 80, a non-ionic surfactant, could adsorb ApoE in serum when it was conjugated onto NPs, and polysorbate 80-coated NPs have been also evaluated as a brain targeting delivery system by many groups [51].

7.7 Nanoparticle Drugs Delivery in Gliomas Treatment

Nanotechnology has notably changed the classical modality in which diagnosis and treatment are achieved mainly due to recent advances in material engineering, drug availability, and the targeting of cancer cells. As already reported, the mechanism of RMT, through binding to specific receptors placed on endothelial and tumor cells, allows the passage of pharmacological compounds through the BBB. Tf-Rs are overexpressed in brain capillary endothelium and glioma cells. Paclitaxel can inhibit cell division via promoting the gathering and stabilization of microtubules. In a recent study, the compound formed by Tf-conjugated magnetic silica PLGA NPs (MNPMSN-PLGA-Tf NPs) loaded with doxorubicin (DOX) and paclitaxel (PTX) (DOX-PTX-NPs-Tf) was evaluated. The cellular uptake of DOX-PTX-NPs-Tf could be enhanced by the presence of magnetic field and the usage of Tf as the targeting ligand; moreover, it showed a significant inhibition of tumor growth [70]. In a recent experimental study, a compound characterized by a transferrin-functionalized PEGylated NPs (Tf-NPs) able to carry TMZ and the bromodomain inhibitor JQ1 was tested. In two intracranial orthotopic mouse models of GBM, the authors demonstrated decreased tumor burden and prolonged survival [45]. It has been reported that low density LRP are overexpressed on BBB and on glioma cells. Conjugation of NPs with angiopep (ANG1005 consists of a 19 amino acid long peptide carrier conjugated to a taxane) is associated with improved delivery of the compound across the BBB via lipoprotein receptor-mediated endocytosis. An interesting research showed a notable penetration and accumulation of ANG-PEG-NP-PTX into 3D glioma spheroids and in gliomas in vivo. The efficacy of PTX loading ANG-PEG-NP was also significantly enhanced [71].

Angiogenesis is a physiological process that plays an important role in various pathological and physiological processes. Alteration in the balance of angiogenesis can cause various disorders including malignancy, ischemic disease, immune disorders, and infectious diseases. The control of the angiogenic process can represent a therapeutic option in gliomas treatment. The brain tumor endothelium, with characteristics of high proliferation, high permeability, and high expression of proangiogenic factors, is a particularly appealing therapeutic target for this strategy. Recently, a nanocarrier system formed by lipid NPs with a VEGF antisense oligonucleotides to down-regulate VEGF expression in an in vitro and in vivo rat glioma model was studied

[72]. The author reported that this compound was highly effective in reducing VEGF expression in both models of rat brain gliomas, allowing quick delivery of antisense oligonucleotides into tumor cells. In another research, a NPs drug delivery system was developed combining enhanced loading capacity of cisplatin and high binding affinity to glioma cells [73]. This was achieved using the highly soluble cisplatin analogue, cis-diamminedinitratoplatinum(II), and antibodies against the native form of VEGF or VEGFR2 conjugated to liposome surfaces. The NPs drug delivery system showed sustained drug release profile, high affinity to antigens, and increased uptake by glioma C6 and U-87 cells. An interesting study evaluated the action of targeted magnetic NPs as MRI agents for in vivo visualization of gliomas [74]. Ferric oxide (Fe_3O_4) cores were coated with bovine serum albumin (BSA) to form NPs. Monoclonal antibodies against VEGF were covalently conjugated to BSA through a polyethylene glycol linker. The results showed an improvement in MRI visualization of the glioma. A significant therapeutic effect was found after convention enhanced delivery (CED) of both iron oxide-NPs (IO-NPs) and EGFRvIIIAb-IO-NPs in mice [75]. Dispersion of the NPs may potentially target infiltrating tumor cells outside the tumor mass that are potentially responsible for tumor recurrence. Use of bioconjugated magnetic NPs may permit the advancement of CED in the treatment of gliomas due to their sensitive imaging qualities on T2-weighted MRI and therapeutic effects [75].

Dendrimers as drug carriers can be utilized in therapeutic applications for drugs and siRNA delivery in various types of cancers. Conjugation of these NPs with D-glucosamine results in increasing their endocytosis and permeability across the BBB and tumor targeting [76]. Methotrexate (MTX)-loaded dendrimers have shown increased distribution in gliomas. Furthermore, glycosylated MTX-loaded dendrimers were able to eradicate MTX-resistance glioma cells [76]. In addition, the poly amidoamine (PAMAM) dendrimer was used to deliver antisense miR-21 oligonucleotide (as-miR-21) and 5-fluorouracil (5-FU) to human GBM cells. The codelivery of as-miR-21 significantly improved the cytotoxicity of 5-FU and remarkably increased the apoptosis of U251 cells [77].

7.8 Nanotoxicity

Nanotoxicity evaluates the interactions of NPs with biological systems and the relationship between the physical and chemical properties of NPs with the induction of toxic biological responses. Currently, a complete evaluation of the size, shape, composition and aggregation-dependent interactions of NPs with biological systems is lacking, and thus it is unclear whether the exposure of humans, animals, and plants to engineered nanostructures could produce harmful biological responses [78]. The same characteristics which make NPs so attractive in medicine may contribute to the toxicological profile of NPs in biological systems. Some NPs contain metals or compounds with known toxicity, and thus the breakdown of these materials could elicit similar toxic responses to the components themselves. NPs seem to stimulate the production of reactive oxygen species (ROS) and oxidative stress. The highly reactive hydroxyl radicals, superoxide anion radicals and hydroxide peroxide [79] are normal products of cellular oxidative metabolism, especially in mitochondria. Nanoparticles, especially metal nanoparticles, can induce overproduction of ROS and oxidative stress that determines changes in cell proteins, lipid peroxidation, modification to nucleic acids and subsequent damage in cell function. All these dysfunctions affect cell operation, through modulation of inflammatory responses and gene expression disorders (activation of redox-sensitive transcription factors) [80]. Nanoparticles that induce the formation of ROS play a primary role in genotoxicity. In fact, DNA damage (such as sugar lesions, DNA-proteins cross-links and single- and double-strand breaks) and unregulated cell signaling are the basis of carcinogenesis. Antioxidants are important to limit and/or prevent oxidative damage due to their role as electron or hydrogen atom donors. As summarized by Fu et al. [78], NPs cause cytotoxicity and genotoxicity by the overproduction of ROS and, as a result, aging based on lipid peroxidation, damage to mitochondria, proteins and the nucleus. Physical and chemical NPs' characteristics which lie at the basis of the production of ROS are size, shape, oxidation status, surface area and coating, solubility, and degree of aggregation [42]. In particular, the most important aspects are size and shape. The small size of NPs allows them to enter cell membranes and other barriers, for example BBB, and cause damage (the smaller the NPs, the greater the inclination to enter cells) [78, 81]. However, it is difficult to explain the molecular mechanisms of nanotoxicity mediated by the ROS, as free radicals have a short half-life since they are very reactive. The best method to demonstrate the toxicity of NPs is the use of cultured cell lines; in fact, this method has high sensitivity [82], as well as high reproducibility and allows concrete measurements of cytotoxicity.

Although targeted NPs have emerged as one strategy to overcome the lack of specificity of conventional chemotherapy, there are other potential risks and challenges associated with this novel strategy. Some cancer cell types develop drug resistance, rendering drugs released from the targeted NPs to be ineffective. Also, the targeted NPs might change the stability, solubility, and pharmacokinetic properties of the carried drugs. The shelf life, aggregation, leakage, and toxicity of materials used to make NPs are other limitations for their use. Some materials such as carbon nanotubes are durable and can persist in the body for months, or even years, making them potentially toxic and limiting their use for repeated treatments. A systematic quantitative analysis of the pharmacokinetics (absorption, distribution, metabolism, and excretion) of NPs, can lead to improvements in the design of NPs for diagnostic and therapeutic applications, a better understanding of nanostructures non-specificity toward tissues and cell types, and assessments of basic distribution and clearance, that serve as the basis in determining their toxicity and future investigative directions.

7.9 Future Developments and Conclusions

The treatment of gliomas is characterized by a high rate of failure due to their infiltrative and aggressive nature, the presence of the BBB, which restricts entry of therapeutic entities to the tumor area, the recurrent nature of the tumor, the paucity of antigen presenting cells, and the immune suppressive nature of the tumor microenvironment. The use of new, multiple, preoperative, and intraoperative techniques such as intra-operative MRI, electrophysiologic monitoring, navigated transcranial magnetic stimulation, neuronavigation, and the use of 5-aminolevulinic acid

(5-ALA), has allowed an improvement of the rates of successful complete tumor resection to 96.2% [14]. Molecular biology studies are certainly more interesting, and allow the identification of many molecular biomarkers that, if affected, could cause a slowing of the tumor progression. However, classical molecular therapy has various limitations. Gliomas show a complex heterogeneity at the genomic and molecular levels. Moreover, the use of EGFR inhibitors, MET and/or PDGFR inhibitors would maintain activation of downstream pathways, which is a theoretical mechanism of therapy resistance [83]. Several studies evaluated the presence of crosstalks between PI3K-MAPK-p53-RB pathways, which can balance any single pathway alteration [84]. Recent research showed the presence of EGFR-VEGF(R) cross-talk in both tumor and tumor-associated endothelial cells which is involved in tumor survival and angiogenesis [85]. Gliomas are also characterized by genomic instability, which favors gene mutations and chromosomal alterations, such that the use of cytotoxic agents and radiotherapy would accelerate the mutagenesis. Another serious obstacle is represented by the fact that the chosen target can be activated by multiple pathways, in different phases, during tumor progression, thus rendering the treatment ineffective. Radio- and chemotherapy treatments have limited action and are burdened by severe side effects.

Nanotechnology can be a valuable and novel approach for the treatment of gliomas. Their small dimensions allow the NPs to cross the BBB and, by acting as carriers, they can convey even more therapeutic compounds able to interact with multiple targets. NP-based drug-delivery systems overcome the BBB with high targeted-cell specificity and selectivity. Thus, NPs permits the delivery of a lower dose of effective drugs both into the central core of the tumor and into the distal foci of tumor cells within areas often characterized by the integrity of the BBB [2]. This feature is of big help in early diagnosis, preoperative histological and grade diagnosis, preoperative treatment planning, and late recurrences. Notwithstanding, there are potential risks related with this novel approach. Some cancer cell types could develop drug resistance making the drugs released from the targeted NPs ineffective. Moreover, NPs might change stability, solubility, and pharmacokinetic properties of the carried drugs. In addition, some materials used to create NPs possess low toxicity but degrade quickly and do not circulate in tissues long enough to warrant a sustained drug/gene delivery.

There is continued debate about the long-term effects of NPs, in particular the effects of interactions between NPs or coating molecules and target cells. In order for this promising field to rapidly progress, focus must be placed on elucidating the safety of these novel materials. Besides new studies on cell and tumor biology are necessary for the rational design of NPs for brain tumor therapy. Identification of multiple targets could be a more viable option, as well as targeting activated molecules in the early stages of tumor progression.

References

1. Ostrom, Q. T., Gittleman, H., Fulop, J., Liu, M., Blanda, R., Kromer, C., Wolinsky, Y., Kruchko, C., Barnholtz-Sloan, J. S. (2015). CBTRUS statistical report: Primary brain and central nervous system tumors diagnosed in the United States in 2008–2012, *Neuro-Oncol.*, **17,** pp. 1–62.
2. Caruso, G., Caffo, M., Alafaci, C., Raudino, G., Cafarella, D., Lucerna, S., Salpietro, F. M., Tomasello, F. (2011). Could nanoparticles systems have a role in the treatment of cerebral gliomas?, *Nanomedicine,* **7**, pp. 744–752.
3. Tzeng, S. Y., Green, J. J. (2013). Therapeutic nanomedicine for brain cancer, *Ther. Deliv.*, **4**, pp. 687–704.
4. Raudino, G., Caffo, M., Caruso, G., Alafaci, C., Raudino, F., Marventano, V., Romano, A., Montemagno, F., Belvedere, M., Salpietro, F. M., Tomasello, F., Schillaci, A. (2013). From gliomagenesis to multimodal therapeutic approaches, ed., Lichtor, T., Chapter 7 *High-Grade Glioma Treatment in Evolution of the Molecular Biology of Brain Tumors and the Therapeutic Implications* (InTech, Rijeka) pp. 203–233.
5. Ostrom, Q. T., Gittleman, H., Stetson, L., Virk, S. M., Barnholtz-Sloan, J. S., (2015). Epidemiology of gliomas, *Cancer Treat. Res.*, **163**, pp. 1–14.
6. Sehedic, D., Cikankowitz, A., Hindre, F., Davodeau, F., Garcion, E. (2015). Nanomedicine to overcome radioresistance in glioblastoma stem-like cells and surviving clones, *Trends Pharmacol. Sci.,* **36,** pp. 236–252.
7. Woodworth, G. F., Dunn, G. P., Nance, E. A., Hanes, J., Brem, H. (2014). Emerging insights into barriers to effective brain tumor therapeutics, *Front. Oncol.,* **4,** p. 126.

8. Ohgaki, H., Kleihues, P. (2013). The definition of primary and secondary glioblastoma, *Clin. Cancer Res.,* **19,** pp. 764–772.
9. Bhujbal, S. V., de Vos, P., Niclou, S. P. (2014). Drug and cell encapsulation: Alternative delivery options for the treatment of malignant brain tumors, *Adv. Drug Deliv. Rev.*, **67–68,** pp. 142–153.
10. Caruso, G., Raudino, G., Caffo, M., Alafaci, C., Granata, F., Lucerna, S., Salpietro, F. M., Tomasello, F. (2010). Nanotechnology platforms in diagnosis and treatments of primary brain tumors. *Recent Pat. Nanotec.*, **4**, pp. 119–122.
11. Auffinger, B., Morshed, R., Tobias, A., Cheng, Y., Ahmed, A. U., Lesniak, M. S. (2013). Drug-loaded nanoparticle systems and adult stem cells: A potential marriage for the treatment of malignant glioma?, *Oncotarget*, **4**, pp. 378–396.
12. Caruso, G., Merlo, L., Caffo, M. (2014). *Innovative Brain Tumor Therapy. Nanoparticle-Based Therapeutic Strategies,* Woodhead Publishing Series in Biomedicine: Number 67. (Woodhead Publishing, Cambridge, UK).
13. Senft, C., Bink, A., Franz, K., Vatter, H., Gasser, T., Seifert, V. (2011). Intraoperative MRI guidance and extent of resection in glioma surgery: A randomised, controlled trial, *Lancet Oncol.*, **12**, pp. 997–1003.
14. Schucht, P., Beck, J., Abu-Isa, J., Andereggen, L., Murek, M., Seidel, K., Stieglitz, L., Raabe, A. (2012). Gross total resection rates in contemporary glioblastoma surgery: Results of an institutional protocol combining 5-aminolevulinic acid intraoperative fluorescence imaging and brain mapping, *Neurosurgery,* **71,** pp. 927–936.
15. Coburger, J., Hagel, V., Wirtz, C. R., König, R. (2015). Surgery for glioblastoma: Impact of the combined use of 5-aminolevulinic acid and intraoperative MRI on extent of resection and survival, *PLoS One*, **10**, p. e0131872.
16. Stupp, R., Mason, W. P., van den Bent, M. J., Weller, M., Fisher, B., et al. (2005). Radiotherapy plus concomitant and adjuvant temozolomide for glioblastoma, N. Engl. J. Med., 352, pp. 987–996.
17. Caffo, M., Barresi, V., Caruso, G., Cutugno, M., La Fata, G., Venza, M., Alafaci, C., Tomasello, F. (2013). Innovative therapeutic strategies in the treatment of brain metastases, *Int. J. Mol. Sci.*, **14**, pp. 2135–2144.
18. Field, K. M., Jordan, J. T., Wen, P. Y., Rosenthal, M. A., Reardon, D. A. (2015). Bevacizumab and glioblastoma: Scientific review, newly reported updates, and ongoing controversies, *Cancer*, **121**, pp. 997–1007.

19. Polyzoidis, S., Ashkan, K. (2014). Dendritic cell immunotherapy for glioblastoma, *Expert Rev. Anticancer Ther.,* **14**, pp. 761–763.

20. Caruso, G., Caffo, M., Raudino, G., Tomasello, F. (2012). Nanoparticles and brain tumor treatment, *Biomedical & Nanomedical Technologies -Concise Monograph Series*. (ASME Press, New York, USA).

21. Caruso, G., Caffo, M., Alafaci, C., Raudino, G., Salpietro, F. M., Tomasello, F. (2010). Antisense oligonucleotides as an innovative therapeutic strategy in the treatment of high-grade gliomas, *Rec. Pat. CNS Drug Discov.*, **5**, pp. 53–69.

22. Cominelli, M., Grisanti, S., Mazzoleni, S., Branca, C., Buttolo, L., et al. (2015). EGFR amplified and overexpressing glioblastomas and association with better response to adjuvant metronomic temozolomide, *J. Natl. Cancer Inst.*, **107**, p. djv041.

23. Kumar, A., Boyle, E. A., Tokita, M., Mikheev, A. M., Sanger, M. C., et al. (2014). Deep sequencing of multiple regions of glial tumors reveals spatial heterogeneity for mutations in clinically relevant genes, *Genome Biol.*, **15**, pp. 530.

24. Westermark, B. (2014). Platelet-derived growth factor in glioblastoma-driver or biomarker?, *Ups. J. Med. Sci.*, **119**, pp. 298–305.

25. Gwak, H. S., Park, M. J., Park, I. C., Woo, S. H., Jin, H. O., Rhee, C. H., Jung, H. W. (2014). Tetraarsenic oxide-induced inhibition of malignant glioma cell invasion in vitro via a decrease in matrix metalloproteinase secretion and protein kinase B phosphorylation, *J. Neurosurg.,* **121,** pp. 1483–1491.

26. Ishida, J., Kurozumi, K., Ichikawa, T., Otani, Y., Onishi, M. (2015). Evaluation of extracellular matrix protein CCN1 as a prognostic factor for glioblastoma, *Brain Tumor Pathol.*, **32**, pp. 245–252.

27. Abbott, N. J., Patabendige, A. A., Dolman, D. E., Yusof, S. R., Begley, D. J. (2010). Structure and function of the blood brain barrier, *Neurobiol. Dis.*, **37**, pp. 13–25.

28. Agarwal, S., Sane, R., Oberoi, R., Ohlfest, J. R., Elmquist, V. F. (2011). Delivery of molecularly targeted therapy to malignant glioma, a disease of the whole brain, *Expert Rev. Mol. Med.*, **13**, p. e17.

29. Cardoso, F. L., Brites, D., Brito, M. A. (2010). Looking at the blood-brain barrier: Molecular anatomy and possible investigation approaches, *Brain Res. Rev.*, **64**, pp. 328–363.

30. Dyrna, F., Hanske, S., Krueger, M., Bechmann, I. (2013). The blood-brain barrier, *J. Neuroimmune Pharmacol.,* **8**, pp. 763–773.

31. Karim, R., Palazzo, C., Evrard, B., Piel, G. (2016). Nanocarriers for the treatment of glioblastoma multiforme: Current state-of-the-art, *J. Control. Release*, **227**, pp. 23–37.

32. Juillerat-Jeanneret, L. (2008). The targeted delivery of cancer drugs across the blood-brain barrier: Chemical modifications of drugs or drug-nanoparticles?, *Drug Discov. Today*, **13**, pp. 1099–1106.

33. Ali, J., Ali, M., Baboota, S., Sahani, J. K., Ramassamy, C., Dao, L. (2010). Potential of nanoparticulate drug delivery systems by intranasal administration, *Curr. Pharm. Des.*, **16**, pp. 1644–1653.

34. Zhang, T. T., Li, W., Meng, G., Wang, P., Liao, W. (2016). Strategies for transporting nanoparticles across the blood-brain barrier, *Biomater. Sci.*, **4**, pp. 219–229.

35. Wohlfart, S., Gelperina, S., Kreuter, J. (2012). Transport of drugs across the blood-brain barrier by nanoparticles, *J. Control. Release*, **161**, pp. 264–273.

36. Birngruber, T., Raml, R., Gladdines, W., Gatschelhofer, C., Gander, E., Ghosh, A., Kroath, T., Gaillard, P. J., Pieber, T. R., Sinner, F. (2014). Enhanced doxorubicin delivery to the brain administered through glutathione PEGylated liposomal doxorubicin (2B3-101) as compared with generic Caelyx(®)/Doxil(®)–a cerebral open flow microperfusion pilot study, *J. Pharm. Sci.*, **103**, pp. 1945–1948.

37. Kim, S. S., Rait, A., Kim, E., Pirollo, K. F., Nishida, M., Farkas., N., Dagata, J. A., Chang, E. H. (2014). A nanoparticle carrying the p53 gene targets tumors including cancer stem cells, sensitizes glioblastoma to chemotherapy and improves survival, *ACS Nano.*, **8**, pp. 5494–5514.

38. Kim, S. S., Rait, A., Kim, E., Pirollo, K. F., Chang, E. H. (2015). A tumor-targeting p53 nanodelivery system limits chemoresistance to temozolomide prolonging survival in a mouse model of glioblastoma multiforme, *Nanomedicine*, **11**, pp. 301–311.

39. Jain, K. K. (2012). Nanobiotechnology-based strategies for crossing the blood-brain barrier, *Nanomedicine*, **7**, pp. 1225–1233.

40. Zhao, J., Castranova, V. (2011). Toxicology of nanomaterials used in nanomedicine, *J. Toxicol. Environ. Heal. B. Crit. Rev.*, **14**, pp. 593–632.

41. Mamot, C., Ritschard, R., Wicki, A., Stehle, G., Dieterle, T., Bubendorf, L., Hilker, C., Deuster, S., Herrmann, R., Rochiltz, C. (2012). Tolerability, safety, pharmacokinetics, and efficacy of doxorubicin-loaded anti-EGFR immunoliposomes in advanced solid tumours: A phase 1 dose-escalation study, *Lancet Oncol.*, **13**, pp. 1234–1241.

42. Wankhede, M., Bouras, A., Kaluzova, M., Hadjipanayis, C. G. (2012). Magnetic nanoparticles: An emerging technology for malignant

brain tumor imaging and therapy, *Expert Rev. Clin. Pharmacol.,* **5**, pp. 173–186.

43. Yamaguchi, S., Kobayashi, H., Narita, T., Kanehira, K., Sonezaki, S., Kubota, Y., Terasaka, S., Iwasaki, Y. (2010). Novel photodynamic therapy using water-dispersed TiO_2-polyethylene glycol compound: Evaluation of antitumor effect on glioma cells and spheroids in vitro, *Photochem. Photobiol.*, **86**, pp. 964–971.
44. Vodyanoy, V., Daniels, Y., Pustovyy, O., MacCrehan, W. A., Muramoto, S., Stan, G. (2016). Engineered metal nanoparticles in the sub-nanomolar levels kill cancer cells, *Int. J. Nanomed.*, **11**, pp. 1567–1576.
45. Lam, F. C., Morton, S. W., Wyckoff, J., Han, T. V., Hwang, M. K., Maffa, A., Yaffe, M. B., Floyd, S. R., Hammond, P. T. (2018). Enhanced efficacy of combined temozolomide and bromodomain inhibitor therapy for gliomas using targeted nanoparticles, *Nat. Commun.*, **9**, p. 1991.
46. Beg, S., Rizwan, M., Sheikh, A. M., Hasnain, M. S., Anwer, K., Kohli, K. (2011). Advancement in carbon nanotubes: Basics, biomedical applications and toxicity, *J. Pharm. Pharmacol.*, **63**, pp. 141–163.
47. Caffo, M., Merlo, L., Marino, D., Caruso, G. (2015). Graphene in neurosurgery: The beginning of a new era. *Nanomedicine (London)*, **10**, pp. 615–625.
48. Yang, K., Feng, L., Shi, X., Liu, Z. (2013). Nanographene in biomedicine: Theranostic applications, *Chem. Soc. Rev.*, **42**, pp. 530–547.
49. Agrawal, A., Min, D. H., Singh, N., Zhu, H., Birjiniuk, A., von Maltzahn, G., Harris, T. J., Xing, D., Woolfenden, S. D., Sharp, P. A., Charest. A., Bhatia, S. (2009). Functional delivery of siRNA in mice using dendriworms, *ACS Nano.*, **3**, pp. 2495–2504.
50. Sundar, S. J., Hsieh, J. K., Manjila, S., Lathia, J. D., Sloan, A. (2014). The role of cancer stem cells in glio- blastoma, *Neurosurg. Focus*, **37**, p. E6.
51. Martins, S. M., Sarmento, B., Nunes, C., Lucio, M., Reis, S., Ferreira, D. C. (2013). Brain targeting effect of camptothecin-loaded solid lipid nanoparticles in rat after intravenous administration, *Eur. J. Pharm. Biopharm.*, **85**, pp. 488–502.
52. Pointer, K. B., Clark, P. A., Zorniak, M., Alrfaei, B. M., Kuo, J. S. (2014). Glioblastoma cancer stem cells: Biomarker and therapeutic advances, *Neurochem. Int.*, **71**, pp. 1–7.
53. Esparza, R., Azad, T. D., Feroze, A. H., Mitra, S. S., Cheshier, S. H. (2015). Glioblastoma stem cells and stem cell-targeting immunotherapies, *J. Neurooncol.*, **123**, pp. 449–457.

54. Jhanwar-Uniyal, M., Labagnara, M., Friedman, M., Kwasnicki, A., Murali, R. (2015). Glioblastoma: Molecular pathways, stem cells and therapeutic targets, *Cancers Basel*, **7**, pp. 538–555.

55. Yang, Y. P., Chien, Y., Chiou, G. Y., Cherng, J. Y., Wang, M. L., Lo, W. L., Chang, Y. L., Huang, P. I., Chen, Y. W., Shih, Y. H., Chen, M. T., Chiou, S. H. (2012). Inhibition of cancer stem cell-like properties and reduced chemoradioresistance of glioblastoma using microRNA145 with cationic polyurethane-short branch PEI, *Biomaterials*, **33**, pp. 1462–1476.

56. Wang, C. H., Chiou, S. H., Chou, C. P., Chen, Y. C., Huang, Y. J., Peng, C. A. (2011). Photothermolysis of glioblastoma stem-like cells targeted by carbon nanotubes conjugated with CD133 monoclonal antibody, *Nanomedicine*, **7**, pp. 69–79.

57. Lo, H. W. (2010). EGFR-targeted therapy in malignant glioma: Novel aspects and mechanisms of drug resistance, *Curr. Mol. Pharmacol.*, **3**, pp. 37–52.

58. Kaluzova, M., Bouras, A., Machaidze, R., Hadjipanayis, C. G. (2015). Targeted therapy of glioblastoma stem-like cells and tumor non-stem cells using cetuximab-conjugated iron-oxide nanoparticles, *Oncotarget*, **6**, pp. 8788–8806.

59. Wang, X., Huang, X., Yang, Z., Gallego-Perez, D., Ma, J., Zhao, X., Xie, J., Nakano, I., Lee, L. J. (2014). Targeted delivery of tumor suppressor microRNA-1 by transferrin-conjugated lipopolyplex nanoparticles to patient-derived glioblastoma stem cells, *Curr. Pharm. Biotechnol.*, **15**, pp. 839–846.

60. Beier, D., Schulz, J. B., Beier, C. P. (2011). Chemoresistance of glioblastoma cancer stem cells–much more complex than expected, *Mol. Cancer*, **10**, pp. 128.

61. Dubin, C. H. (2004). Special delivery: Pharmaceutical companies aim to target their drugs with nano precision, *Mech. Eng.*, **126**, pp. S10–S12.

62. Suka, J. S., Xua, Q., Kima, N., Hanesa, J., Ensigna, L. M. (2016). PEGylation as a strategy for improving nanoparticle-based drug and gene delivery, *Adv. Drug Deliv. Rev.*, **99**, pp. 28–51.

63. Byrne, J. D., Betancourt, T., Brannon-Peppas, L. (2008). Active targeting schemes for nanoparticle systems in cancer therapeutics, *Adv. Drug Deliv. Rev.*, **60**, pp. 1615–1626.

64. Rapoport, N., Marin, A. P., Timoshin, A. A. (2000). Effect of a polymeric surfactant on electron transport in HL-60 cells, *Arch. Biochem. Biophys.*, **384**, pp. 100–108.

65. Wei, X., Chen, X., Ying, M., Lu, W. (2014). Brain tumor-targeted drug delivery strategies, *Acta Pharm. Sin. B*, **4**, pp. 193–201.

66. Qin, Y., Fan, W., Chen, H., Yao, N., Tang, W., Yuan, W., Kuai, R., Zhang, Z., Wu. Y., He, Q. (2010). In vitro and in vivo investigation of glucose-mediated brain-targeting liposomes, *J. Drug Target*, **18**, pp. 536–549.

67. Kumar, P., Wu, H., McBride, J. L., Jung, K. E., Kim, M. H., Davidson, B. L., Lee, S. K., Shankar, P., Manjunath, N. (2009). Transvascular delivery of small interfering RNA to the central nervous system, *Nature*, **448**, pp. 39–43.

68. Ulbrich, K., Hekmatara, T., Herbert, E., Kreuter, J. (2009). Transferrin- and transferrin-receptor-antibody- modified nanoparticles enable drug delivery across the blood-brain barrier (BBB), *Eur. J. Pharm. Biopharm.*, **71**, pp. 251–256.

69. Caruso, G., Merlo, L., Tot, E., Pignataro, C., Caffo, M. (2017). Nanotechnology and the new frontiers of drug delivery in cerebral gliomas, ed., Grumezescu, A. M., Chapter 6 Nano- and Microscale Drug Delivery Systems (Elsevier, Amsterdam, NL) pp. 95–107.

70. Cui, Y., Xu, Q., Chow, P. K., Wang, D., Wang, C. (2013). Transferrin-conjugated magnetic silica PLGA nanoparticles loaded with doxorubicin and paclitaxel for brain glioma treatment, *Biomaterials*, **34**, pp. 8511–8520.

71. Xin, H., Sha, X., Jiang, X., Zhang, W., Chen, L., Fang, X. (2012). Anti-glioblastoma efficacy and safety of paclitaxel-loading Angiopep-conjugated dual targeting PEG-PCL nanoparticles, *Biomaterials*, **33**, pp. 8167–8176.

72. Brioschi, M., Calderoni, S., Pradotto, L., Guido, M., Strada, A. (2009). Solid lipid nanoparticles carrying oligonucleotides inhibit vascular endothelial growth factor expression in rat glioma models, *J. Nanoneurosci.*, **1**, pp. 65–74.

73. Shein, S. A., Kuznetsov, I. I., Abakumova, T. O., Chelushkin, P. S., Melnikov, P. A., Korchagina, A. A., Bichov, D. A., Seregina, I. F., Nukulova, N. V. (2016). VEGF- and VEGFR2-targeted liposomes for cisplatin delivery to glioma cells, *Mol. Pharm.*, **13**, pp. 3712–3723.

74. Abakumov, M. A., Nukolova, N. V., Sokolsky-Papkov, M., Shein, S. A., Sandalova, T. O. (2015). VEGF-targeted magnetic nanoparticles for MRI visualization of brain tumor, *Nanomedicine*, **11**, pp. 825–833.

75. Hadjipanayis, C. G., Machaidze, R., Kaluzova, M., Wang, L., Schuette, A. J., Chen, H., Wu, X., Mao, H. (2010). EGFRvIII antibody conjugated

iron oxide nanoparticles for MRI guided convection-enhanced delivery and targeted therapy of glioblastoma, *Cancer Res.*, **70**, pp. 6303–6312.

76. Yan, H., Wang, J., Yi, P., Lei, H., Zhan, C., Xie, C., Feng, L., Qian, J., Zhu, J., Lu. W., Li, C. (2011). Imaging brain tumor by dendrimer-based optical/paramagnetic nanoprobe across the blood-brain barrier, *Chem. Commun.*, **47**, pp. 8130–8132.

77. Ren, Y., Kang, C. S., Yuan, X. B., Zhou, X., Xu, P., Han, L., Wang, G. X., Jia, Z., Zhong, Y., Yu, S., Sheng, J., Pu, P. Y. (2019). Co-delivery of as-miR-21 and 5-FU by poly(amidoamine) dendrimer attenuates human glioma cell growth in vitro, *J. Biomater. Sci. Polym. Ed.*, **21**, pp. 303–314.

78. Fu, P. P., Xia, Q., Hwang, H. M., Ray, P. C., Yu, H. (2014). Mechanisms of nanotoxicity: Generation of reactive oxygen species, *Food Drug Anal.*, **22**, pp. 64–75.

79. Yin, J. J., Liu, J., Ehrenshaft, M., Roberts, J. E., Fu, P. P., Mason, R. P., Zhao, B. (2012). Phototoxicity of nanotitanium dioxides in HaCaT keratinocytes e generation of reactive oxygen species and cell damage, *Toxicol. Appl. Pharmacol.*, **263**, p. 81e8.

80. Crawford, D. R. (1999) Regulation of mammalian gene expression by reactive oxygen species, eds., Gilbert, D. L., Colton, C. A., *Reactive Oxygen Species in Biological Systems: An Interdisciplinary Approach* (Kluwer Academic Publishers, New York, USA) p. 155e71.

81. Yoshida, T., Yoshikawa, T., Nabeshi, H., Tsutsumi, Y. (2012). Relation analysis between intracellular distribution of nanomaterials, ROS generation and DNA damage, *Yakugaku Zasshi*, **132**, pp. 295–300.

82. Nel, A., Xia, T., Madler, L., Li, N. (2006). Toxic potential of materials at the nanolevel, *Science*, **311**, p. 622e7.

83. Snuderl, M., Fazlollahi, L., Le, L. P., Nitta, M., Zhelyazkova, B. H., Davidson, C. J., Cahill, D. P., Aldape, K. D., Betensky, R. A., Louis, D. N., Iafrate, A. J. (2011). Mosaic amplification of multiple receptor tyrosine kinase genes in glioblastoma, *Cancer Cell*, **20**, pp. 810–817.

84. Furnari, F. B., Fenton, T., Bachoo, R. M., Mukasa, A., Stommel, J. M., Stegh, A., Hahn, W. C., Ligon, K. L., louis, D. N., Brennan, C., Chin, L., DePinho, R. A., Cavenee, W. K. (2007). Malignant astrocytic glioma: Genetics, biology, and paths to treatment, *Genes Dev.*, **21**, pp. 2683–2710.

85. Larsen, A. K., Ouaret, D., El Ouadrani, K., Petitprez, A. (2011). Targeting EGFR and VEGF(R) pathway cross-talk in tumor survival and angiogenesis, *Pharmacol. Ther.*, **131**, pp. 80–90.

Chapter 8

Plant-Derived Edible Nanoparticles in Cancer Drug Delivery

Siavash Iravani[a] and Ghazaleh Jamalipour Soufi[b]

[a]*Faculty of Pharmacy and Pharmaceutical Sciences, Isfahan University of Medical Sciences, Isfahan, Iran*
[b]*Radiology Department, Isfahan University of Medical Sciences, Isfahan, Iran*

siavashira@gmail.com

There have been several innovative scientific investigations to develop natural nanomedicines. In this regard, plant-derived edible nanoparticles have shown great potential to be applied as cancer drug delivery systems due to their unique properties. These include their relative safety, low immunogenicity, high stability, and industrial production capability. In this chapter, we present recent literature and discuss some applications of plant-derived edible nanoparticles in cancer drug delivery.

Nanoparticle Drug Delivery Systems for Cancer Treatment
Edited by Hala Gali-Muhtasib and Racha Chouaib

ISBN 978-981-4800-54-9 (Hardcover), 978-0-429-34125-0 (eBook)
www.jennystanford.com

8.1 Introduction

Plant-derived edible nanoparticles (PDENs) are nano-sized membrane vesicles released by the edible plants (such as, ginger, grapefruit, broccoli, and lemon). These NPs might have intrinsic therapeutic activities against specific diseases, for example, in the treatment of inflammatory bowel disease (IBD) and cancer; moreover, they are non-toxic and can be mass-produced. Nowadays, nano-researchers are looking for selecting and characterizing the specific populations of PDENs which have these activities and are analyzing their intrinsic targeting. Plant-derived lipid nanoparticles (PDLNs) might act as natural drug carriers which can effectively target drug delivery to a specific location of the human body [1].

PDENs and PDLNs show unique benefits, including suitable nanoparticle size, low toxicity, good tissue-specific targeting, excellent environmental safety, and significant potential for large-scale production. It was reported that NPs from the juice of grapefruits, grapes, ginger, carrots, and tomatoes might be isolated by using an environmentally friendly process [2]. Various characterization procedures have demonstrated that these nano-sized vesicles have structures similar to mammalian-derived exosomes. It was reported that PDENs might be implicated in plant cell–cell communication, could potentially regulate the innate immunity of plants, and might also transport mRNAs, miRNAs, bioactive lipids, and proteins into animal cells [3, 4].

For the isolation of PDENs, differential ultracentrifugation plus density gradient centrifugation can be applied. Plants are usually ground to juice in a mixer; low-speed centrifugation is applied to remove large particles and plant fibers; medium-speed centrifugation is used to remove large debris and intact organelles; and high-speed centrifugation is used to pellet the exosome-like NPs. The quantity and quality of the obtained NPs are sensitive to the g force rotor type, and angle of rotor sedimentation. Moreover, this method sediments other vesicles, proteins and/or protein/RNA aggregates. Therefore, a subsequent sucrose density gradient step is used to separate the PDENs from contaminants of different densities. Gradient ultracentrifugation requires an extended centrifugation time (about 1–5 h), but provides a more purified edible plant nanoparticle isolate than ultracentrifugation alone [5–7].

The RNA, lipid, and protein contents in PDENs differ from those found in mammalian exosomes [8]. For instance, grape exosome-like NPs (GELNs) contained 96 miRNAs and 28 identified proteins [9]. In comparison, mammalian exosomes typically contain 100–300 miRNAs and more than 1,000 proteins. In terms of the lipid profile, mammalian exosomes are generally rich in cholesterol and sphingomyelin but have only low levels of phosphatidylethanolamine and the mitogenic compound, phosphatidic acid (PA). In contrast, GELNs were found to have 98% phospholipids (approximately, 50% of which was PA) and 2% typical plant lipids (for example, galactolipids). Phosphatidic acid was reported to interact with mammalian target of rapamycin (mTOR), and was shown to trigger cell growth and proliferation [10]. In addition, the mitogenic phospholipid PA is highly fusogenic in the presence of calcium, and therefore, it has been postulated to induce inter-vesicular fusion [11].

Grapefruit-derived exosome-like NPs (GDNs) were isolated. As a result, it was reported that GDNs contained about 137 proteins and had a lipid profile that differed from that of GELNs [12]. Mass spectrometry-based protein profiling of GDNs revealed that a number of the proteins were responsible for regulating carbohydrate/lipid metabolism. The lipidomic data indicated that GDNs were enriched for phosphatidylethanolamine (45.52%) and phosphatidylcholine (28.53%), with particularly high levels of phosphatidylethanolamine (34:2) and phosphatidylcholine (34:2). These differences of plant-derived exosome-like NPs might play critical roles in interspecies communication, such as long distance communication throughout the entire mammalian digestive tract.

Transmission electron microscopy (TEM) has been used to characterize PDENs. It seems that this characterization method does not allow researchers to obtain additional biochemical and mechanical information, but it is the main method used for ultrastructural analysis at the subcellular level [13–15]. For TEM analysis, exosome-like NPs could be resuspended in a water-soluble solution (such as deionized water), spotted onto a carbon-coated grid, fixed with 1% glutaraldehyde, stained with 2% phosphotungstic acid, and visualized. The structure of various

PDENs (such as from grapefruit, carrots, ginger, and tomatoes) have been reported [16]. Atomic force microscopy (AFM) can be used to study the sizes and structures of individual PDENs. Atomic force microscopy is a very high-resolution type of scanning probe microscopy (SPM) which has a demonstrated resolution on the order of fractions of a nanometer, which is more than 1,000 times better than the diffraction limit of optical microscopy [17–19].

Dynamic light scattering (DLS, also known as photon correlation spectroscopy or quasi-elastic light scattering) is used for studying the size-distribution profiles of small particles present in suspension, and to determine the sizes and zeta potentials of PDENs [16]. In DLS, fluctuations of the scattered light intensity (which are due to the Brownian movement of the particles) are correlated to get information about particle size. Since DLS is noninvasive and highly sensitive, and requires only very small sample volumes, it has been widely adopted in many scientific fields. It is currently the gold standard for quickly and accurately determining the size distribution of suspensions [20–22]. The median diameter of a nanoparticle is considered to be a key parameter for the therapeutic efficacy of a nanoparticle. PDENs obtained from grapefruit, ginger, tomatoes, and garlic reportedly have plant-specific monodispersed size distributions, with median diameters of about 400 nm for edible grape NPs, and about 250 nm for edible grapefruit and ginger NPs. In contrast, edible carrot NPs has two such distributions, one about 100 nm and the other about 1,000 nm. PDENs also show negative zeta potentials (about 49.2 to 1.52 mV) of relatively low value at pH of about 6, indicating that they exhibit mutual repulsion and no tendency towards aggregation [23–26].

In this chapter, some important applications of PDENs in cancer drug delivery are highlighted.

8.2 Plant-Derived Edible Nanoparticles in Cancer Drug Delivery

In general, the current chemotherapeutics used against cancer have some disadvantages, including toxicity and a lack of specificity. Actually, various polymeric NPs such as micelles, and other organic and inorganic biomaterials have been used as nanoscale drug carriers in the fight against cancer [27]. In one research, grapefruit-derived lipid NPs (GDNVs) was coated with inflammatory-related receptor enriched membranes of activated leukocytes (IGNVs) [28]. It was demonstrated that the resulting particles could deliver the anti-cancer drug doxorubicin to an inflamed tumor site [28]. Using different mouse models of inflammation-driven disease, they demonstrated that IGNVs showed better targeting of inflamed tumor tissues compared with GDNVs. Moreover, this homing to inflamed tissues was significantly inhibited by the blockade of LFA-1 or CXCR1/CXCR2 on the IGNV membranes.

Indeed, lipids are used as preferred carriers for drug delivery, but synthetic liposomes and lipid NPs can generate adverse effects including cell stress, inflammasome activation and apoptosis [29]. Plant-derived lipid nanoparticles are applied as suitable candidates for delivering various therapeutic agents. The use of nontoxic "natural" nanovectors assembled from PDLNs has been suggested as a feasible alternate strategy for in vivo drug delivery [30]. The feasibility of generating "natural" nanovectors from ginger NPs derived lipids was reported [30]. The generated lipid NPs had no cytotoxicity or negative effect on intestinal barrier function, suggesting that they may be useful for efficient in vivo drug delivery [30]. Nanoparticles made of grapefruit-derived lipids, and grapefruit-derived nanovectors, could deliver chemotherapeutic agents, siRNAs, DNA expression vectors, and proteins to different cell types [30]. Grapefruit-derived nanovectors was used to code liver therapeutic agents plus folic acid, and reported that this strategy significantly increased the targeting efficiency to cells that also express folate receptors. These nanovectors were further demonstrated to enhance the chemotherapy-induced inhibition of tumor growth in CT26- and SW620-cell-derived tumors in mice. The authors showed that grapefruit derived nanovectors were less toxic than NPs made of synthetic lipids and did not cross the placental barrier when injected intravenously into pregnant mice, suggesting that they could prove useful for drug delivery. In another study, intravenously injected grapefruit derived nanovectors with folic acid (FA)-enriched surfaces were used to deliver paclitaxel (PTX) to the tumor site. FA-enriched NPs take advantage of the striking elevation of FA receptors in a variety of tumor cells; they showed significantly increased targeting efficiency and were able to reduce the tumor volumes in mice xenografted with SW620 or CT26 cells [30].

It has been shown that oral administration of grape exosome-like NPs to mice reportedly triggered proliferation of cells in the intestinal epithelium and was associated with intestinal stem cell proliferation throughout the intestine and colon [9]. Furthermore, orally administered grape exosome-like NPs appeared to be resistant to degradation by saliva, the acidic environment of the stomach, and the highly active proteolytic enzymes present along the intestinal tract. These recent observations suggested that PDENs can be orally delivered to the intestine, where they may be taken up by intestinal cells and exert functions such as intestinal regeneration. PDENs, which naturally target colonic tissues and have anti-inflammatory properties, might therefore represent a novel natural and nontoxic delivery system that could easily be scaled up for the treatment of digestive tract diseases such as IBD. Ju et al. [9] reported interspecies communication between plant and gut host cells by demonstrating that grape exosome-like NPs have unique transport properties and biological functions. They showed that these NPs can penetrate the intestinal mucus barrier, be taken up by mouse intestinal stem cells, and significantly induce the activation of Lgr5C intestinal stem cells through the Wnt/β-catenin pathway. It was demonstrated that oral administration of grape exosome-like NPs had protective effects on dextran sulfate sodium induced colitis in mice by triggering the growth and proliferation of intestinal stem cells which have a crucial role in regulating intestinal epithelial cells differentiation and are required for intestinal tissue homeostasis and repair.

In one study, grapefruit-derived nanovesicles were selectively taken up by intestinal macrophages and ameliorated dextran sulfate sodium-induced mouse colitis [12]. Their anti-inflammatory effects were mediated by up-regulating the expression of heme oxygenase-1 (HO-1) and inhibiting the productions of IL-1b and TNF-α in intestinal macrophages. It was demonstrated that these nanovesicles are biocompatible, biodegradable, and stable across a wide range of pH values, suggesting that they could be developed as an oral drug delivery system. Therefore, the anti-inflammatory drug methotrexate (MTX) was incorporated into grapefruit-derived nanovesicles. Consequently, it was found that the encapsulated drug showed a lower toxicity than free

MTX and exhibited remarkably higher therapeutic effects against dextran sulfate sodium-induced colitis in mice. It was suggested that grapefruit-derived nanovesicles could serve as immune modulators in the intestine and promote homeostasis of intestinal macrophages. These NPs showed a potential for development as an oral delivery system for small molecule drugs aimed at attenuating inflammatory responses in human disease.

In one research, PDENs from ginger, carrots, grape and grapefruit were isolated and characterized [16]. Consequently, they were similar to mammalian-derived exosomes in terms of their sizes and structures. They contained proteins, lipids, and miRNAs, and were found to be taken up by intestinal macrophages and stem cells. The results obtained from PDEN-transfected macrophages indicated that ginger PDENs preferentially induced the expression levels of the antioxidant, heme oxygenase-1 (HO-1), and the anti-inflammatory cytokine, IL-10. Nuclear factor-like 2 (nrf2) as a key regulator of HO-1 gene, has an important role in anti-inflammation and anti-oxidation, which can be used as an indicator for regulating the cellular anti-inflammatory response. The authors also confirmed grapefruit, ginger, and carrot PDENs could activate nuclear translocation of nrf2 [16]. Furthermore, analysis of the intestines of β-catenin/TCF/LEF reporter transgenic mice treated with the various PDENs revealed increased numbers of activated β-catenin in intestinal crypts. This suggests that PDENs have significant effects on modulating Wnt/TCF/LEF activity and play an important role in gut homeostasis [31]. Therefore, PDENs-mediated interspecies communication might involve the activation of anti-inflammatory cytokines, antioxidants, and Wnt signaling, which are crucial for maintaining intestinal homeostasis.

Plant-derived exosome-like NPs might affect cancer progression, and inhibit cell proliferation in various tumor cell lines [32]. In one study, vesicles from *Citrus limon* L. were isolated by differential ultracentrifugation; this fraction had a density of 1.15 to 1.19 g/ml and was characterized by a specific proteomic profile [33]. It was demonstrated that these NPs inhibited cancer cell proliferation in different tumor cell lines in vitro by activating TRAIL-mediated apoptotic cell death. Furthermore, lemon nanovesicles suppressed the growth of chronic myeloid leukemia in vivo via activating TRAIL-mediated apoptosis, suggesting that plant-edible nanovesicles could be used for cancer treatment [33].

NPs were isolated from ginger (GDNP 2), and their lipids were reassembled. Consequently, these natural molecules were applied to prepare ginger-derived nano-lipids (as nanovectors). For targeted therapy of tumor tissues, nanoscientists altered these nanovectors with folic acid (FA) to prepare FA-modified nanovectors (FA nanovectors). FA demonstrated accelerated affinity binding to the folate receptors which were highly expressed on various tumors, and almost undetectable on non-tumor cells. The FA nanovectors were evaluated as delivery systems for doxorubicin, a chemotherapeutic drug applied for treatment of colon cancers. It was found that doxorubicin was completely loaded into the FA nanovectors, and then, these nanovectors were efficiently taken up by colon cancer cells, showed great biocompatibility and successfully hindered tumor growth. Compared to commercially available approaches for doxorubicin delivery, the mentioned nanovectors release the drug more quickly in an acidic pH that resembles the tumor surroundings. It appears that this method might reduce the severe side effects of doxorubicin. It was reported that FA nanovectors made of edible ginger-derived lipids could shift the current paradigm of drug delivery away from artificially synthesized NPs toward the use of nature-derived nanovectors from edible plants. Moreover, researchers have shown that they are nontoxic and can be produced on a large scale; FA nanovectors derived from edible plants could represent one of the safest targeted therapeutic delivery platforms [34]. Zhang et al. [34] reported a nanovector made from ginger-derived lipids which could serve as a delivery platform for the therapeutic agent doxorubicin to treat colon cancer. Consequently, viability and apoptosis assays and electric cell-substrate impedance-sensing technology revealed that ginger-derived nanovectors exhibited excellent biocompatibility; by contrast, cationic liposomes at the same concentrations decreased cell proliferation and increased apoptosis. These nanovectors were capable of loading doxorubicin with high efficiency and showed a better pH-dependent drug-release profile than commercially available liposomal-doxorubicin. Modified GDNVs conjugated with the targeting ligand folic acid mediated the targeted delivery of doxorubicin to Colon-26 tumors in vivo and enhanced the chemotherapeutic inhibition of tumor growth compared with free drug [34]. Findings from these investigations showed that PDENs can be good candidates for the delivery of anti-cancer drugs (summarized in Table 8.1).

Table 8.1 Plant-derived edible nanoparticles in cancer drug delivery

PDENs sources	Target cells	Pharmaceutical agents	Refs
Grapefruit	Colon cancer	Dox, curcumin	[28]
Grapefruit	Brain tumor	JSI-124 (Cucurbitacin I)	[30]
Grapefruit	Colon cancer	Paclitaxel	[30]
Grapefruit	Colon cancer	Luciferase gene siRNA	[30]
Lemon	Chronic myeloid leukemia (CML)	Proteins	[33]
Ginger	Colon	Doxorubicin	[34]
Broccoli	Colon	Sulforaphane	[35]

8.3 Conclusion

In conclusion, in the field of nanomedicine, PDENs have shown suitable properties including relatively high internalization rate, low immunogenicity, proven stability in the gastrointestinal tract, and the ability to overcome the blood-brain barrier but not cross the placental barrier. Therefore, they are good candidates for cancer drug delivery. In fact, non-targeted chemotherapy normally applied for treatment of cancers, but this strategy is not able to differentiate cancerous and normal cells, leading to inadequate therapeutic outcomes, and unwanted and harsh toxic side effects on normal cells. Enabling chemotherapeutic drugs to target cancer cells would be an important improvement in the treatment of cancer. Application of nanotechnology for drug delivery has shown great promise for improving cancer treatment. However, potential toxicity, hazardous environmental effects, issues with large-scale production, and potential excessive costs are important challenges which should be addressed for further clinical applications. PDENs, which are released from edible plant cells, are not cytotoxic and are free from immunotoxicity; for example, in the case of ginger-derived NPs. These NPs have shown the capacity to target the colon, reduce colorectal tumorigenesis by decreasing the pro-inflammatory cytokine levels, inhibit intestinal epithelial cell proliferation, and induce apoptosis. The surface modification of PDENs or the combination of plant-derived nanolipids with other nano-sized biomaterials could potentially be applied to tune the PDENs to target different cancer sites without generating adverse immunotoxicity. It appears that PDENs can be applied as next-generation therapeutic delivery systems for the treatment of cancers.

References

1. Yang, C., Zhang, M., Merlin, D. (2018). Advances in plant-derived edible nanoparticlebased lipid nano-drug delivery systems as therapeutic nanomedicines. *J. Mater. Chem. B*, **6**, pp. 1312–1321.
2. Quesenberry, P. J., Aliotta, J., Camussi, G., Abdel-Mageed, A. B., Wen, S., Goldberg, L., et al. (2015). Potential functional applications of extracellular vesicles: A report by the NIH common fund extracellular RNA communication consortium. *J. Extracell. Vesicles*, **4**, p. 27575.
3. Nielsen, M. E., Feechan, A., Bohlenius, H., Ueda, T., Thordal-Christensen, H. (2012). Arabidopsis ARF-GTP exchange factor, GNOM, mediates transport required for innate immunity and focal accumulation of syntaxin PEN1. *Proc. Natl. Acad. Sci. U. S. A.*, **109**, pp. 11443–11448.
4. Wei, T., Hibino, H., Omura, T. (2009). Release of Rice dwarf virus from insect vector cells involves secretory exosomes derived from multivesicular bodies. *Commun. Integr. Biol.*, **2**, pp. 324–326.
5. Taylor, D. D., Gercel-Taylor, C. (2008). MicroRNA signatures of tumor-derived exosomes as diagnostic biomarkers of ovarian cancer. *Gynecol. Oncol.*, **110**, pp. 13–21.
6. Lakhal, S., Wood, M. J. A. (2011). Exosome nanotechnology: An emerging paradigm shift in drug delivery Exploitation of exosome nanovesicles for systemic in vivo delivery of RNAi heralds new horizons for drug delivery across biological barriers. *Bioessays*, **33**, pp. 737–741.
7. Zhou, H., Yuen, P. S. T., Pisitkun, T., Gonzales, P. A., Yasuda, H., Dear, J. W., et al. (2006). Collection, storage, preservation, and normalization of human urinary exosomes for biomarker discovery. *Kidney Int.*, **69**, pp. 1471–1476.

8. Xiao, J., Feng, S., Wang, X., Long, K., Luo, Y., Wang, Y., et al. (2018). Identification of exosome-like nanoparticle-derived microRNAs. *Peer. J.*, **6**, p. 5186.

9. Ju, S., Mu, J., Dokland, T., Zhuang, X., Wang, Q., Jiang, H., et al. (2013). Grape exosome-like nanoparticles induce intestinal stem cells and protect mice from DSS-induced colitis. *Mol. Ther.*, **21**, pp. 1345–1357.

10. Wiczer, B. M., Thomas, G. (2012). Phospholipase D and mTORC1: Nutrients are what bring them together. *Sci. Signal.*, **5**, 13; PMID:22457329.

11. Blackwood, R. A., Smolen, J. E., Transue, A., Hessler, R. J., Harsh, D. M., Brower, R. C., et al. (1997). Phospholipase D activity facilitates Ca2C-induced aggregation and fusion of complex liposomes. *Am. J. Physiol. Cell. Ph.*, **272**, pp. C1279–C1285.

12. Wang, B., Zhuang, X., Deng, Z. B., Jiang, H., Mu, J., Wang, Q., et al. (2014). Targeted drug delivery to intestinal macrophages by bioactive nanovesicles released from grapefruit. *Mol. Ther.*, **22**, pp. 522–534.

13. Chevillet, J. R., Kang, Q., Ruf, I. K., Briggs, H. A., Vojtech, L. N., Hughes, S. M., et al. (2014). Quantitative and stoichiometric analysis of the microRNA content of exosomes. *P. Natl. Acad. Sci. U. S. A.*, **111**, pp. 14888–14893.

14. Gangalum, R. K., Atanasov, I. C., Zhou, Z. H., Bhat, S.-P. (2011). α,β-crystallin is found in detergent-resistant membrane micro-domains and is secreted via exosomes from human retinal pigment epithelial cells. *J. Biol. Chem.*, **286**, pp. 3261–3269.

15. Van der Pol, E., Hoekstra, A. G., Sturk, A., Otto, C., van Leeuwen, T. G., Nieuwland, R. (2010). Optical and non-optical methods for detection and characterization of microparticles and exosomes. *J. Thromb. Haemost.*, **8**, pp. 2596–2607.

16. Mu, J. Y., Zhuang, X. Y., Wang, Q. L., Jiang, H., Deng, Z. B., Wang, B. M., et al. (2014). Interspecies communication between plant and mouse gut host cells through edible plant derived exosome-like nanoparticles. *Mol. Nutr. Food Res.*, **58**, pp. 1561–1573.

17. Sharma, S., Rasool, H. I., Palanisamy, V., Mathisen, C., Schmidt, M., Wong, D. T., et al. (2010). Structural mechanical characterization of nanoparticle exosomes in human saliva, using correlative AFM, FESEM, and force spectroscopy. *ACS. Nano*, **4**, pp. 1921–1926.

18. Palanisamy, V., Sharma, S., Deshpande, A., Zhou, H., Gimzewski, J., Wong, D.-T. (2010). Nanostructural and transcriptomic analyses of human saliva derived exosomes. *PLoS One*, **5**, p. e8577.

19. Thery, C., Amigorena, S., Raposo, G., Clayton, A. (2006). Isolation and characterization of exosomes from cell culture supernatants and biological fluids. *Curr. Protoc. Cell Biol.* Chapter 3: pp. 3.22.1–3.22.29; PMID:18228490.

20. Zhang, M. Z., Yu, R. N., Chen, J., Ma, Z. Y., Zhao, Y.-D. (2012). Targeted quantum dots fluorescence probes functionalized with aptamer and peptide for transferrin receptor on tumor cells. *Nanotechnology*, **23**, p. 485104.

21. Zhang, M. Z., Yu, Y., Yu, R. N., Wan, M., Zhang, R. Y., Zhao, Y.-D. (2013). Tracking the down-regulation of folate receptor-a in cancer cells through target specific delivery of quantum dots coupled with antisense oligonucleotide and targeted peptide. *Small*, **9**, pp. 4183–4193.

22. Yang, J., Yao, M. H., Wen, L., Song, J. T., Zhang, M. Z., Zhao, Y. D., et al. (2014). Multifunctional quantum dot-polypeptide hybrid nanogel for targeted imaging and drug delivery. *Nanoscale*, **6**, pp. 11282–11292.

23. Song, J. T., Yang, X. Q., Zhang, X. S., Yan, D. M., Wang, Z. Y., Zhao, Y.-D. (2015). Facile synthesis of gold nanospheres modified by positively charged mesoporous silica, loaded with near-infrared fluorescent dye, for in vivo X-ray computed tomography and fluorescence dual mode imaging. *ACS Appl. Mater. Interfaces*, **7**, pp. 17287–17297.

24. Mohammadinejad, R., Karimi, S., Iravani, S. Varma, R.-S. (2016). Plant-derived nanostructures: Types and applications. *Green. Chem.*, **18,** pp. 20–52.

25. Gadad, A. P., Vijay, Kuma, S. V., Dandagi, P. M., Bolm, U. B., Pallavi, N. P. (2014). Nanoparticles and their therapeutic applications in pharmacy. *Int. J. Pharm. Sci. Nanotechnol.*, **7**, pp. 2509–2519.

26. Nehete, J. Y., Bhambar, R. S., Narkhede, M. R., Gawali, S.-R. (2013). Natural proteins: Sources, isolation, characterization and applications. *Pharmacogn. Rev.*, **7**, pp. 107–116.

27. Lu, M., Xing, H., Yang, Z., Sun, Y., Yang, T., Zhao, X., et al. (2017). Recent advances on extracellular vesicles in therapeutic delivery: Challenges, solutions, and opportunities. *Eur. J. Pharm. Biopharm.*, **119**, pp. 381–395.

28. Wang, Q. L., Ren, Y., Mu, J. Y., Egilmez, N. K., Zhuang, X. Y., Deng, Z. B., et al. (2015). Grapefruit derived nanovectors use an activated leukocyte trafficking pathway to deliver therapeutic agents to inflammatory tumor sites. *Cancer. Res.*, **75**, pp. 2520–2529.

29. Allen, T. M., Cullis, P.-R. (2013). Liposomal drug delivery systems: From concept to clinical applications. *Adv. Drug. Deliv. Rev.*, **65**, pp. 36–48.

30. Wang, Q. L., Zhuang, X. Y., Mu, J. Y., Deng, Z. B., Jiang, H., Zhang, L. F., et al. (2013). Delivery of therapeutic agents by nanoparticles made of grapefruit-derived lipids. *Nat. Commun.*, **4**, p. 1867.

31. Kim, B. M., Mao, J., Taketo, M. M., Shivdasani, R.-A. (2007). Phases of canonical Wnt signaling during the development of mouse intestinal epithelium. *Gastroenterology*, **133**, pp. 529–538.

32. Urbanelli, L., Buratta, S., Sagini, K., Leonardi, L., Emiliani, C. (2017). Nanovesicles from plants as edible carriers of bioactive compounds. *AgroLife. Sci. J.*, **6**, pp. 167–171.

33. Raimondo, S., Naselli, F., Fontana, S., Monteleone, F., Lo Dico, A., Saieva, L., et al. (2015). Citrus limon-derived nanovesicles inhibit cancer cell proliferation and suppress CML xenograft growth by inducing TRAIL-mediated cell death. *Oncotarget*, **6**, pp. 19514–19527.

34. Zhang, M., Xiao, B., Wang, H., Han, M. K., Zhang, Z., Viennois, E., et al. (2016). Edible ginger-derived nano-lipids loaded with doxorubicin as a novel drug-delivery approach for colon cancer therapy. *Mol. Ther.*, **24**, pp. 1783–1796.

35. Deng, Z., Rong, Y., Teng, Y., Mu, J., Zhuang, X., Tseng, M., et al. (2017). Broccoli-derived nanoparticle inhibits mouse colitis by activating dendritic cell AMP-activated protein kinase. *Mol. Ther.*, **25**, pp. 1641–1654.

Chapter 9

Gold Nanoparticles as Promising Agents for Cancer Therapy

Nadine Karaki,[a,b] **Hassan Hajj Ali,**[c] **and Assem El Kak**[a]

[a]*Department of Medical Biology, Faculty of Pharmacy, Lebanese University, Hadath, Beirut, Lebanon*
[b]*Biological and Chemical Sciences Department, Faculty of Arts and Sciences, Lebanese International University, Mousaytbeh, Beirut, Lebanon*
[c]*Faculty of Sciences, Ecole Nationale Supérieure d'Agronomie et des Industries Alimentaires, The Biomolecular Engineering Laboratory (LIBio), Lorraine University, Vandoeuvre-les-Nancy, Nancy, France*

Nadine_karaki@hotmail.com

This chapter is intended for authors interested in the use of gold nanoparticles (GNPs) as promising agents for cancer therapy. Gold nanoparticles have been the subject of many studies because of their unique physicochemical properties including their size, shape, optical and electronic properties. This chapter reviews the different preparation techniques reported for the synthesis and functionalization of gold nanoparticles with different molecules and highlight a variety of applications of these nanosystems in cancer nanotechnology. Finally, the toxicity of gold nanoparticles in vitro and in vivo is discussed briefly.

Nanoparticle Drug Delivery Systems for Cancer Treatment
Edited by Hala Gali-Muhtasib and Racha Chouaib

ISBN 978-981-4800-54-9 (Hardcover), 978-0-429-34125-0 (eBook)
www.jennystanford.com

9.1 Introduction

Cancer is a generic term for a broad group of diseases that can affect any parts of the body and cause abnormal and uncontrolled cell division, associated with malignancy, such as invasion and metastasis. Metastasis is the process by which tumor cells proliferate and spread throughout the body to reach other organs and is the principal reason for the mortality of patients with cancer. Despite the enormous progress made in the modern medical sciences, cancer remains one of the leading causes of death in the world [1]. Nowadays, surgery is the first treatment used for the removal of the tumor; it is often followed by intrusive processes such as chemotherapy, radiotherapy, hormone therapy [2], targeted therapy such as immunotherapy [3], or combined therapy [4, 5]. Although all these intrusive treatments have some limitations and side effects, the main limitation of combination therapy is the cumulative toxicity of normal tissues caused by the combined modalities used [6]. Hence, scientists have made remarkable efforts to minimize adverse effects by using selective methods to target cancer cells. The field of nanomedicine research has evolved rapidly and has been integrated with existing therapies and imaging to destroy cancer cells while causing much less damage to normal cells. Researchers have developed organic (e.g., polymers, liposomes, dendrimers) and inorganic nanosystems (e.g., quantum dot, magnetic, polystyrene, and metallic) for the treatment of cancer [7]. Among the inorganic metallic nanoparticles (NPs), tremendous progress has been made in the synthesis and functionalization of spherical gold nanoparticles (GNPs) and their use in cancer therapy and biomedical applications due to their excellent biocompatibility and stability, and their enhanced properties. Recent studies in cancer research show that the incorporation of GNPs into different treatment modalities such as radiation therapy, chemotherapy, photodynamic therapy (PDT), photothermal therapy (PTT) and targeted drug delivery has improved tumor cell destruction [8, 9].

This chapter reviews the properties of GNPs, their synthesis methods, and the surface functionalization with a wide range of molecules, expanding and improving GNPs use in biomedical applications and reducing their cytotoxicity. Overall, GNPs are a promising carrier for drug delivery, in addition to gene and cancer therapy.

9.2 Characteristics of Gold Nanoparticles

Gold nanoparticles, also known as colloidal gold, are a suspension of sub-micrometer-sized gold particle in a fluid. They have gained increasing interest due to their special properties, such as controllable size and shape dispersion, composition, optical and electronic properties, high stability, chemical inertness, high electron density, encapsulation, low toxicity, and easy surface functionalization due to the negative reactive groups on the surface [10]. These properties determine the applications of GNPs.

9.2.1 Morphology (Size and Shape)

Gold nanoparticles are synthesized in different sizes that vary from 3 nm to 200 nm [11]. GNPs of very small size (<100 nm) offer a large surface/volume ratio, so that the physical and chemical properties are very different from those of the same bulk material. These properties include enhanced electrical and heat conductivity, and improved photoemission and surface catalytic activity [12]. Based on their size, GNPs can accumulate passively at the tumor sites by enhanced permeability and retention effect (EPR) [13]. They enter the body cells most likely by non-specific receptor mediated endocytosis mechanism. They extravasate through these large leaky vascular pores in the tumor site. GNPs remain trapped at this site, where they exert their effects, due to inefficient lymphatic drainage and to their poor diffusion [14].

Gold nanoparticles exhibit different shapes like spherical, sub-octahedral, octahedral, decahedral, polyhedron multiple twined, multiple twisted, irregular form, tetrahedral, nanotriangles, nano-prisms, hexagonal platelets, nanoflowers, and nanorods [15, 16]. Spherical and gold nanorods nanoparticles attract great attention due to their significant nature, and their tunable intrinsic properties [15]. In fact, the shape and size of GNPs affect the viability of cells, due to their different geometries as well as physical and chemical interactions with cell membranes [17].

9.2.2 Optical Properties

The optical properties of GNPs are significant because absorption and emission are within the visible range of light [18]. The optical properties of GNPs depend on their size and shape. When the size of GNPs is bigger than 10 nm, they can also absorb and scatter specific wavelengths of light across the visible and the near-infrared spectra (800 nm) and have a maximum absorption at 520 nm [19]. As for the shape, gold nanospheres exhibit a unique absorption peak between 510 and 550 nm in the visible range, while the gold nanorods have 2 bands at 550–600 nm and at 520 nm due to their rod-like shape [20].

9.2.3 Electrical Properties

Gold nanoparticles exhibit size-dependent electrical properties. These properties depend on the type of motion of GNPs electrons, which depends on the space available for them [18]. GNPs conduct only electricity when their size is bigger than 10 nm [21]. Studies have shown that the conductivity of the GNPs solution increases with light brightness or thermal heating [22].

9.2.4 Thermal Properties

The thermal properties of GNPs are also size-dependent. Decreasing the GNPs size causes a decrease in the melting point, this is known as "melting point depression" [23]. At the bulk state, gold has a melting point of 1064°C. Whereas, at the colloidal state, the decrease in GNPs size from 100 to 10 nm diameter provokes a significant drop in their melting temperatures of about 100°C. As the size is reduced to about 2 nm, the melting point decreases to 532°C which is about half of the melting point at the macro scale level [24].

9.2.5 Photothermal Properties

Gold nanoparticles have a high atomic number, which leads to greater absorption of kilovoltage X-rays and provides greater contrast than standard agents. They resonate when exposed to light of specific energies, thus producing heat. The light to heat conversion is especially strong if the nanoparticles are excited by a laser frequency corresponding to their surface plasmon resonance position. GNPs with sizes above 2 nm have a larger extinction cross-section, and a greater chance of reaching 100% of light-to-heat conversion efficiency and a high photostability [25].

9.2.6 Localized Surface Plasmon Resonance

Gold nanoparticles have a strong interaction with light because the conduction of electrons on the surface of gold undergoes a collective oscillation when excited by light at specific wavelengths [26]. This oscillation is known as localized surface plasmon resonance (LSPR) and causes absorption and diffusion intensities of gold nanoparticles much higher than those of non-plasmonic nanoparticles of identical size. The LSPR phenomenon is influenced by the size, the shape of GNPs and by the dielectric properties of the environment and the proximity of other nanoparticles [27]. The LSPR adds to the benefits of the enhanced local heating effect.

9.2.7 Surface Chemistry

The small size of GNPs and their large surface area can lead to their aggregation and could limit the loading of functional constituents and burst liberation. For this reason, it is important to synthesize GNPs with appropriate sizes and surface chemistry that prevent their immediate recognition by the immune system and increase their circulation times [28]. Surface chemistry affects the stability of GNPs, their size uniformity in a dispersed phase and their biocompatibility. It also provides functional groups (carboxyl, amine, etc.) or charges that can be exploited for biofunctionalization or ligand exchange [29].

9.2.8 Biological Properties

GNPs are biologically inert and potentially biodegradable in vivo [30, 31]. Hence, the clinical perspectives of gold-based multifunctional NPs are promising.

9.3 Synthesis Methods of Gold Nanoparticles

The colloidal state of GNPs can be evidently observed by color change. Depending on the size of the nanoparticles, small to medium-sized GNPs presents either an intense red color while larger particles are purple and aggregates are blue [32]. Since the properties and the applications of GNPs rely on their size and shape, the tracking of color change of GNPs provides effective signs to the scientist at each step of the nanoparticle preparation.

The first experiment for the synthesis of GNPs was reported by Faraday in the 1850s [33]. He reported the formation of deep red gold colloidal solutions by reduction of an aqueous chloroaurate ($AuCl_4^-$) solution using a phosphorus reducing agent in a two-phase system (carbon disulfide). He referred to this solution as "activated gold." Faraday further investigated the optical properties of thin films prepared from dried colloidal solutions. Because of their easy preparation routes, high stability and high density per particle, GNPs have likewise been utilized in the improvement of physical partition techniques, like ultracentrifugation and electron microscopy dyes as well [34]. In this part, we will discuss the different synthesis strategies of GNPs.

9.3.1 Direct Reduction: The Turkevich Method

A wide cluster of arrangement-based methodologies have been created to control the physicochemical properties of GNPs by adjusting their size, shape, and surface chemistry (functionality of grafted molecules) [35–37]. The chemical reduction of $AuCl_4$ by citric acid in a hot aqueous solution was first investigated by Turkevich et al. in 1951 [38]. In this approach, citric acid acts as both a reducing agent and stabilizing agent that stabilizes the GNPs through electrostatic interactions. This method allows the growth of GNPs with many different shapes and sizes ranging from 10 to 50 nm (Fig. 9.1).

More recently, a substitutional procedure demonstrated that a high grouping of Cl^- ions additionally prompts larger sizes for GNPs reduced by citric acid, since it diminishes their surface charge and thereby promotes coarser particle size and even aggregation [39]. A similar result was obtained by increasing the pH of the NP solution just after initiating the reducing step. This increase in NP size is due to the limitation of the nucleation process which decreases the final number of grown GNPs [40].

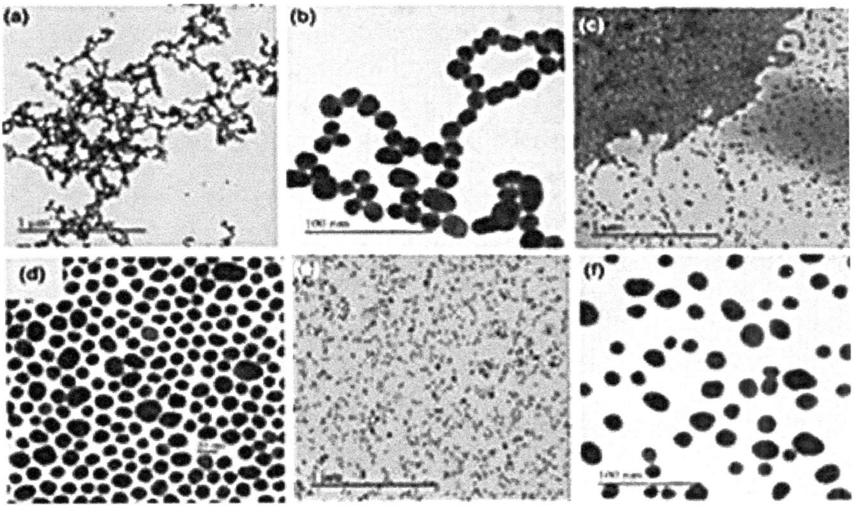

Figure 9.1 Gold nanoparticles synthesized by the Turkevitch method in citrate: (a, b) as-synthesized; (c, d) grafted with polyethylene glycol PEG 1000 Mw; and (e, f) grafted with polyethylene glycol PEG 5000 Mw. The longer the chain, the larger the spacing between the particles. Adapted from [56].

9.3.2 Seed-Mediated Growth

The seed-mediated growth technique is one of the most promising approaches to obtain highly monodisperse GNPs with controlled-size distribution critical for many applications [41]. Particles with different morphologies can be prepared by controlling the ratio of crystal to growth solution, such as spherical and triangular [42], but this method is mainly established for synthesis of gold nanorods [43]. The small NPs seeds prepared first by water-phase oxidation-reduction, act as a catalytic nucleation center for the development of bigger NPs [44]. The nucleation of very small NPs seeds is usually made through the reduction of $HAuCl_4$, used in the growth solution, into gold atom by the toxic reducing agent sodium borohydride ($NaBH_4$), which is tolerated as an initiator of GNPs seeds. The main limitation of this method is that an important population of smaller NPs frequently remains in the solution in addition to the grown seeds, requiring further purification to achieve a narrow NPs size distribution. However, using hydroxylamine [45], ascorbic acid [46] and hydroquinone [47] as a selective reducing agent for gold ions helped solve this issue.

9.3.3 Gold Nanoparticles Synthesis in Organic Solvents

The synthesis methods of GNPs in an aqueous medium are easy to achieve and they have the advantage of not using toxic solvents. Unfortunately, it is relatively difficult to maintain the size distribution of GNPs at small and narrow values. However, it is well known that organic synthesis often results in much smaller and narrower particle size distributions [34].

9.3.3.1 The Brust–Schiffrin method

The Brust–Schiffrin method is one of the first reported to synthesize GNPs in organic media, and it is until now one of the most widely performed to prepare uniform small particles of narrow distributions [48]. It is a classic sulfhydryl ligand method that consists of two biphasic reduction procedures. To prepare GNPs, phase transfer agent and stabilizer of GNPs such as tetraoctylammonium bromide (TOABr) or tetraoctylammonium chloride (TOACl), are used in addition to a reducing agent such as $NaBH_4$ and a thiol ligand (Fig. 9.2). Generally, the gold salts are transferred from a water phase into an organic phase (e.g., toluene, ethanol and ethylene glycol) by a phase transfer agent. Then, NPs are nucleated through a reduction step through thiols ligands and $NaBH_4$. A recent study showed that the species [$AuBr_4^-$] is a better precursor in the Brust–Schiffrin method to synthesize GNPs as it is more resistant to the formation of Au(I) thiolate species than $[AuCl_4]$ [49], knowing that the formation of Au(I) thiolate must be avoided to achieve a good yield of NPs [50].

This methodology produces small and stable GNPs that can be easily functionalized by other ligands [51].

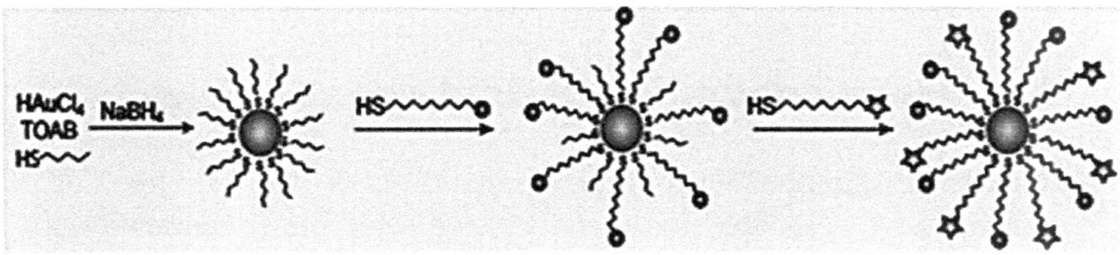

Figure 9.2 Two-phase synthesis of GNPs by reduction of $HAuCl_4$ in presence of alkanethiols as stabilizing ligands and $NaBH_4$ as reducing agent. Adapted from [52]. TOAB: TetraOctylAmmonium Bromide.

9.3.3.2 The polyol method

This method is a robust strategy for preparation of monodisperse GNPs with a variety of well-controlled morphologies such as octahedron, polyhedron, sheet hexagon, nano-rods and icosahedron [53, 54]. The synthesis consists of reducing the gold salt by a polyol solvent in the presence of a suitable capping agent such as the polyvinylpyrrolidone (PVP), at a high temperature. The capping agent is an essential factor for stabilization of the formed nuclei (via adsorption on the metal surface) at the preliminary stage of reaction and for directing the morphology of the GNPs [55].

In summary, the polyol process is a simple and inexpensive technology used for the industrial production of symmetric narrow size distribution of GNPs [56]. Nevertheless, the presence of toxic chemical compounds and solvent residues requires the use of tedious and time-consuming techniques of filtration and purification [34].

9.3.4 The Biological Synthesis Method

The green synthesis of GNPs has attracted attention as a low-cost, eco-friendly and easily scaled-up technology in comparison to the expensive and toxic physical and chemical methods [57].

Apart from the chemical synthetic methods, GNPs can be also synthesized by a green biological route using some microorganisms such as bacteria [58], fungi [59], and yeast [60] or from plant biomass or extracts [61]. This new safe technology might overcome the toxicity issues regarding the preparation of GNPs from chemical and toxic compounds, which will limit their usage in biomedical applications [62].

9.4 Functionalization of Gold Nanoparticles

Nowadays, nanoparticles functionalization is the subject of concentrated research, to introduce required bio-functionalities to widen the range of potential biomedical applications. The functionalized gold nanoparticles have good biocompatibility and controllable biodistribution patterns, making them particularly suitable candidates for new therapies [63].

Several parameters were shown to strongly affect the final surface coverage of GNPs by biomolecules and their effectiveness; such as surface chemistry, stabilizing agents, pH as well as the functionalization procedure [64].

Different stabilizing agents such as ligands, polymers, surfactants, and biomolecules must be used to prevent the uncontrollable growth and aggregation of NPs [65]. Thiolate is the most frequently used stabilizer or capping ligand because it forms more compact nanoparticles by strong gold-thiol bonds (chemisorption) knowing that there are other less used ligand motifs such as amines or phosphines, among others [66].

9.4.1 Functionalization Strategies

The stabilizers of gold nanoparticles such as thiols, citrates, or other adsorbed ligands may be substituted with external thiols via a ligand exchange reaction in order to synthesize monolayer-protected mixed GNPs. These monolayer-protected mixed-GNPs can be used for synergistic applications when two or more functional ligands are introduced during the ligand exchange reaction. Non-covalent conjugation allows for organic molecules or biomolecules to be attached on the surface of GNPs by different physical interactions, such as specific binding affinity [67], hydrophobic interactions [68] and electrostatic interactions known as ionic coupling [69]. The advantages of these strategies are their ease of functionalization, rapid binding and that neither the biomolecules nor the GNPs must be modified in case of electrostatic and hydrophobic interactions. However, it is difficult to control the amount and orientation of bound molecules [11]. The non-covalent functionalization strategies are widely used in the delivery of molecules because of their ease of release and their reversible nature.

In contrast, the covalent functionalization of GNPs stabilizes the conjugates, which is necessary in the case of stable constructs used in imaging, for example. There are two main strategies for covalent conjugation: the direct attachment of the thiolate molecule to the surface of GNPs by chemisorption and the formation of covalent bonds, and exploiting functional groups on particles and biomolecules (1-ethyl-3-(3-dimethylaminopropyl)-carbodiimide coupling known as EDC coupling [70], Malemide coupling [71], and click reaction [72]. These covalent conjugations are usually made between the functional group of the molecule and the pre-grafted thiolate ligands on the GNPs surface. Nonetheless, some biomolecules such as antibodies, oligonucleotides or peptides do not contain functional groups and must be modified before conjugation [73]. However, since chemical modification may compromise the activity of biomolecules, linker molecules are used as one of the most common ways of conjugating molecules to GNPs. These linker molecules can be modified to contain a thiol group on one terminal end, which is used to attach to the gold surface and an amine or carboxylic acid functional group on the other end [32].

9.4.2 Types of Molecules Functionalized to GNPs Surfaces

The functionalization of GNPs with different molecules offers the possibility to load and specifically transfer large amounts of materials with therapeutic effect. Thus, functionalized GNPs are used in cancer diagnosis and therapy.

GNPs can be easily functionalized with chemical molecules and biomolecules such as nucleic acids (DNA and RNA), drugs, biomarkers, antibodies, and peptides (Fig. 9.3).

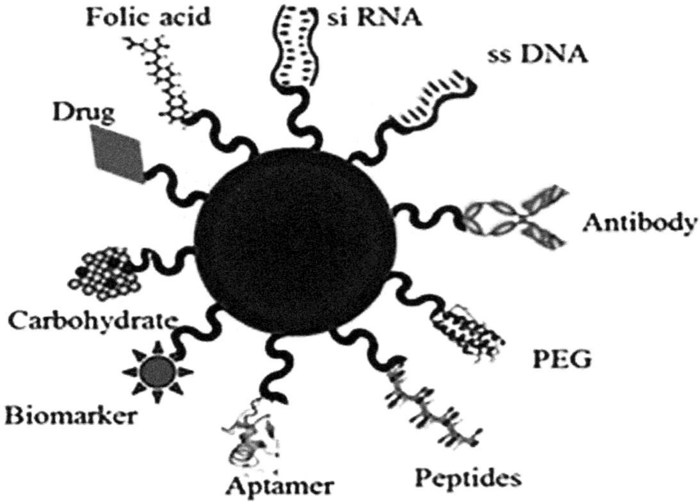

Figure 9.3 Schematic representation of GNPs functionalized with various molecules for cancer therapy. ss DNA: single strand DNA.

9.4.2.1 Chemical functional group

The sequential functionalization of tyrosine-reduced GNPs with anionic polyoxometalates (POMs) followed by the cationic lysine was found to exhibit a strong antibacterial activity against a gram-negative bacterium *Escherichia coli* and an antitumoral property when tested against human lung carcinoma cells [74].

Poly(ethylene glycol) (PEG) is another chemical molecule used for the functionalization of GNPs to provide them with the ability to circulate for a prolonged period and target a particular organ, to deliver proteins, nucleic acids, peptides and drugs. In fact, PEGs have a strong effect on nanoparticle structure, stabilization and biodistribution both in vitro and in vivo [75].

9.4.2.2 Biological molecules

9.4.2.2.1 Peptides

Functionalization of GNPs with peptides was an effective way to improve specificity and efficacy of nanoparticle-based delivery systems. For systemic applications, peptide surface functionalized and long-circulating GNPs coated with PEG were developed as delivery agents and were highly desired for passive targeting to tumors and inflammatory sites [76]. Cysteine-containing peptides were linked to GNPs using PEG as a spacer. This would allow the modified GNPs to remain in the blood for a longer time to provide a more efficient interaction with the target [69]. PEG also generates a steric repulsion between adjacent nanoparticles to prevent their aggregation when exposed to oppositely charged peptides.

GNPs functionalized with peptides have also been used as effective cell-targeting agents depending on the type of peptide used. Cysteine-modified arginine-glycine-aspartic acid peptide and nuclear localization signal peptide that derive from large T antigen were used to sequentially functionalize GNPs being conjugated with thiolated PEG first. This sequential conjugation allows the selective and effective delivery of GNPs to the nucleus of cancer cells [77].

Similarly, the surface modification of GNPs with cell penetrating peptides and a peptide acting as a nuclear localization signal, led to their uptake by human fibroblast cells and their successful nuclear targeting [78].

Peptide functionalized GNPs are also known to induce cancer cell death. The conjugation of an oncogene peptide to GNPs using 3-mercaptopropionic acid as a linker [66] was found to reduce the cell proliferation and to increase apoptosis in retinoblastoma cells [79].

Moreover, the reaction of the immune system with reference to the GNPs can be modulated by the peptide coating. A homogeneous conjugation of peptides onto the GNPs surface activates the immune system by enhancing the macrophage response [80].

9.4.2.2.2 Peptide–drug conjugates

Gold nanoparticles stabilized with a monolayer of L-aspartate and several cytostatic drugs (doxorubicin, cisplatin, and capecitabine), were successfully used as a complex tumor-targeting drug-delivery system. The proliferation of hepatic cancer cells was statistically lower than those of cells exposed to the cytostatic drugs alone, indicating that GNPs facilitated an increased susceptibility of cancer cells to drugs with minimal toxicity [81].

Peptide–drug conjugates (PDCs) are used as an effective strategy for specific delivery of cytotoxic drugs to cancer cells. They have relatively low stability in the blood, liver and kidneys. Since GNPs have a longer plasma half-life than PDCs, the latter could be conjugated to GNPs, as they have demonstrated favorable physicochemical and safety properties for drug delivery systems.

PDC conjugated GNPs have facilitated the effective delivery of anticancer drugs at low concentrations and can cause more DNA damage and more cell death than the free PDC when irradiated [82]. Thus, it can be used in combined therapy with chemo-radiation [83].

9.4.2.2.3 Folic acid

The surface of GNPs has been functionalized with folic acid, which is used as a recognition molecule to perform a selective internalization in cancer cells. Recent studies have functionalized GNPs with folic acid using linkers (4-aminothiophenol, 6-mercapto-1-hexanol) [84] or incorporating other molecules amine-terminated PEG-folic acid [85]. These functionalized GNPs decreased significantly the viability of human cervical carcinoma cells. In addition, modifying GNPs with folic acid without using linkers or other molecules, could complicate, according to some researchers, the GNPs action mechanism or affect their biocompatibility [86]. However, other researchers have reported that folic acid-conjugated GNPs, without using other molecules or linkers for the surface functionalization, could be used in the cancer therapy modality known as PTT, either alone [87] or in combination with radiotherapy [88]. Its use with radiotherapy significantly induced apoptosis in mouth epidermal carcinoma cells [88].

9.4.2.2.4 Proteins/antibodies

The attachment of antibodies to the surface of GNPs can prevent the binding to antigens if the antigen binding sites are sterically blocked upon conjugation.

These functionalized GNPs can be used for different medical applications such as contrast agents in biomedical imaging of cancer cells. The non-covalent functionalization of GNPs with monoclonal antibody specific to human epidermal growth factor receptor 2 (anti-EGFR) was used for imaging of breast cancer cells [68], ovarian cancer cells [89] and for the in vivo and in vitro diagnosis of oral cancer cells [90]. Furthermore, antibody-functionalized GNPs loaded with drugs can be used to actively target cancer cells [91].

9.4.2.2.5 Nucleic acids

Nucleic acids are promising tools for therapeutic targets, but they require a delivery vector to protect against nucleases and other environmental agents and to facilitate their entry into the cell. GNPs are exploited as useful synthetic vectors for the creation of transfection agents in gene therapy to cure cancer and genetic disorders [92] and for enhanced molecular recognition [93].

GNPs functionalized with small silencing RNA (siRNA) can be used as therapeutic polymeric molecules which bind via covalent [69, 94] or electrostatic interactions [137] to targets and stop the growth of the cancer. In addition, GNPs modified with DNAs are applied in biosensing or as DNA probes for diagnosis [143]. Moreover, drugs can be efficiently conjugated with DNA-GNPs to further enhance the intracellular uptake of GNPs to achieve their efficient and targeted delivery while minimizing drug toxicity [95].

9.5 Applications of Gold Nanoparticles

The small sized GNPs are biologically inert and have several significant physicochemical properties convenient for their use in a variety of fields such as medicine, catalysis, molecular sensors, surfactants, renewable and sustainable energy production and storage, and environmental applications [96]. In medicine, GNPs are mainly used in cancer nanotechnology, including molecular diagnosis, imaging, therapy, and in vitro assays (Fig. 9.4). In the following section, we will review the use of GNPs in cancer therapy modalities, drug delivery, and gene therapy.

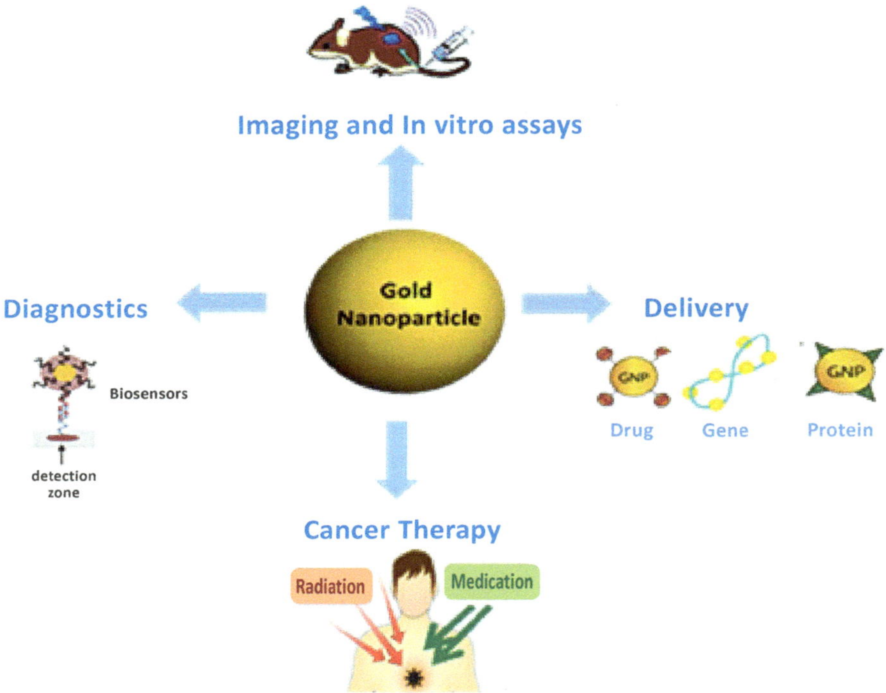

Figure 9.4 General scheme for cancer nanotechnology applications of GNPs.

9.5.1 Gold Nanoparticles in Cancer Therapy

Gold nanoparticles have emerged as successful candidates for cancer therapy. They are used in chemotherapy, radiation therapy, PTT and PDT, in combination modalities as well as in drug therapeutics and in gene therapy (Fig. 9.5).

The following section provides a summary of GNPs-based treatment modalities developed for fighting cancer.

9.5.1.1 Gold nanoparticles in cancer radiation therapy

Unlike chemotherapy, radiation therapy is a local treatment that affects only the part of the body being treated and it is effective in treating approximately 50% of all cancer patients. It relies on the deposition of energy directly into tumor tissue usually by photon radiation (gamma or X-rays) that scatter the electrons of

NPs, or by energy beams of ions enough to damage the cancer cells or their vascular system and thus induce tumor death or nutrient starvation. However, photon radiation therapy is nonspecific since healthy tissues along the path of the photons are affected [97]. Due to their high X-ray absorption, GNPs have emerged as novel radiosensitizers used to enhance radiotherapeutic performance [98, 99]. Radiation works by directly damaging the cells by ionizing fragmentations of the DNA and in the mitochondria, which triggers cancer cell necrosis [100].

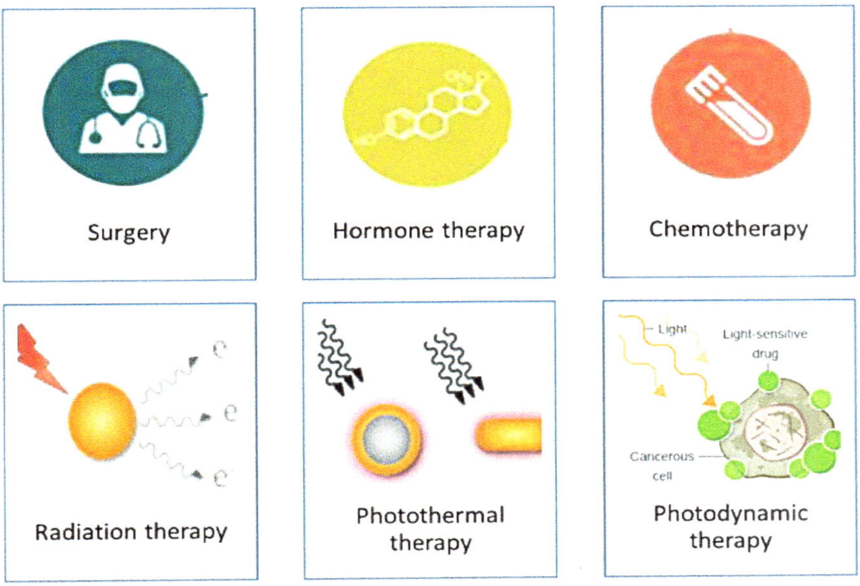

Figure 9.5 Different modalities of cancer treatment using gold nanoparticles.

Studies conducted by using GNPs in combination with X-ray radiotherapy showed that radiation beam will deposit a lower dose after having passed through the GNPs containing region, thus increasing the therapeutic ratio. GNPs were found to increase the production of free radicals thus killing the cells.

The radiotherapy-combined chemotherapy treatment strategy is commonly used in cancer therapy. GNPs can be used as radiation dose enhancers and as anti-cancer drug vectors which minimizes the side effects of treatments [83].

9.5.1.2 Gold nanoparticles in photothermal therapy

Photothermal therapy is a physical method of cancer treatment for those tumors that cannot be removed by surgery. It is most efficient in the form of three modes: local, regional, and whole-body hyperthermia [101].

Thermotherapy is an invasive technique, which uses hyperthermia, i.e. a small rise in tumor temperature between 41–47°C for tens of minutes. Such condition will cause irreversible damage to the cells due to denaturing of proteins and/or destruction of cell membranes generated by photothermal agents to destroy cancer cells. The heating sources can come from radiation (infrared, radio frequency, and microwaves) and ultrasounds [102].

More precise thermal damage effects could be enhanced using strong absorption agents, such as metal nanoparticles. Gold nanoparticles, due to their LSPR, have been explored as photothermal agents that absorb the external energy of electromagnetic radiation and convert it into heat energy due to electron excitation and relaxation, and dissipate the heat near cancer cells without harming the surrounding healthy tissues [103]. The photoexcitation of GNPs at their surface plasmon resonance maximum wavelength results in an efficient light-to-heat conversion [104].

The LSPR peaks of GNPs can be easily turned to the near-infrared (NIR) region for better treatment due to its tissue penetration ability of several centimeters in biological tissues. Thus, after the accumulation of GNPs at the tumor site, the NIR laser is focused on this site. This light energy excites the GNPs and overheats them, thus ultimately killing cancer cells [105]. Three mechanisms explain the cell death process: cell membrane destruction, tumoral DNA denaturation, and blocking of angiogenesis [106].

This PTT method of tumor reduction has few side effects and is of low toxicity except in the case of prolonged exposure to the IR laser which can potentially start to burn surrounding tissues. No harmful chemicals are used in GNPs; so long-term damage is not a problem. This form of energy also emits no kind of toxic radiation to the human body, making it safer and healthier for patients than other targeted radiation techniques [107]. However, one weak feature of PTT is that hyperthermia leads to thermotolerance in cancer cells, mediated by heat shock proteins [108]. Thus, a synergistic approach in which PTT is combined with another therapeutic method is needed to overcome this limitation.

Combination therapy is a treatment modality that combines two or more therapeutic approaches to enhance cancer treatment efficacy in comparison to the mono-therapy approach [109]. Studies have shown how GNP-induced hyperthermia associated with more common treatments can yield positive treatment results [110]. A hyperthermia makes cancer cells more susceptible to radiation and chemotherapy. GNPs-mediated hyperthermia coupled with radiation therapy resulted in not only a greater reduction of tumor volume but also a reduction of the proportion of stem cells in the residual tumor [111].

A study combining the effects of chemotherapy and the GNPs-mediated PTT in the visible light was found more effective to fight drug resistance in cancer cells than individual therapy approaches [112]. In addition, another approach combined the effect of radiation therapy and a conjugated GNPs-mediated PTT to selectively induce apoptosis in cancer cells [88].

9.5.1.3 Gold nanoparticles in photodynamic therapy

Photodynamic therapy is increasingly being recognized as an attractive, alternative treatment modality for superficial cancer. It is a minimally invasive procedure that can be used repeatedly at the same site, with minimum functional disturbance; it is normally an outpatient therapy. PDT is a clinical method to damage diseased cells by highly toxic singlet oxygen generated by a photosensitizer, which is excited by external light source [113]. Most photosensitizers are highly hydrophobic and require delivery systems.

After accumulation of a photosensitizer in cancer cells, illumination with a light induces cell death with two mechanisms: direct mechanism (necrosis and apoptosis) and indirect mechanism (microvascular damage and antitumor immune responses) [106]. So, the combination of GNPs-mediated PTT and photosensitizers-mediated PDT was reported to achieve synergistic effects on destroying cancer cells [114, 115].

A study highlighted the unique advantages of nanomedicine-based PTT among these three modalities in cancer treatment in terms of killing tumor cells located far from tumor blood vessels [116].

9.5.1.4 Gold nanoparticles in chemotherapy

Chemotherapy is based on the inhibition of the division of rapidly dividing cells, which is a characteristic not only of cancerous cells, but of some normal cells (hair follicles, bone marrow and gastrointestinal tract cells). Chemotherapy uses cytotoxic agents to treat many malignancies. However, these agents have many problems, including drug resistance, low water solubility (most anticancer drugs), non-specific targeting and distribution, systemic toxicity, and low therapeutic index.

To solve these issues, GNPs are engineered to treat cancer more effectively due to their small sizes and unique physical, chemical and biological properties [101].

9.5.2 Gold Nanoparticles in Targeted Delivery

9.5.2.1 Gold nanoparticles as drug carriers

Gold nanoparticles themselves are used as therapeutic agents to destroy tumor cells due to their anti-angiogenic properties. They have the ability to bind to vascular endothelium growth factor and, consequently lead to a decrease in macrophage infiltration and inflammation [117].

They can also be employed to trigger controlled drug release by changing the pH, temperature, and redox potential [118].

Drugs are loaded in GNPs to increase their solubility and stability, their therapeutic efficiency, biodistribution, controlled release and site-specific delivery and to reduce their toxicity [119–121].

GNPs, drug carriers, are used in passive and active targeted cancer therapy. In passive targeting, GNPs, administered systemically, passively accumulate inside tumors due to EPR effect, which is a consequence of the leaky tumor vasculature and the poorly developed lymphatic system that limits drainage of NPs [122].

The blood circulation time of GNPs can be increased by coating their surface with PEG, improving their efficiency for drug delivery [123]. While in active targeting, the GNPs are functionalized with specific molecules that interact with receptors known to be selectively present in tumor cells [124]. This has been achieved with antibodies [125], peptides [126], glycoproteins [127], folic acid [128], aptamers [129], and glucose molecules [130].

It is also possible to combine the two targeting strategies. To utilize PEG for increased circulation time, the ratio of PEG to targeting ligand has to be optimized, and the length of PEG molecules should not exceed the length of the targeting ligands in order to prevent PEG molecules from blocking the receptor–ligand interaction [131].

In addition, loading of drugs onto GNPs can be performed through either non-covalent interactions or covalent conjugation. GNPs capped with thiolated cyclodextrins conjugated with a ferrocene derivative were found to be effective phase transfer agents to chloroform solution [132]. Ferrocene derivatives have attracted significant interest as anticancer, antibacterial, antifungal and antiparasitic drug candidates [133].

GNPs were also functionalized to achieve high levels of accumulation in tumors. Different sizes of GNPs from 2 to 15 nm were prepared and their surfaces were coated with tiopronin (thiol drug). Results proved that ultra-small GNPs (smaller than 10 nm) displayed unique advantages over nanoparticles larger than 10 nm in terms of localization, and penetration of breast cancer cells, multicellular tumor spheroids, and tumors in mice [134].

GNPs have also solved the problem of the lack of selectivity of PDT against tumor cells; because they could act both as a radical oxygen species (ROS) enhancer and as a photosensitizer carrier of porphyrin and its derivatives, for example the chlorin E6. However, due to the limited surface area of GNPs, the carrying capacity of the photosensitizer is relatively low and it can leak rapidly during NPs circulation in biological systems. To solve this problem, researchers have coated the GNPs surface with silica. Silica provides a large surface area and a high loading capacity for the photosensitizer chlorin E6, thereby improving the efficiency of ROS generated after cellular uptake for PDT. In addition, the functionalized GNPs can be used as a chemotherapeutic drug carrier for loading doxorubicin. Thus, functionalized GNPs combine cancer treatment: PDT and chemotherapy, simultaneously loading different types of therapeutic agents, chlorin and doxorubicin, which show synergistic effects of destruction of cancer cells [135].

9.5.2.2 Gold nanoparticles in gene therapy

Gene therapy is a medical therapeutic approach used to fix a genetic defect by introducing new genes into cells, repairing or replacing existing abnormal genes, a normal version of a defective or missing gene, or regulating the expression of specific genes, thus fixing the underlying disorder [136].

In tumor cells, siRNA is an effective method for regulating gene expression in vitro and in vivo. However, their delivery to the cell and then to the cytosol is the main requisite for their therapeutic potential [137].

For this reason, GNPs were functionalized with siRNA by chemisorption allowing the effective delivery of gene-silencing oligonucleotides (siRNA) at the cellular level without being degraded by nucleases due to the low immunogenic response of GNPs [138].

These functionalized GNPs help the clinical translation of siRNA treatments for genetic diseases [138]. Researchers have reported the use of folic acid-conjugated GNPs to deliver functional siRNA to treat prostate cancer [139]. Moreover, dendrimer-entrapped GNPs were used as a platform for the delivery of siRNA into a human glioblastoma cell line [140] and human cervical carcinoma cell line, to induce gene silencing of cancer cells in vitro and in vivo [141]. The use of β-cyclodextrin-modified dendrimer-entrapped gold nanoparticles to enhance the delivery siRNA to glioblastoma cells was also reported [142].

In conclusion, GNPs provide a useful platform to integrate many therapeutic modalities leading to effective combination therapy in the fight against cancer.

9.6 Toxicity of Gold Nanoparticles

Nanoparticles-based drug delivery systems were primarily created to ensure the delivery of the drug to its specific target site without affecting the healthy tissues or organs. Most anticancer drugs are highly toxic; thus site-specific targeting would surely reduce their side effects. It is well known that the drug biodistribution and toxicity can be affected by many factors, including particle size and shape, types, functionalization techniques, particle administration methods and doses, among others [143]. Furthermore, nanocarriers can cause biodistribution and toxicity problems as they are immediately cleared from the blood by the reticuloendothelial system and can stay in organs for a long span of time.

Despite the huge number of reported studies on the synthesis, characterization and functionalization of GNPs, there are few GNPs-based pharmaceuticals in the market [144]. There are many studies on GNPs toxicity in vitro; those reports cannot be used as good indicators of possible toxicity in vivo. Moreover, the studies on GNPs toxicity remain incomplete and at times even contradictory [145]. For example, Chen et al. found that the toxicity of the same GNPs was inconsistent in vivo and in vitro [146]. It should be affirmed that GNPs are biodegradable but since the excretion of accumulated GNPs from spleen and liver can take up to 3–4 months, the question about the injected doses and possible inflammation processes is of great importance. It should be emphasized that bioaccumulated GNPs in reticulo-endothelial system organs can interfere with different diagnostic techniques. All these challenges, along with their potential toxicity are the major barriers for the successful clinical translation of GNPs.

9.7 Conclusion

Recent advances in nanotechnology result from the development of engineered nanoparticles. Efficiently, metallic nanoparticles have been widely exploited for biomedical applications, and among them, GNPs are highly remarkable. Their intrinsic properties such as optical, electrical, physicochemical and, surface plasmon resonance; can be altered by changing the GNPs shape and size.

These GNPs could be synthesized and functionalized with many agents, such as polymers, oligonucleotides, ligands, drugs, proteins, and peptides. Their ease of synthesis and functionalization properties have resulted in their use in different fields of biomedicine such as diagnosis, in vitro assays, cancer therapy, targeted drug delivery, and gene therapy.

References

1. Bray, F., Ferlay, J., Soerjomataram, I., Siegel, R. L., Torre, L. A., Jemal, A. (2018). Global cancer statistics 2018: GLOBOCAN estimates of incidence and mortality worldwide for 36 cancers in 185 countries, *CA Cancer J. Clin.*, **68**, pp. 394–424.
2. Tremont, A., Lu, J., Cole, J. T. (2017). Endocrine therapy for early breast cancer: Updated review, *Ochsner.* J., **17**, pp. 405–411.
3. Zhang, H., Chen, J. (2018). Current status and future directions of cancer immunotherapy, *J. Cancer*, **9**, pp. 1773–1781.
4. Bhalla, N., Brooker, R., Brada, M. (2018). Combining immunotherapy and radiotherapy in lung cancer, *J. Thorac. Dis.*, **10**, pp. S1447–S1460.
5. Wirsdörfer, F., de Leve, S., Jendrossek, V. (2019). Combining radiotherapy and immunotherapy in lung cancer: Can we expect limitations due to altered normal tissue toxicity? *Int. J. Mol. Sci.*, **20**, p. 24.
6. Barnett, G. C., West, C. M. L., Dunning, A. M., Elliott, R. M., Coles, C. E., Pharoah, P. D. P., Burnet, N. G. (2009). Normal tissue reactions to radiotherapy, *Nat. Rev. Cancer*, **9**, pp. 134–142.
7. Grigore, M. E. (2017). Organic and inorganic nano-systems used in cancer treatment, *J. Med. Res. Health Educ.*, **1**, pp. 1–8.
8. Harish, K. K., Nagasamy, V., Himangshu, B., Anuttam, K. (2018). Metallic nanoparticle: A review, *Biomed. J. Sci. Tech. Res.*, **4**.
9. Sharma, A., Goyal, A. K., Rath, G. (2018). Recent advances in metal nanoparticles in cancer therapy, *J. Drug Target.*, **26**, pp. 617–632.
10. Kim, D., Jon, S. (2012). Gold nanoparticles in image-guided cancer therapy, *Inorganica Chim. Acta*, **393**, pp. 154–164.
11. Conde, J., Dias, J. T., Grazú, V., Moros, M., Baptista, P. V., de la Fuente, J. M. (2014). Revisiting 30 years of biofunctionalization and surface chemistry of inorganic nanoparticles for nanomedicine, *Front. Chem.*, **2**, p. 48.
12. Madkour, L. H. (2018). Applications of gold nanoparticles in medicine and therapy, *Pharm. Pharmacol. Int. J.*, **6**, pp. 157–174.
13. Grumezescu, A. M. (2017). *Nano- and Microscale Drug Delivery Systems: Design and Fabrication*, William Andrew.

14. Wang, A. Z., Tepper, J. E. (2014). Nanotechnology in radiation oncology, *J. Clin. Oncol.*, **32**, pp. 2879–2885.

15. Elahi, N., Kamali, M., Baghersad, M. H. (2018). Recent biomedical applications of gold nanoparticles: A review, *Talanta*, **184**, pp. 537–556.

16. Murali, K., Neelakandan, M. S., Thomas, S. (2018). Biomedical applications of gold nanoparticles, *JSM Nanotechnol. Nanomed.*, **6**, p. 1064.

17. Woźniak, A., Malankowska, A., Nowaczyk, G., Grześkowiak, B. F., Tuśnio, K., Słomski, R., Zaleska-Medynska, A., Jurga, S. (2017). Size and shape-dependent cytotoxicity profile of gold nanoparticles for biomedical applications, *J. Mater. Sci. Mater. Med.*, **28**, p. 92.

18. El-Sayed, M. A. (2001). Some interesting properties of metals confined in time and nanometer space of different shapes, *Acc. Chem. Res.*, **34**, pp. 257–264.

19. Kumar, A., Boruah, B. M., Liang, X.-J. (2011). Gold nanoparticles: promising nanomaterials for the diagnosis of cancer and HIV/AIDS, *J. Nanomater.*, **2011**, pp. 1–17.

20. Ma, Z., Xia, H., Liu, Y., Liu, B., Chen, W., Zhao, Y. (2013). Applications of gold nanorods in biomedical imaging and related fields, *Chin. Sci. Bull.*, **58**, pp. 2530–2536.

21. Chouhan, N. (2018). Silver nanoparticles: Synthesis, characterization and applications, in *Silver Nanoparticles-Fabrication, Characterization and Applications*, eds., Maaz, K. (InTechOpen).

22. Zhang, Y., Schwartzberg, A. M., Xu, K., Gu, C., Zhang, J. Z. (2005). Electrical and thermal conductivities of gold and silver nanoparticles in solutions and films and electrical field enhanced surface-enhanced Raman scattering (SERS), *Physical Chemistry of Interfaces and Nanomaterials IV, SPIE Proc.*, **5929**, pp. 121–129.

23. Quidant, R. (2017). Optical and thermal properties of gold nanoparticles for biology and medicine, in *Gold Nanoparticles for Physics, Chemistry and Biology*, eds., Louis, C., Pluchery, O. (World scientific, Europe) pp. 483–508.

24. Bonner, J. T. (2009). Why size matters, in *The Big Ideas of Nanoscale Science and Engineering*, eds., Stevens, S. Y., Sutherland, L. M., Krajcik, J. S. (NSTA Press, Princeton) pp. 4–5 (80p).

25. Amendola, V., Pilot, R., Frasconi, M., Maragò, O. M., Iatì, M. A. (2017). Surface plasmon resonance in gold nanoparticles: A review, *J. Phys. Condens Matter*, **29**, p. 203002.

26. Maier, S. A. (2007). Localized surface plasmons, in *Plasmonics: Fundamentals and Applications*, ed., Maier, S. A. (Springer US, New York) pp. 65–88.
27. Mayer, K. M., Hafner, J. H. (2011). Localized surface plasmon resonance sensors, *Chem. Rev.,* **111,** pp. 3828–3857.
28. Shvedova, A. A., Kagan, V. E., Fadeel, B. (2010). Close encounters of the small kind: Adverse effects of man-made materials interfacing with the nano-cosmos of biological systems, *Annu. Rev. Pharmacol. Toxicol.,* **50**, pp. 63–88.
29. Chen, Y., Xianyu, Y., Jiang, X. (2017). Surface modification of gold nanoparticles with small molecules for biochemical analysis, *Acc. Chem. Res.,* **50**, pp. 310–319.
30. Bhowmik, T., Saha, P. P., Sarkar, A., Gomes, A. (2017). Evaluation of cytotoxicity of a purified venom protein from Naja kaouthia (NKCT1) using gold nanoparticles for targeted delivery to cancer cell, *Chem. Biol. Interact.,* **261**, pp. 35–49.
31. Abadeer, N. S., Murphy, C. J. (2016). Recent progress in cancer thermal therapy using gold nanoparticles, *J. Phys. Chem. C,* **120**, pp. 4691–4716.
32. Dreaden, E. C., Alkilany, A. M., Huang, X., Murphy, C. J., El-Sayed, M. A. (2012). The golden age: Gold nanoparticles for biomedicine, *Chem. Soc. Rev.,* **41**, pp. 2740–2779.
33. Faraday, M. (1857). X. The Bakerian Lecture.—Experimental relations of gold (and other metals) to light, *Philos. Trans. R. Soc. Lond.,* **147**, pp. 145–181.
34. Fortin, M.-A., Simão, T., Laprise-Pelletier, M. (2018). Gold nanoparticles for imaging and cancer therapy, in *Nanooncology: Engineering Nanomaterials for Cancer Therapy and Diagnosis,* eds., Gonçalves, G., Tobias, G. (Springer, Cham) pp. 1–50.
35. Sardar, R., Shumaker-Parry, J. S. (2011). Spectroscopic and microscopic investigation of gold nanoparticle formation: Ligand and temperature effects on rate and particle size, *J. Am. Chem. Soc.,* **133**, pp. 8179–8190.
36. Wilton-Ely, J. D. (2008). The surface functionalisation of gold nanoparticles with metal complexes, *Dalton Trans.,* **1**, pp. 25–29.
37. Grzelczak, M., Perez-Juste, J., Mulvaney, P., Liz-Marzan, L. M. (2008). Shape control in gold nanoparticle synthesis, *Chem. Soc. Rev.,* **37**, pp. 1783–1791.

38. Turkevich, J., Stevenson, P. C., Hillier, J. (1951). A study of the nucleation and growth processes in the synthesis of colloidal gold, *Discuss. Faraday Soc.*, **11**, pp. 55–75.

39. Zhao, L., Jiang, D., Cai, Y., Ji, X., Xie, R., Yang, W. (2012). Tuning the size of gold nanoparticles in the citrate reduction by chloride ions, *Nanoscale*, **4**, pp. 5071–5076.

40. Shiba, F. (2013). Size control of monodisperse Au nanoparticles synthesized via a citrate reduction process associated with a pH-shifting procedure, *CrystEngComm*, **15**, pp. 8412–8415.

41. Bastús, N. G., Comenge, J., Puntes, V. (2011). Kinetically controlled seeded growth synthesis of citrate-stabilized gold nanoparticles of up to 200 nm: Size focusing versus Ostwald ripening, *Langmuir*, **27**, pp. 11098–11105.

42. Liu, G., Lu, M., Huang, X., Li, T., Xu, D. (2018). Application of gold-nanoparticle colorimetric sensing to rapid food safety screening, *Sensors*, **18**, p. 4166.

43. Yadav, H. K. S., Raizaday, A. (2016). Inorganic nanobiomaterials for medical imaging, in *Nanobiomaterials in Medical Imaging*, eds., Grumezescu, A. M., Chapter 12 (Elsevier) pp. 365–401.

44. Jana, N. R., Gearheart, L., Murphy, C. J. (2001). Evidence for seed-mediated nucleation in the chemical reduction of gold salts to gold nanoparticles, *Chem. Mat.*, **13**, pp. 2313–2322.

45. Brown, K. R., Natan, M. J. (1998). Hydroxylamine seeding of colloidal Au nanoparticles in solution and on surfaces, *Langmuir*, **14**, pp. 726–728.

46. Ziegler, C., Eychmüller, A. (2011). Seeded growth synthesis of uniform gold nanoparticles with diameters of 15–300 nm, *J. Phys. Chem. C*, **115**, pp. 4502–4506.

47. Kumar, D., Mutreja, I., Sykes, P. (2016). Seed mediated synthesis of highly mono-dispersed gold nanoparticles in the presence of hydroquinone, *Nanotechnology*, **27**, p. 355601.

48. Brust, M., Walker, M., Bethell, D., Schiffrin, D. J., Whyman, R. (1994). Synthesis of thiol-derivatised gold nanoparticles in a two-phase liquid–liquid system, *J. Chem. Soc. Chem. Commun.*, **0**, pp. 801–802.

49. Booth, S. G., Uehara, A., Chang, S.-Y., La Fontaine, C., Fujii, T., Okamoto, Y., Imai, T., Schroeder, S. L., Dryfe, R. (2017). The significance of bromide in the Brust–Schiffrin synthesis of thiol protected gold nanoparticles, *Chem. Sci.*, **8**, pp. 7954–7962.

50. Uehara, A., Booth, S. G., Chang, S. Y., Schroeder, S. L., Imai, T., Hashimoto, T., Mosselmans, J. F. W., Dryfe, R. A. (2015). Electrochemical insight into the Brust–Schiffrin synthesis of Au nanoparticles, *J. Am. Chem. Soc.*, **137**, pp. 15135–15144.
51. Nicol, J. R., Dixon, D., Coulter, J. A. (2015). Gold nanoparticle surface functionalization: A necessary requirement in the development of novel nanotherapeutics, *Nanomedicine*, **10**, pp. 1315–1326.
52. Yeh, Y.-C., Creran, B., Rotello, V. M. (2012). Gold nanoparticles: Preparation, properties, and applications in bionanotechnology, *Nanoscale*, **4**, pp. 1871–1880.
53. Fu, Y. Z., Lai, G. C., Li, J. R. (2009). Synthesis of different morphologies of mostly mono-disperse gold nanoparticles in a polyol process, *J. Disper. Sci. Technol.*, **30**, pp. 514–519.
54. Li, C., Shuford, K. L., Chen, M., Lee, E. J., Cho, S. O. (2008). A facile polyol route to uniform gold octahedra with tailorable size and their optical properties, *ACS Nano*, **2**, pp. 1760–1769.
55. Eid, K., Wang, H., Wang, L. (2017). Nanoarchitectonic metals, in *Supra-Materials Nanoarchitectonics,* eds., Ariga, K., Aono, M., Chapter 6 (Elsevier) pp. 135–171.
56. Fiévet, F., Ammar-Merah, S., Brayner, R., Chau, F., Giraud, M., Mammeri, F., Peron, J., Piquemal, J.-Y., Sicard, L., Viau, G. (2018). The polyol process: A unique method for easy access to metal nanoparticles with tailored sizes, shapes and compositions, *Chem. Soc. Rev.*, **47**, pp. 5187–5233.
57. Sheoran, N., Kaur, P. (2018). Biosynthesis of nanoparticles using eco-friendly factories and their role in plant pathogenicity: A review, *Biotechnol. Res. Innovation*, **2**, pp. 63–73.
58. Li, J., Li, Q., Ma, X., Tian, B., Li, T., Yu, J., Dai, S., Weng, Y., Hua, Y. (2016). Biosynthesis of gold nanoparticles by the extreme bacterium Deinococcus radiodurans and an evaluation of their antibacterial properties, *Int. J. Nanomed.*, **11**, p. 5931.
59. Vala, A. K. (2015). Exploration on green synthesis of gold nanoparticles by a marine-derived fungus Aspergillus sydowii, *Environ. Prog. Sustain. Energy*, **34**, pp. 194–197.
60. Zhang, X., Qu, Y., Shen, W., Wang, J., Li, H., Zhang, Z., Li, S., Zhou, J. (2016). Biogenic synthesis of gold nanoparticles by yeast Magnusiomyces ingens LH-F1 for catalytic reduction of nitrophenols, *Colloids Surf. A Physicochem. Eng. Asp.*, **497**, pp. 280–285.

61. Mata, R., Bhaskaran, A., Sadras, S. R. (2016). Green-synthesized gold nanoparticles from Plumeria alba flower extract to augment catalytic degradation of organic dyes and inhibit bacterial growth, *Particuology*, **24**, pp. 78–86.

62. Kitching, M., Ramani, M., Marsili, E. (2015). Fungal biosynthesis of gold nanoparticles: Mechanism and scale up, *Microb. Biotechnol.*, **8**, pp. 904–917.

63. Sztandera, K., Gorzkiewicz, M., Klajnert-Maculewicz, B. (2018). Gold nanoparticles in cancer treatment, *Mol. Pharm.*, **16**, pp 1–23.

64. Ciaurriz, P., Bravo, E., Hamad-Schifferli, K. (2014). Effect of architecture on the activity of glucose oxidase/horseradish peroxidase/carbon nanoparticle conjugates, *J. Colloid Interface Sci.*, **414**, pp. 73–81.

65. Sperling, R. A., Parak, W. J. (2010). Surface modification, functionalization and bioconjugation of colloidal inorganic nanoparticles, *Philos. Trans. A Math. Phys. Eng. Sci.*, **368**, pp. 1333–1383.

66. Karni-Schmidt, O., Lokshin, M., Prives, C. (2016). The roles of MDM2 and MDMX in cancer, *Annu. Rev. Pathol.*, **11**, pp. 617–644.

67. Jain, A., Cheng, K. (2017). The principles and applications of avidin-based nanoparticles in drug delivery and diagnosis, *J. Control. Release*, **245**, pp. 27–40.

68. Rayavarapu, R. G., Petersen, W., Ungureanu, C., Post, J. N., van Leeuwen, T. G., Manohar, S. (2007). Synthesis and bioconjugation of gold nanoparticles as potential molecular probes for light-based imaging techniques, *Int. J. Biomed. Imaging*, **2007**, pp. 1–10.

69. Conde, J., Ambrosone, A., Sanz, V., Hernandez, Y., Marchesano, V., Tian, F., Child, H., Berry, C. C., Ibarra, M. R., Baptista, P. V., Tortiglione, C., de la Fuente, J. M. (2012). Design of multifunctional gold nanoparticles for in vitro and in vivo gene silencing, *ACS Nano*, **6**, pp. 8316–8324.

70. Dhar, S., Daniel, W. L., Giljohann, D. A., Mirkin, C. A., Lippard, S. J. (2009). Polyvalent oligonucleotide gold nanoparticle conjugates as delivery vehicles for platinum(IV) warheads, *J. Am. Chem. Soc.*, **131**, pp. 14652–14653.

71. Ravi, S., Krishnamurthy, V. R., Caves, J. M., Haller, C. A., Chaikof, E. L. (2012). Maleimide–thiol coupling of a bioactive peptide to an elastin-like protein polymer, *Acta Biomater.*, **8**, pp. 627–635.

72. Li, X., Guo, J., Asong, J., Wolfert, M. A., Boons, G.-J. (2011). Multifunctional surface modification of gold-stabilized nanoparticles by bioorthogonal reactions, *J. Am. Chem. Soc.*, **133**, pp. 11147–11153.

73. Nobs, L., Buchegger, F., Gurny, R., Allémann, E. (2004). Current methods for attaching targeting ligands to liposomes and nanoparticles, *J. Pharm. Sci.*, **93**, pp. 1980–1992.

74. Daima, H. K., Selvakannan, P. R., Shukla, R., Bhargava, S. K., Bansal, V. (2013). Fine-tuning the antimicrobial profile of biocompatible gold nanoparticles by sequential surface functionalization using polyoxometalates and lysine, *PLoS One*, **8**, p. e79676.

75. Maldiney, T., Richard, C., Seguin, J., Wattier, N., Bessodes, M., Scherman, D. (2011). Effect of core diameter, surface coating, and PEG chain length on the biodistribution of persistent luminescence nanoparticles in mice, *ACS Nano*, **5**, pp. 854–862.

76. Liu, Y., Shipton, M. K., Ryan, J., Kaufman, E. D., Franzen, S., Feldheim, D. L. (2007). Synthesis, stability, and cellular internalization of gold nanoparticles containing mixed peptide–poly(ethylene glycol) monolayers, *Anal. Chem.*, **79**, pp. 2221–2229.

77. Qian, W., Curry, T., Che, Y., Kopelman, R., Targeted delivery of peptide-conjugated biocompatible gold nanoparticles into cancer cell nucleus, *Colloidal Nanocrystals for Biomedical Applications VIII, SPIE Proc.,* **2013**, pp. 85951D.

78. Nativo, P., Prior, I. A., Brust, M. (2008). Uptake and intracellular fate of surface-modified gold nanoparticles, *ACS Nano*, **2**, pp. 1639–1644.

79. Kalmodia, S., Parameswaran, S., Ganapathy, K., Yang, W., Barrow, C. J., Kanwar, J. R., Roy, K., Vasudevan, M., Kulkarni, K., Elchuri, S. V., Krishnakumar, S. (2017). Characterization and molecular mechanism of peptide-conjugated gold nanoparticle inhibiting p53-HDM2 interaction in retinoblastoma, *Mol. Ther. Nucleic Acids*, **9**, pp. 349–364.

80. Bastús, N. G., Sánchez-Tilló, E., Pujals, S., Farrera, C., López, C., Giralt, E., Celada, A., Lloberas, J., Puntes, V. (2009). Homogeneous conjugation of peptides onto gold nanoparticles enhances macrophage response, *ACS Nano*, **3**, pp. 1335–1344.

81. Tomuleasa, C., Soritau, O., Orza, A., Dudea, M., Petrushev, B., Mosteanu, O., Susman, S., Florea, A., Pall, E., Aldea, M., Kacso, G., Cristea, V., Berindan-Neagoe, I., Irimie, A. (2012). Gold nanoparticles conjugated with cisplatin/doxorubicin/capecitabine lower the chemoresistance of hepatocellular carcinoma-derived cancer cells, *J. Gastrointestin Liver Dis.*, **21**, pp. 187–196.

82. Kalimuthu, K., Lubin, B.-C., Bazylevich, A., Gellerman, G., Shpilberg, O., Luboshits, G., Firer, M. A. (2018). Gold nanoparticles stabilize peptide-drug-conjugates for sustained targeted drug delivery to cancer cells, *J. Nanobiotechnol.*, **16**, p. 34.

83. Yang, C., Bromma, K., Di Ciano-Oliveira, C., Zafarana, G., van Prooijen, M., Chithrani, D. B. (2018). Gold nanoparticle mediated combined cancer therapy, *Cancer Nano*, **9**, p. 4.

84. Mansoori, G. A., Brandenburg, K. S., Shakeri-Zadeh, A. (2010). A comparative study of two folate-conjugated gold nanoparticles for cancer nanotechnology applications, *Cancers*, **2**, pp. 1911–1928.

85. Xiao, Z., Ji, C., Shi, J., Pridgen, E. M., Frieder, J., Wu, J., Farokhzad, O. C. (2012). DNA self-assembly of targeted near-infrared-responsive gold nanoparticles for cancer thermo-chemotherapy, *Angew. Chem.*, **124**, pp. 12023–12027.

86. Bertel Garay, L., Martínez Ortega, F., Méndez-Sanchez, S. C. (2018). Use in vitro of gold nanoparticles functionalized with folic acid as a photothermal agent on treatment of HeLa cells, *J. Mex. Chem. Soc.*, **62**, , p. 00012.

87. Zeinizade, E., Tabei, M., Shakeri-Zadeh, A., Ghaznavi, H., Attaran, N., Komeili, A., Ghalandari, B., Maleki, S., Kamrava, S. K. (2018). Selective apoptosis induction in cancer cells using folate-conjugated gold nanoparticles and controlling the laser irradiation conditions, *Artif. Cells Nanomed. Biotechnol.*, **46**, pp. 1026–1038.

88. Neshastehriz, A., Tabei, M., Maleki, S., Eynali, S., Shakeri-Zadeh, A. (2017). Photothermal therapy using folate conjugated gold nanoparticles enhances the effects of 6 MV X-ray on mouth epidermal carcinoma cells, *J. Photochem. Photobiol.*, **172**, pp. 52–60.

89. Sokolov, K., Follen, M., Aaron, J., Pavlova, I., Malpica, A., Lotan, R., Richards-Kortum, R. (2003). Real-time vital optical imaging of precancer using anti-epidermal growth factor receptor antibodies conjugated to gold nanoparticles, *Cancer Res.*, **63**, p. 1999.

90. El-Sayed, I. H., Huang, X., El-Sayed, M. A. (2005). Surface plasmon resonance scattering and absorption of anti-EGFR antibody conjugated gold nanoparticles in cancer diagnostics: Applications in oral cancer, *Nano Lett.*, **5**, pp. 829–834.

91. Fernandes, A. R., Jesus, J., Martins, P., Figueiredo, S., Rosa, D., Martins, L. M. R. D. R. S., Corvo, M. L., Carvalheiro, M. C., Costa, P. M., Baptista, P. V. (2017). Multifunctional gold-nanoparticles: A nanovectorization tool for the targeted delivery of novel chemotherapeutic agents, *J. Control. Release*, **245**, pp. 52–61.

92. Ding, Y., Jiang, Z., Saha, K., Kim, C. S., Kim, S. T., Landis, R. F., Rotello, V. M. (2014). Gold nanoparticles for nucleic acid delivery, *Mol. Ther.*, **22**, pp. 1075–1083.

93. Wang, H., Yang, R., Yang, L., Tan, W. (2009). Nucleic acid conjugated nanomaterials for enhanced molecular recognition, *ACS Nano*, **3**, pp. 2451–2460.

94. Daraee, H., Pourhassanmoghadam, M., Akbarzadeh, A., Zarghami, N., Rahmati-Yamchi, M. (2016). Gold nanoparticle–oligonucleotide conjugate to detect the sequence of lung cancer biomarker, *Artif. Cells Nanomed. Biotechnol.*, **44**, pp. 1417–1423.

95. Gossai, N. P., Naumann, J. A., Li, N.-S., Zamora, E. A., Gordon, D. J., Piccirilli, J. A., Gordon, P. M. (2016). Drug conjugated nanoparticles activated by cancer cell specific mRNA, *Oncotarget*, **7**, pp. 38243–38256.

96. Muzibur Rahman, M., Mohamed Asiri, A. (2019). Introductory chapter: Basic concept of gold nanoparticles, in *Gold Nanoparticles–Reaching New Heights*, eds., Rahman, M., Mohammed Asiri, A. (IntechOpen).

97. Hainfeld, J. F., Dilmanian, F. A., Slatkin, D. N., Smilowitz, H. M. (2008). Radiotherapy enhancement with gold nanoparticles, *J. Pharm. Pharmacol.*, **60**, pp. 977–985.

98. Nagi, N. M. S., Khair, Y. A. M., Abdalla, A. M. E. (2017). Capacity of gold nanoparticles in cancer radiotherapy, *Jpn. J. Radiol.*, **35**, pp. 555–561.

99. Her, S., Jaffray, D. A., Allen, C. (2017). Gold nanoparticles for applications in cancer radiotherapy: Mechanisms and recent advancements, *Adv. Drug Deliv. Rev.*, **109**, pp. 84–101.

100. Alizadeh, E., Sanz, A. G., García, G., Sanche, L. (2013). Radiation damage to DNA: The indirect effect of low energy electrons, *J. Phys. Chem. Lett.*, **4**, pp. 820–825.

101. Ranganathan, R., Madanmohan, S., Kesavan, A., Baskar, G., Krishnamoorthy, Y. R., Santosham, R., Ponraju, D., Rayala, S. K., Venkatraman, G. (2012). Nanomedicine: Towards development of patient-friendly drug-delivery systems for oncological applications, *Int. J. Nanomed.*, **7**, pp. 1043–1060.

102. Norouzi, H., Khoshgard, K., Akbarzadeh, F. (2018). In vitro outlook of gold nanoparticles in photo-thermal therapy: A literature review, *Lasers Med. Sci.*, **33**, pp. 917–926.

103. Adnan, N. N. M., Cheng, Y. Y., Ong, N. M. N., Kamaruddin, T. T., Rozlan, E., Schmidt, T. W., Duong, H. T. T., Boyer, C. (2016). Effect of gold nanoparticle shapes for phototherapy and drug delivery, *Polym. Chem.*, **7**, pp. 2888–2903.

104. Cabral, R. M., Baptista, P. V. (2014). Anti-cancer precision theranostics: A focus on multifunctional gold nanoparticles, *Expert Rev. Mol. Diagn.*, **14**, pp. 1041–1052.

105. Ou, Y.-C., Webb, J. A., Faley, S., Shae, D., Talbert, E. M., Lin, S., Cutright, C. C., Wilson, J. T., Bellan, L. M., Bardhan, R. (2016). Gold nanoantenna-mediated photothermal drug delivery from thermosensitive liposomes in breast cancer, *ACS Omega*, **1**, pp. 234–243.

106. Pinto, A., Pocard, M. (2018). Photodynamic therapy and photothermal therapy for the treatment of peritoneal metastasis: A systematic review, *Pleura and Peritoneum*, **3**, p. 20180124.

107. Ajnai, G., Chiu, A., Kan, T., Cheng, C.-C., Tsai, T.-H., Chang, J. (2014). Trends of gold nanoparticle-based drug delivery system in cancer therapy, *J. Exp. Clin. Cancer Res.*, **6**, pp. 172–178.

108. Lepock, J. R. (2003). Cellular effects of hyperthermia: Relevance to the minimum dose for thermal damage, *Int. J. Hyperthermia*, **19**, pp. 252–266.

109. Mokhtari, R. B., Homayouni, T. S., Baluch, N., Morgatskaya, E., Kumar, S., Das, B., Yeger, H. (2017). Combination therapy in combating cancer, *Oncotarget*, **8**, pp. 38022–38043.

110. Ren, Y., Qi, H., Chen, Q., Ruan, L. (2017). Thermal dosage investigation for optimal temperature distribution in gold nanoparticle enhanced photothermal therapy, *Int. J. Heat Mass Transf.*, **106**, pp. 212–221.

111. Chatterjee, D. K., Diagaradjane, P., Krishnan, S. (2011). Nanoparticle-mediated hyperthermia in cancer therapy, *Ther. Deliv.*, **2**, pp. 1001–1014.

112. Pedrosa, P., Mendes, R., Cabral, R., Martins, L. M. D. R. S., Baptista, P. V., Fernandes, A. R. (2018). Combination of chemotherapy and Au-nanoparticle photothermy in the visible light to tackle doxorubicin resistance in cancer cells, *Sci. Rep.*, **8**, p. 11429.

113. Yao, J., Feng, J., Chen, J. (2016). External-stimuli responsive systems for cancer theranostic, *Asian J. Pharm. Sci.*, **11**, pp. 585–595.

114. Wei, X., Chen, H., Tham, H. P., Zhang, N., Xing, P., Zhang, G., Zhao, Y. (2018). Combined photodynamic and photothermal therapy using cross-linked polyphosphazene nanospheres decorated with gold nanoparticles, *ACS Appl. Nano Mater.*, **1**, pp. 3663–3672.

115. Kim, H., Lee, D. (2018). Near-infrared-responsive cancer photothermal and photodynamic therapy using gold nanoparticles, *Polymers*, **10**, p. 961.

116. Liu, J., Liu, K., Feng, L., Liu, Z., Xu, L. (2017). Comparison of nanomedicine-based chemotherapy, photodynamic therapy and photothermal therapy using reduced graphene oxide for the model system, *Biomater. Sci.*, **5**, pp. 331–340.

117. Arvizo, R. R., Rana, S., Miranda, O. R., Bhattacharya, R., Rotello, V. M., Mukherjee, P. (2011). Mechanism of anti-angiogenic property of gold nanoparticles: Role of nanoparticle size and surface charge, *Nanomedicine*, **7**, pp. 580–587.

118. Cheng, W., Kumar, J. N., Zhang, Y., Liu, Y. (2015). pH- and redox-responsive self-assembly of amphiphilic hyperbranched poly(amido amine)s for controlled doxorubicin delivery, *Biomater. Sci.*, **3**, pp. 597–607.

119. Khutale, G. V., Casey, A. (2017). Synthesis and characterization of a multifunctional gold-doxorubicin nanoparticle system for pH triggered intracellular anticancer drug release, *Eur. J. Pharm. Biopharm.*, **119**, pp. 372–380.

120. Zhao, X., Pan, J., Li, W., Yang, W., Qin, L., Pan, Y. (2018). Gold nanoparticles enhance cisplatin delivery and potentiate chemotherapy by decompressing colorectal cancer vessels, *Int. J. Nanomed.*, **13**, pp. 6207–6221.

121. Tamarkin, L., Kingston, D. G. I. (2017). Exposing the tumor microenvironment: How gold nanoparticles enhance and refine drug delivery, *Ther. Deliv.*, **8**, pp. 363–366.

122. Maeda, H. (2015). Toward a full understanding of the EPR effect in primary and metastatic tumors as well as issues related to its heterogeneity, *Adv. Drug Deliv. Rev.*, **91**, pp. 3–6.

123. Suk, J. S., Xu, Q., Kim, N., Hanes, J., Ensign, L. M. (2016). PEGylation as a strategy for improving nanoparticle-based drug and gene delivery, *Adv. Drug Deliv. Rev.*, **99**, pp. 28–51.

124. Salvati, A., Pitek, A. S., Monopoli, M. P., Prapainop, K., Bombelli, F. B., Hristov, D. R., Kelly, P. M., Åberg, C., Mahon, E., Dawson, K. A. (2013). Transferrin-functionalized nanoparticles lose their targeting capabilities when a biomolecule corona adsorbs on the surface, *Nat. Nanotechnol.*, **8**, pp. 137–143.

125. Shmeeda, H., Tzemach, D., Mak, L., Gabizon, A. (2009). Her2-targeted pegylated liposomal doxorubicin: Retention of target-specific binding and cytotoxicity after in vivo passage, *J. Control. Release*, **136**, pp. 155–160.

126. Kumar, A., Ma, H., Zhang, X., Huang, K., Jin, S., Liu, J., Wei, T., Cao, W., Zou, G., Liang, X.-J. (2012). Gold nanoparticles functionalized with therapeutic and targeted peptides for cancer treatment, *Biomaterials*, **33**, pp. 1180–1189.

127. Chanda, N., Kan, P., Watkinson, L. D., Shukla, R., Zambre, A., Carmack, T. L., Engelbrecht, H., Lever, J. R., Katti, K., Fent, G. M., Casteel, S. W.,

Smith, C. J., Miller, W. H., Jurisson, S., Boote, E., Robertson, J. D., Cutler, C., Dobrovolskaia, M., Kannan, R., Katti, K. V. (2010). Radioactive gold nanoparticles in cancer therapy: Therapeutic efficacy studies of GA-198AuNP nanoconstruct in prostate tumor-bearing mice, *Nanomedicine*, **6**, pp. 201–209.

128. Samadian, H., Hosseini-Nami, S., Kamrava, S. K., Ghaznavi, H., Shakeri-Zadeh, A. (2016). Folate-conjugated gold nanoparticle as a new nanoplatform for targeted cancer therapy, *J. Cancer Res. Clin. Oncol.*, **142**, pp. 2217–2229.

129. Hong, E. J., Kim, Y.-S., Choi, D. G., Shim, M. S. (2018). Cancer-targeted photothermal therapy using aptamer-conjugated gold nanoparticles, *J. Ind. Eng. Chem.*, **67**, pp. 429–436.

130. Suvarna, S., Das, U., Kc, S., Mishra, S., Sudarshan, M., Saha, K. D., Dey, S., Chakraborty, A., Narayana, Y. (2017). Synthesis of a novel glucose capped gold nanoparticle as a better theranostic candidate, *PLoS One*, **12**, p. e0178202.

131. Dai, Q., Walkey, C., Chan, W. C. W. (2014). Polyethylene glycol backfilling mitigates the negative impact of the protein corona on nanoparticle cell targeting, *Angew. Chem. Int. Ed. Engl.*, **53**, pp. 5093–5096.

132. Liu, J., Alvarez, J., Ong, W., Román, E., Kaifer, A. E. (2001). Phase transfer of hydrophilic, cyclodextrin-modified gold nanoparticles to chloroform solutions, *J. Am. Chem. Soc.*, **123**, pp. 11148–11154.

133. Patra, M., Gasser, G. (2017). The medicinal chemistry of ferrocene and its derivatives, *Nat. Rev. Chem.*, **1**, p. 66.

134. Huang, K., Ma, H., Liu, J., Huo, S., Kumar, A., Wei, T., Zhang, X., Jin, S., Gan, Y., Wang, P. C., He, S., Zhang, X., Liang, X.-J. (2012). Size-dependent localization and penetration of ultrasmall gold nanoparticles in cancer cells, multicellular spheroids, and tumors in vivo, *ACS Nano*, **6**, pp. 4483–4493.

135. Li, C.-H., Kuo, T.-R., Su, H.-J., Lai, W.-Y., Yang, P.-C., Chen, J.-S., Wang, D.-Y., Wu, Y.-C., Chen, C.-C. (2015). Fluorescence-guided probes of aptamer-targeted gold nanoparticles with computed tomography imaging accesses for in vivo tumor resection, *Sci. Rep.*, **5**, p. 15675.

136. Soleimani, M., Merheb, M., Matar, R. (2015). Human gene therapy: the future of health care, *Hamdan Med. J.*, **212**, pp. 1–10.

137. Mendes, R., Fernandes, A. R., Baptista, P. V. (2017). Gold nanoparticle approach to the selective delivery of gene silencing in cancer—the case for combined delivery? *Genes (Basel)*, **8**, p. 94.

138. Barnaby, S. N., Lee, A., Mirkin, C. A. (2014). Probing the inherent stability of siRNA immobilized on nanoparticle constructs, *Proc. Natl. Acad. Sci. U. S. A.*, **111**, pp. 9739–9744.

139. Guo, J., O'Driscoll, C. M., Holmes, J. D., Rahme, K. (2016). Bioconjugated gold nanoparticles enhance cellular uptake: A proof of concept study for siRNA delivery in prostate cancer cells, *Int. J. Pharm.*, **509**, pp. 16–27.

140. Kong, L., Wu, Y., Alves, C. S., Shi, X. (2016). Efficient delivery of therapeutic siRNA into glioblastoma cells using multifunctional dendrimer-entrapped gold nanoparticles, *Nanomedicine (Lond)*, **11**, pp. 3103–3115.

141. Hou, W., Wei, P., Kong, L., Guo, R., Wang, S., Shi, X. (2016). Partially PEGylated dendrimer-entrapped gold nanoparticles: A promising nanoplatform for highly efficient DNA and siRNA delivery, *J. Mater. Chem. B*, **4**, pp. 2933–2943.

142. Qiu, J., Kong, L., Cao, X., Li, A., Wei, P., Wang, L., Mignani, S., Caminade, A.-M., Majoral, J.-P., Shi, X. (2018). Enhanced delivery of therapeutic siRNA into glioblastoma cells using dendrimer-entrapped gold nanoparticles conjugated with β-cyclodextrin, *Nanomaterials (Basel)*, **8**, p. 131.

143. Fratoddi, I., Venditti, I., Cametti, C., Russo, M. V. (2015). How toxic are gold nanoparticles? The state-of-the-art, *Nano Res.*, **8**, pp. 1771–1799.

144. Gil, P. R., Hühn, D., Loretta, L., Sasse, D., Parak, W. J. (2010). Nanopharmacy: Inorganic nanoscale devices as vectors and active compounds, *Pharmacol. Res.*, **62**, pp. 115–125.

145. Jia, Y.-P., Ma, B.-Y., Wei, X.-W., Qian, Z.-Y. (2017). The in vitro and in vivo toxicity of gold nanoparticles, *Chinese Chem. Lett.*, **28**, pp. 691–702.

146. Chen, Y.-S., Hung, Y.-C., Liau, I., Huang, G. S. (2009). Assessment of the in vivo toxicity of gold nanoparticles, *Nanoscale Res. Lett.*, **4**, pp. 858.

Chapter 10

Anti-Tumor Activity of Verbascoside-Loaded Noble Metal Nanoparticles

Yihai Liu,[a] Mubashir Hussain,[b] Xixi Wang,[a] Bin Liu,[a] and Yi Yuan[c]

[a]*Key Laboratory of Clinical and Medical Engineering, Department of Biomedical Engineering, School of Biomedical Engineering and Informatics, Nanjing Medical University, Nanjing, China*
[b]*State Key Laboratory of Bioelectronics, School of Biological Science and Medical Engineering, Southeast University, Nanjing, China*
[c]*Jiangsu Key Laboratory of Oral Diseases, Nanjing Medical University, Nanjing, China*

sslb_112@hotmail.com, xpyywy@126.com

Nanoparticles (NPs) have played a significant role in the biomedical field by enhancing the bioavailability of drugs and reducing their cytotoxicity. Among the NPs used are noble metal compounds such as silver, gold, and platinum that are particularly unreactive and stable at high temperatures, thus providing more enhanced signaling, and facilitating their use as site-specific vehicles. There

Nanoparticle Drug Delivery Systems for Cancer Treatment
Edited by Hala Gali-Muhtasib and Racha Chouaib

ISBN 978-981-4800-54-9 (Hardcover), 978-0-429-34125-0 (eBook)
www.jennystanford.com

has been extensive interest to utilize noble metal nanoparticles as drug carriers. Verbascoside (VB) is a natural medical plant that has promising antitumor activities. The combination of VB and noble metal nanoparticles has been shown to enhance the antitumor activity of VB. In this chapter, we introduce our previous work on the VB-loaded Au and Ni nanoparticles as anti-tumor agents and propose future directions and challenges in this field.

10.1 Noble Metal Nanoparticles

Conventional methodologies for drug delivery are oral and infusion, which are limited by not just systemic side effects but also inefficient therapeutic efficacy. Accordingly, it is deemed as quite pivotal to develop alternative drug delivery systems through the incorporation of one or more of the innovative materials that include biological materials, polymers, silicon and carbon materials, besides metals. Currently, materials are structured in nanoscale formats (Table 10.1). Among them, nanoparticles are playing a significant role in the biomedicine field. Nanoparticles are capable of offering appropriate form for all routes of administration, together with enhancing the bioavailability of the drug. In the circulatory system, they have the ability to augment the aqueous solubility of the drug, in addition to safeguarding the drug against degradation. More significantly, they provide targeted delivery of the drug, together with producing a prolonged release of the drug at the site of tumors. Another advantage involves lowering the toxic adverse effects of the drug.

10.1.1 Noble Metal-Based Nanoparticles

Metal nanoparticles are characterized by small sizes of ~50 nm and large surface area. They have the potential of efficiently penetrating across barriers through small capillaries into tissues and cells because of their small size. They are capable of carrying more drugs because of their large surface. As a result, metal NP-loaded drugs are capable of accumulating in the target site. Accordingly, this leads to reduced systemic side effects as well as enhanced therapeutic efficacy [1]. There are numerous types of metal-based NPs that serve as drug-delivery vehicles for loading drugs by different kinds of methodologies; for instance, encapsulation, surface attachment or entrapment.

Table 10.1 The well-known used nanoscale formats

Drug delivery technology	Materials	Nanostructure forms
Biologicals	Lipids	Vesicles, nanotubes, rings
	Peptides	Nanoparticles
	Nucleic acids	
	Polysaccharides	
	Viruses	
Polymeric	Poly(lactic acid)	Vesicles, spheres, nanoparticles
	Poly(glycolic acid)	Micelles, dendrimers
	Poly(alkyl cyanoacrylate)	
	Poly(3-hydroxybutanoic acid)	
	Poly(organophosphazene)	
	Poly(ethylene glycol)	
	Poly(caprolactone)	
	Poly(ethylene oxide)	
	Poly(amidoamine)	
	Poly(L-glutamic acid)	
	Poly(ethyleneimine)	
	Poly(propylene imine)	
Silicon based	Silicon	Porous, nanoparticles
	Silicon dioxide	Nanoneedles
Carbon-based	Carbon	Nanotubes, fullerenes
Metallic	Gold	Nanoparticles, nanoshells
	Silver	
	Palladium	
	Platinum	

Source: Data from Lajos P. Balogh, *Nanomedicine in Cancer*, 2017 Pan Stanford Publishing Pte. Ltd, Chapter 2, Page 23.

Among the metal nanoparticles, noble metal NPs have manifested potential applications in biomedical fields. Nobel metal NPs include silver, gold, platinum, palladium, rhodium, rhenium, ruthenium, iridium and osmium [2]. The difference from common metal NPs indicates that noble metal NPs are particularly unreactive and stable in the high temperature; accordingly, they are not conveniently impacted by oxidation by surrounding liquor and gas, thus providing more enhanced signaling, and facilitating their use as site-specific vehicles for carrying several chemotherapeutic agents. For their exceptional role in diagnostic and therapeutic fields, noble metal NPs, in particular, silver, gold, and platinum, have garnered extensive attention from researchers.

10.1.2 Surface Modifications

Surface modification is quite pivotal for NPs since it not only lowers the toxicity of initial stabilizing agents and NPs, but also enhances their targeting potential for tumor cells. The surface modification of noble metal NPs is comparatively easier, in the form of non-covalent and covalent interactions. In general, the non-covalent modification includes electrostatic interactions, hydrophobic entrapment, and van der Waals forces. Besides that, covalent modifications primarily make use of direct chemical attachment, linker molecules, or click chemistry [3] (Fig. 10.1).

For the unique surface stability. Noble metal NPs can be coated with different kinds of biological molecules, which include peptides, antibodies, nucleic acid, and tumor markers, enhancing the efficacy of delivery, besides being suitable for various applications [4] (Fig. 10.2).

10.1.3 Drug Carrier

Considering the benefits associated with noble metal NPs, they are designed to be effective and safe drug-loaded carriers (Table 10.2). NPs make use of the characteristics of newly formed vascularity around the tumor and are capable of specifically targeting tumors by either passively accumulating or actively anchoring targeting moieties to the overexpressed biomarkers in tumor cells, thus eventually

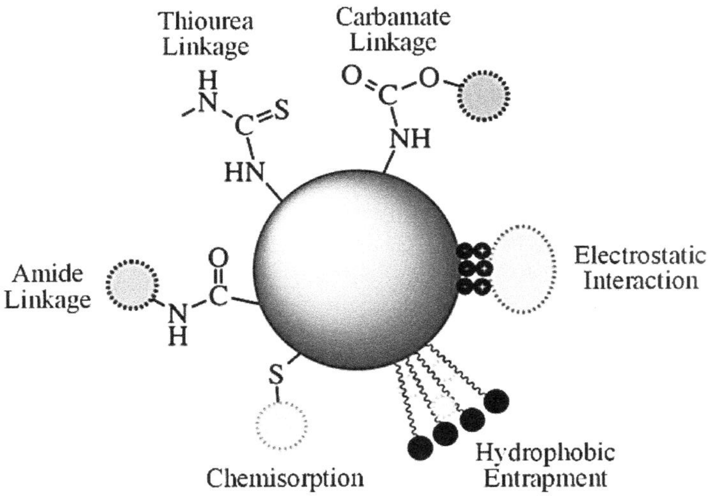

Figure 10.1 Surface modification of common nanoparticles by covalent and non-covalent interactions.

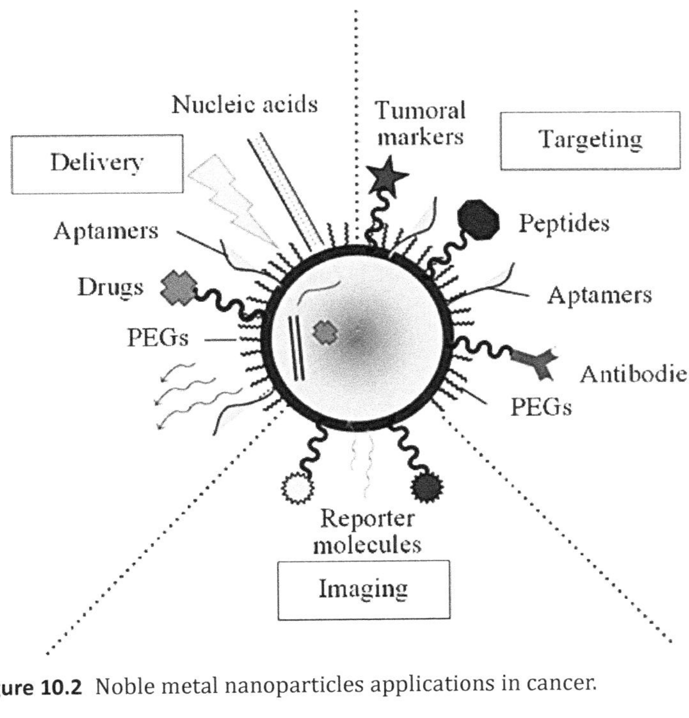

Figure 10.2 Noble metal nanoparticles applications in cancer.

Table 10.2 The anti-tumor activity of different types of noble NPs

Type of noble NPs	Shape	Size (nm)	Model	Reference
Pt	Spherical	4–12	Lungs cancer cells line (A549)	[11]
TPP@Pt	Cubic	30–60	Cervical cancer cells (SiHa)	[12]
Pt-bp	Spherical	200	HeLa, A549 and MCF-7; S180 bearing mice	[13]
Pt NPs-C	Cuboidal	5–6	Neuro 2A	[14]
Au@Pt	17 ± 1.6	17 ± 1.6	SKOV3 cells, U87-MG cells and LO2 cells	[15]
Pt	Spherical	20–50	MCF-7	[16]
Pt	Spherical	54.3	HCT 116	[17]
Au@Pt	Spherical	5–20	SiHa	[18]
PVP@Ag	Spherical	3–30	SHI-1, THP-1, Acute Myeloid Leukemia cell lines	[19]
Ag	Spherical	5–8	MCF-7	[20]
TAT@Ag	Spherical	8	Malignant Melanoma	[21]
Ag	Spherical	12–40	Human Cervical Carcinoma cell	[22]
Ag	Oval, Triangle	50	MCF-7	[23]
Ag	Spherical	15	Lung Epithelial Adenocarcinoma cell A549	[24]
PVP@Ag	Spherical	30–50	L5178Y murine Lymphoma tumor	[25]
Ag	Spherical	2–6	SiHa	[26]
Au@ tiopronin	Spherical	50	MCF-7 cells	[27]
gold@ DTDTPA	NA	2.4 ± 0.5	PNT2-C2, DU145; PC-3 Xenograft Model	[28]
Au	Spherical	5–60	SW579 Human Thyroid Carcinoma Cell line	[29]
Au@CB	Spherical	8.4±2.2	HT-29 and Caco-2	[30]
PLGA@Au	Round	180	U87MG	[31]
Au@ polymer/ MB	Cubic	610	HeLa	[32]

releasing carried drugs for killing cancer cells. Nevertheless, an efficient drug delivery carrier must have a number of requirements. First, it should be specific, and should have an affinity to bind with the antitumor drug and conveniently release the drug at the site of the tumor. In addition to being target specific, the NP-drug complex must be stable in the serum. Even then NPs should be safely degraded, besides having the least number of side effects [5] (Fig. 10.3).

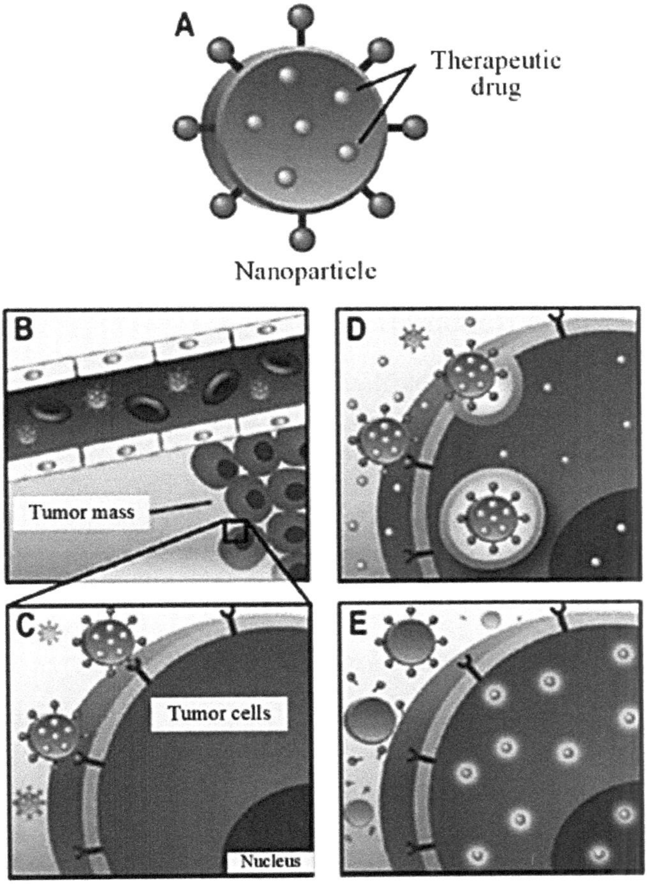

Figure 10.3 (A) An affinity to bind with the antitumor drug; (B) The NP-drug complex must remain stable in the serum; (C) Complex must be target specific; (D) Easily release the drug at the site of the tumor; (E) Safely degraded.

10.1.4 Gold Nanoparticles

The gold NPs (AuNPs) are the most frequently used, besides being among the important noble metal NPs, since they are easily synthesized and are nontoxic. Furthermore, AuNPs have exclusive physicochemical characteristics. These include an ultra-small size, large surface-area-to-mass ratio, and high surface reactivity.

AuNPs have a variety of applications in the biomedical field. With regard to cancer diagnosis and treatment, they are used not just in biological imaging but also in clinical sensing, radiotherapy and drug delivery.

Our team earlier revealed that VB-loaded NPs had an inhibitory effect on the growth of immortalized myelogenous leukemia cell line chronic K562 cells both in vitro and in vivo. We have shed light on the fact that, in comparison with AuNPs or VB alone, VB-AuNP substantially lowered the proliferation of K562 cells, and induced of tumor cells [6].

10.1.5 Silver Nanoparticles

Silver NPs are termed as the most common nanomaterials put to use in medical and research fields, in particular, acting as antimicrobial agents [7]. They have been shown to possess antifungal, antiviral, antiprotozoal, and anti-arthropod effects, which suggests that they could be used to potentially treat infectious diseases [8]. There is still a need for extensive research in the field of therapeutic application of silver nanoparticles.

10.1.6 Platinum Nanoparticles

Among various noble metal NPs, platinum is a rare metal. Also, commonly utilized platinum NPs have a diameter that ranges between 10 and 50 nm. A platinum NP that is highly effective against cancer is known as cisplatin. For the capacity of entering the cells, platinum NPs have an advantage over platinum when it comes to cancer treatment. A report shed light on the fact that platinum could exert an impact on the DNA integrity and redox systems of human colon carcinoma cell line. Besides that, the toxic effects have an inverse correlation with the diameter of platinum NPs [9]. Mohammadi et al. investigated the cytotoxic effect of PtNPs on MCF-7 and HepG-2 cell line by MTT assay, and confirmed that the IC_{50} was 2.904 and 6.829 mg/ml in HepG-2 and MCF-7, respectively [10].

10.2 Verbascoside

10.2.1 Discovery of Verbascoside

Verbascoside (VB) is a type of phenylethanoid glycoside, which is extracted from medicinal plants. Scientists reported VB in the year 1963, and naturally, VB is a phenylethanoid glycoside that is isolated from mullein (Scrophulariaceae) [33]. In the following years, it was also discovered in common lilac (Oleaceae) [34], glory tree (Lamiaceae) and greater broomrape (Orobanchaceae), besides over 200 plant species [35]. Structurally, it is characterized by phenylethanoid and caffeic acid moieties attached to a β-glucopyranose [36] (Fig. 10.4).

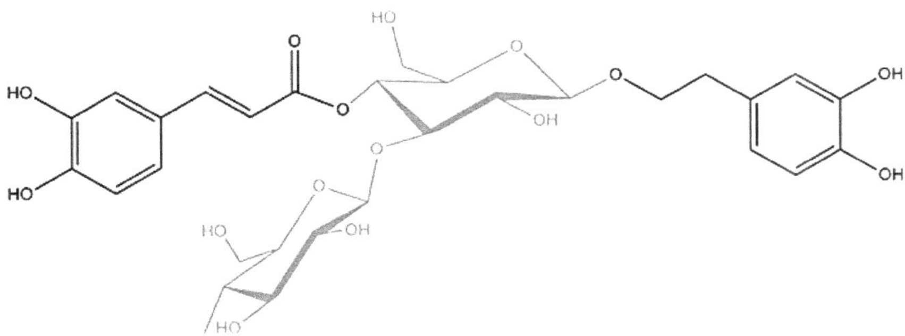

Figure 10.4 The structure of verbascoside.

10.2.2 Pharmacological Effect

There are numerous research works, which reported that VB manifests anti-inflammatory [37], anti-oxidative [38], antibacterial [39], anti-fatigue [40], anti-anxiety and hypnotic characteristics [41]. Additionally, VB has a specific anti-tumor activity, which has the potential for treating not just colorectal cancer, but also gastric cancer, leukemia and oral squamous cell carcinoma [42, 43].

10.2.3 Anti-Tumor Activity and Mechanistic Insight

Traditional chemotherapy drugs, for instance, doxorubicin, cisplatin, vinblastine, bleomycin, and daunorubicin, have limited and poor specificity, high toxicity, and have resulted in the emergence of drug-resistant diseases. Traditional plants, on the other hand, that are relatively nontoxic to normal cells and less likely to cause resistance, have attracted great attention from researchers. The proposed study sheds light on the fact that verbascoside has selective cytotoxicity against tumor cells (Fig. 10.5) [44]. VB augmented the oxidative stress, besides suppressing MMPs and modulating the immune response in tumor cell lines, while it was not cytotoxic in normal physiological cells. In addition, it had the ability to enhance the cytotoxic effect of various chemotherapy agents through the sensitization of chemoresistant cancer cells (Table 10.3).

Table 10.3 Anti-tumor activity and mechanistic insight of verbascoside

Tumor	Cell line	Effect	Mechanism	Ref.
Colorectal cancer	HCT-116, Caco-2 HT29 HCT-116	Sensitizing CRC to 5-FU Anti-metastasis and anti-invasion inducing apoptosis	Targeting PI3K/AKT pathway suppressing Rac-1, HIF-1alpha, and Zeb-1 regulating HIPK2-p53 signaling	[45–47]
Glioblastoma	U87	Inhibiting cell proliferation, migration, and invasion	Activating SHP-1, inactivating STAT3	[48]

Tumor	Cell line	Effect	Mechanism	Ref.
Hepatocellular carcinoma	Hepa-1c1c7 Hep3B	Blocking Carcinogenesis against menadione cytotoxicity Preventing from Diethynitrosamine-induced carcinogenesis	Inducing NQO1 expression reversing the expression of STAT-3	[38, 49]
Oral squamous cell carcinoma	HN4	Inhibiting metastasis and promoting apoptosis	Inactivating NF-kappaB pathway	[43]
Colon adenocarcinoma	WiDr HT-29	Anti-inflammation Anti-oxidative stress	Suppressing IL-8 scavenging Superoxide radical	[50, 51]
Non-small lung cancer	A549	Antagonizing epithelial-mesenchymal transition	Inhibiting TGF-bata1 pathway	[52]
Glioma	Glioma primary cells C6	Cell cycle arrest Regulating inflammatory response	Alterating Bax/Bcl-2-p53 axis Suppressing NF-kappaB and MAPK	[53]
Breast cancer	MCF-7	Anti-metastasis	Inhibiting NF-kappaB/AP-1/MMP9 activation	[54]
Leukemia	HL-60	Enhancing the anticancer potential of AP-cd	Anti-oxidative load	[54]
Melanoma	B16	Inhibiting metastasis	Inactivating adenyl cyclase	[55]

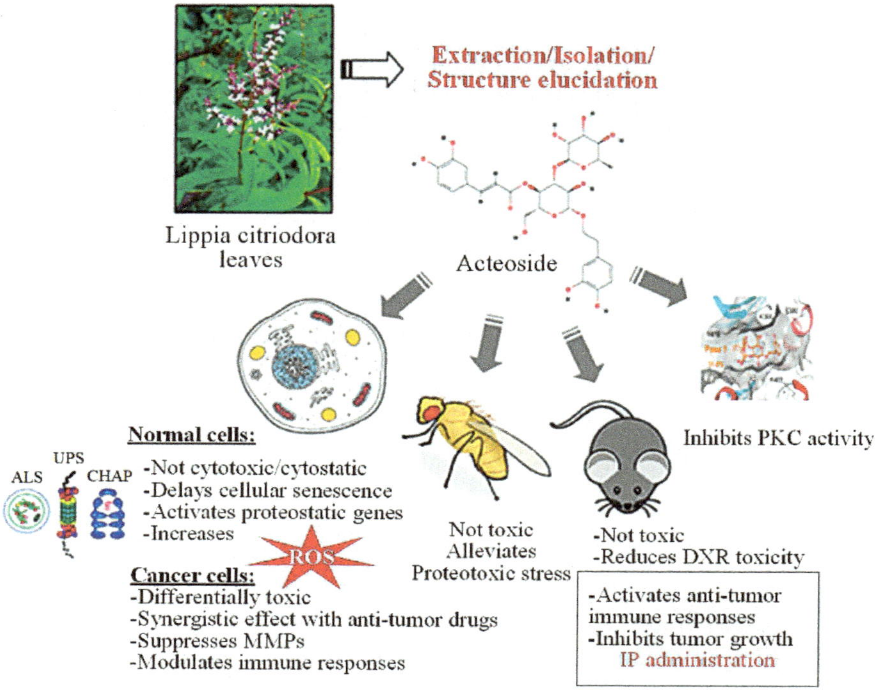

Figure 10.5 Verbascoside has anti-tumor activity.

10.3 Anti-Tumor Activity of Verbascoside-Loaded Noble Metal Nanoparticles

The combination of the antitumor pharmacological action of verbascoside and the advantage of noble metal nanoparticles as drug carriers would result in an enhanced antitumor activity of verbascoside-loaded noble metal nanoparticles.

In earlier research works, we have synthesized VB-loaded Au nanoparticles [6]. Even though there exists no significant difference in size between VB-Au and unmodified Au, the zeta potential analysis showed that the zeta potential of VB-Au was less than –20 mv, which indicated their stable binding (Fig. 10.6). The in vitro assay illustrated that VB-Au induced K562 cell line apoptosis which was mediated by caspace-3 as evidenced by

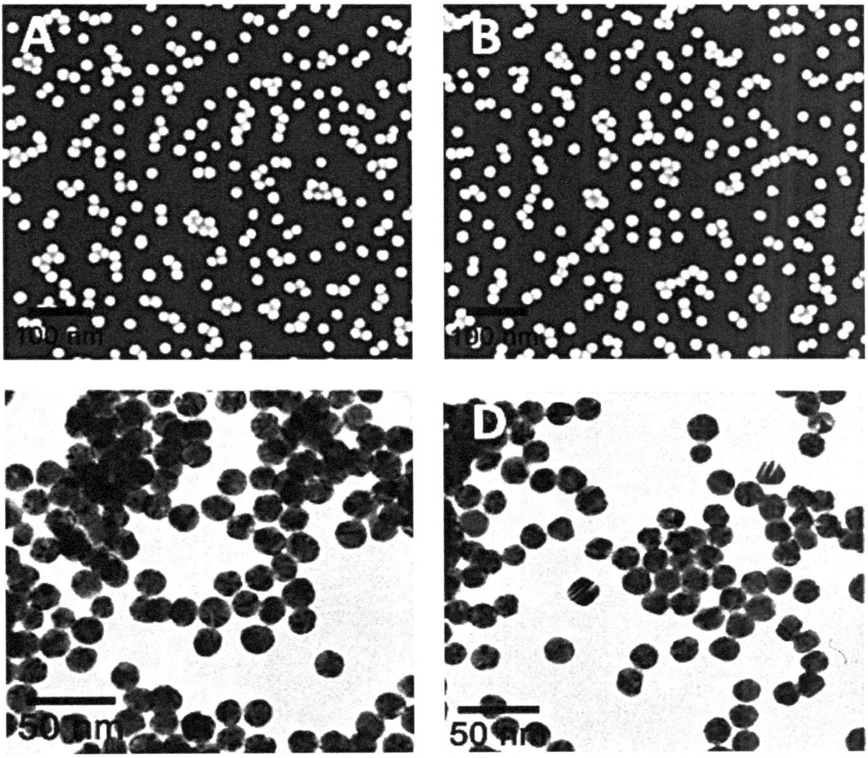

Figure 10.6 SEM and TEM images of VB-Au nanoparticles. (A) In addition, (B) SEM images of unmodified gold NPs (Au) (A) and VB-modified gold NPs (VB-Au) (B); (C) and (D) TEM images of Au (C) and VB-Au (D).

electron microscopy, viability assay, DNA fragmentation and western blot analysis, the effect of which was stronger than Au and VB alone (Fig. 10.7). In vivo experiments also shed light on the fact that, in comparison with Au and VB alone, VB-Au had the potential of inhibiting the growth of the tumor, together with promoting the survival of tumor-bearing mice more effectively (Fig. 10.8).

In vivo, female C57 nude mice (6 weeks old) were induced as tumor-bearing mice by tail vein injection of K562 tumor cells every two days until the 20th day. Each group had 7 mice in Fig. 10.8 while 5 mice in Fig. 10.11.

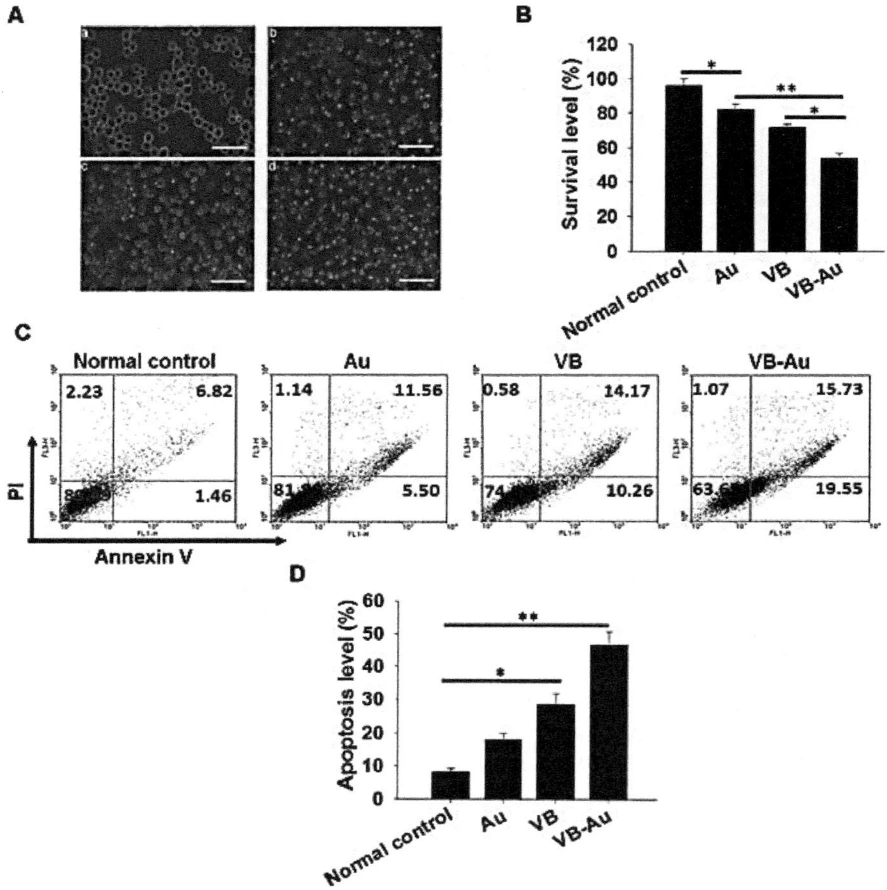

Figure 10.7 The effects of VB-Au in immortalized myelogenous leukemia cell line K562 cells. (A) morphological assay; (B) MTT assay; (C) Flow cytometric analysis; (D) the quantitative analysis of flow cytometry data.

Also, a VB-Ni NP was prepared by mixing 2 mg VB with 30 mg Ni in solution. The average diameter of VB-Ni was approximately 10 nm in TEM (Fig. 10.9). With the use of a microscopic fluorescence assay, we are able to observe the proapoptotic impact of VB-Ni on K562 cells (Fig. 10.10). Furthermore, the tumor volumes and weights of Adriamycin (ADM) resistant cell line K562 (K652/ADM) mice were inhibited more significantly by VB-Ni as compared with Ni alone (Fig. 10.11).

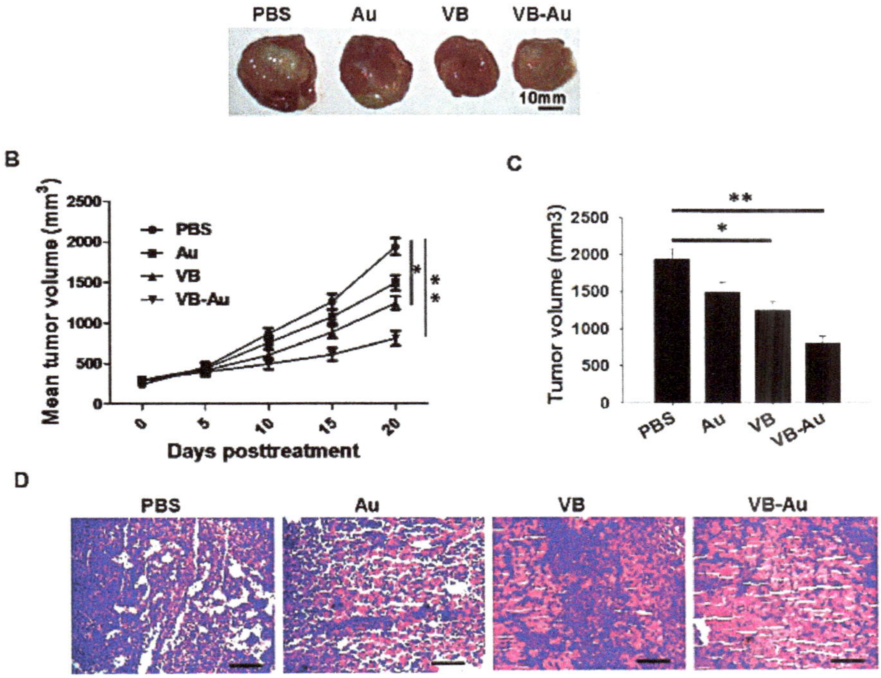

Figure 10.8 The effects of VB-Au in K562 nude mice. (A) the images of tumors; (B) and (C) Tumor volumes in mice; (D) Tumor tissue sections via HE staining. Each group had 7 mice.

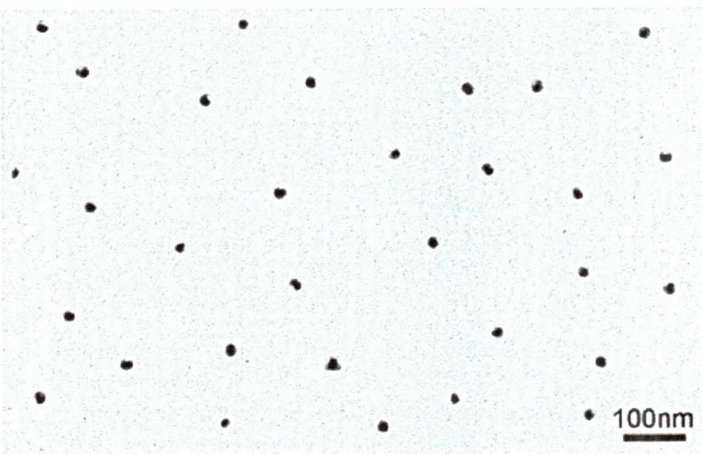

Figure 10.9 TEM images of VB-Ni NP.

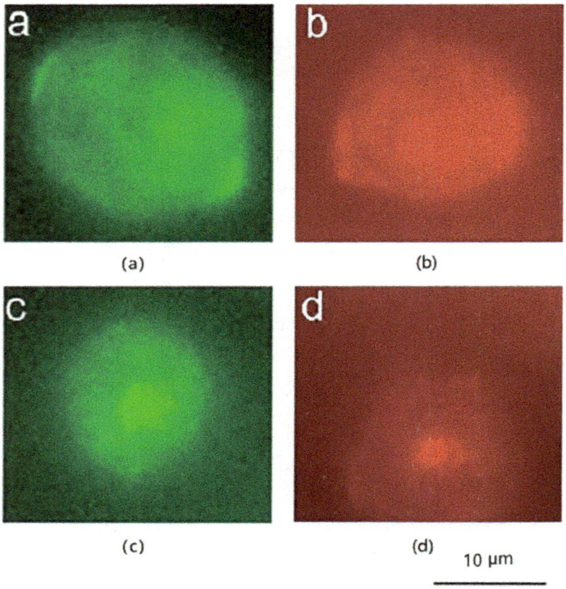

Figure 10.10 Fluorescence detection of the apoptosis of K562 cells. (A) Early and (B) later apoptotic nuclei of normal control K562 cells; (C) early and (D) later apoptotic nuclei of K562 cells treated with VB-Ni.

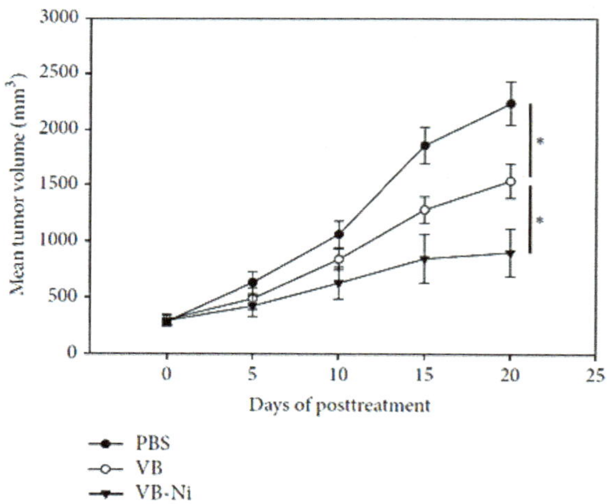

Figure 10.11 Inhibited tumor growth in K562 adriamycin resistant cell line (K652/ADM) mice with VB-Ni treatments. Female C57 nude mice were induced as tumor-bearing mice by tail vein injection of K562/ADM tumor cells every two days until the 20th day. Each group had 5 mice.

10.4 Challenges

There were major constraints for using NPs in biomedical applications because of human and environmental toxicity caused by their uncontrolled release. Thereafter, numerous research works have focused on the development of strategies for the purpose of enhancing their advantages and lowering adverse effects [56]. At first, lowering their toxicity could be accomplished through the addition of some nontoxic materials such as capping agents. Chen et al. lowered the toxicity of AgNPs by coating them with PVP and gum Arabic [57]. Another investigation highlighted the fact that the PEG-coated gold nanorods also lowered their toxicity [58]. Preventing nanoparticle aggregation is also a means of improving the efficiency of NPs, since the aggregation decreased the surface sizes of NPs, which resulted in their low activity. Dominguez et al. discovered that if AuNPs were coated using bovine serum albumin, their capability of aggregating became low and tetradecyltrimethylammonium bromide had the potential to inhibit PtNPs ability to form aggregations [59].

Additionally, PEG, PCP and natural polymers all manifested their role in the prevention of nanoparticle aggregation to some levels. It is fully known that the reproducible production of NPs with the controlled size and shape is quite difficult, which constitutes a challenge for nanotechnology researchers. Sau et al. carried out the successful synthesis of 5–20 nm-sized AuNPs at a temperature of 28°C by UV irradiation for few minutes and AgNPs at the size of 25–110 nm through the use of ascorbic acid as a reducing agent [60]. A report illustrated the synthesis of size-controlled AgNPs by an autoclaving mixture of silver-containing glass powder and glucose at 121°C and 200 kPa for a period of 20 min [61].

10.5 Conclusion

Nanotechnology plays a pivotal role in biomedicine. In particular, nanoparticles could be used as drug carriers for treating fatal diseases such as cancer. The increasing need for sustainable and stable nanoparticles triggers the development of NP with surface modification, manifesting their long-term stability. Surface materials do not just manifest their exclusive biological activity but also result in a synergistic effect with nanoparticles. The advantage of surface modification of NPs indicates a significant prospect in the delivery of anti-tumor drugs.

References

1. Yih, T. C., and Al-Fandi, M. (2006). Engineered nanoparticles as precise drug delivery systems, *J. Cell Biochem.*, **97**, pp. 1184–1190.
2. Chatterjee, K. K. (2007). *Uses of Metals and Metallic Minerals*, New Age International.
3. Austin, L. A., Mackey, M. A., Dreaden, E. C., and El-Sayed, M. A. (2014). The optical, photothermal, and facile surface chemical properties of gold and silver nanoparticles in biodiagnostics, therapy, and drug delivery, *Arch. Toxicol.*, **88**, pp. 1391–1417.
4. Conde, J., Doria, G., and Baptista, P. (2012). Noble metal nanoparticles applications in cancer, *J. Drug Deliv.*, **2012**, p. 751075.
5. Thakor, A. S., and Gambhir, S. S. (2013). Nanooncology: The future of cancer diagnosis and therapy, *CA Cancer J. Clin.*, **63**, pp. 395–418.
6. Zhang, Y., Liu, B., Wu, H., Li, B., Xu, J., Duan, L., Jiang, C., Zhao, X., Yuan, Y., Zhang, G., and Zeng, X. (2014). Anti-tumor activity of verbascoside loaded gold nanoparticles, *J. Biomed. Nanotechnol.*, **10**, pp. 3638–3646.
7. Mijnendonckx, K., Leys, N., Mahillon, J., Silver, S., and Van Houdt, R. (2013). Antimicrobial silver: Uses, toxicity and potential for resistance, *Biometals*, **26**, pp. 609–621.
8. Rai, M., Kon, K., Ingle, A., Duran, N., Galdiero, S., and Galdiero, M. (2014). Broad-spectrum bioactivities of silver nanoparticles: The emerging trends and future prospects, *Appl. Microbiol. Biotechnol.*, **98**, pp. 1951–1961.
9. Pelka, J., Gehrke, H., Esselen, M., Turk, M., Crone, M., Brase, S., Muller, T., Blank, H., Send, W., Zibat, V., Brenner, P., Schneider, R., Gerthsen, D., and Marko, D. (2009). Cellular uptake of platinum nanoparticles in human colon carcinoma cells and their impact on cellular redox systems and DNA integrity, *Chem. Res. Toxicol.*, **22**, pp. 649–659.
10. Mohammadi, H., Abedi, A., Akbarzadeh, A., Mokhtari, M. J., Shahmabadi, H. E., Mehrabi, M. R., Javadian, S., and Chiani, M. (2013). Evaluation

of synthesized platinum nanoparticles on the MCF-7 and HepG-2 cancer cell lines, *Int. Nano Lett.*, **3**, p. 28.

11. Ullah, S., Ahmad, A., Wang, A., Raza, M., Jan, A. U., Tahir, K., Rahman, A. U., and Qipeng, Y. (2017). Bio-fabrication of catalytic platinum nanoparticles and their in vitro efficacy against lungs cancer cells line (A549), *J. Photochem. Photobiol. B*, **173**, pp. 368–375.
12. Alshatwi, A. A., Athinarayanan, J., and Vaiyapuri, S. P. (2015). Green synthesis of platinum nanoparticles that induce cell death and G2/M-phase cell cycle arrest in human cervical cancer cells, *J. Mater. Sci. Mater. Med.*, **26**, p. 5330.
13. Huang, Y., He, Y., Huang, Z., Jiang, Y., Chu, W., Sun, X., Huang, L., and Zhao, C. (2017). Coordination self-assembly of platinum-bisphosphonate polymer-metal complex nanoparticles for cisplatin delivery and effective cancer therapy, *Nanoscale*, **9**, pp. 10002–10019.
14. Manikandan, M., Hasan, N., and Wu, H. F. (2013). Platinum nanoparticles for the photothermal treatment of Neuro 2A cancer cells, *Biomaterials*, **34**, pp. 5833–5842.
15. Liang, H., Wu, Y., Ou, X. Y., Li, J. Y., and Li, J. (2017). Au@Pt nanoparticles as catalase mimics to attenuate tumor hypoxia and enhance immune cell-mediated cytotoxicity, *Nanotechnology*, **28**, p. 465702.
16. Baskaran, B., Muthukumarasamy, A., Chidambaram, S., Sugumaran, A., Ramachandran, K., and Rasu, M. T. (2017). Cytotoxic potentials of biologically fabricated platinum nanoparticles from Streptomyces sp. on MCF-7 breast cancer cells, *IET Nanobiotechnol.*, **11**, pp. 241–246.
17. Yang, C., Wang, M., Zhou, J., and Chi, Q. (2017). Bio-synthesis of peppermint leaf extract polyphenols capped nano-platinum and their in-vitro cytotoxicity towards colon cancer cell lines (HCT 116), *Mater. Sci. Eng. C Mater. Biol. Appl.*, **77**, pp. 1012–1016.
18. Alshatwi, A. A., Athinarayanan, J., and Periasamy, V. S. (2015). Green synthesis of bimetallic Au@Pt nanostructures and their application for proliferation inhibition and apoptosis induction in human cervical cancer cell, *J. Mater. Sci. Mater. Med.*, **26**, p. 148.
19. Guo, D., Zhu, L., Huang, Z., Zhou, H., Ge, Y., Ma, W., Wu, J., Zhang, X., Zhou, X., Zhang, Y., Zhao, Y., and Gu, N. (2013). Anti-leukemia activity of PVP-coated silver nanoparticles via generation of reactive oxygen species and release of silver ions, *Biomaterials*, **34**, pp. 7884–7894.

20. Gurunathan, S., Han, J. W., Dayem, A. A., Eppakayala, V., Park, J. H., Cho, S. G., Lee, K. J., and Kim, J. H. (2013). Green synthesis of anisotropic silver nanoparticles and its potential cytotoxicity in human breast cancer cells (MCF-7), *J. Ind. Eng. Chem.*, **19**, pp. 1600–1605.

21. Liu, J., Zhao, Y., Guo, Q., Wang, Z., Wang, H., Yang, Y., and Huang, Y. (2012). TAT-modified nanosilver for combating multidrug-resistant cancer, *Biomaterials.*, **33**, pp. 6155–6161.

22. Jeyaraj, M., Rajesh, M., Arun, R., Mubarak Ali, D., Sathishkumar, G., Sivanandhan, G., Dev, G. K., Manickavasagam, M., Premkumar, K., Thajuddin, N., and Ganapathi, A. (2013). An investigation on the cytotoxicity and caspase-mediated apoptotic effect of biologically synthesized silver nanoparticles using Podophyllum hexandrum on human cervical carcinoma cells, *Colloids Surf. B Biointerfaces*, **102**, pp. 708–717.

23. Gurunathan, S., Raman, J., Abd, M. S., John, P. A., and Vikineswary, S. (2013). Green synthesis of silver nanoparticles using Ganoderma neo-japonicum Imazeki: A potential cytotoxic agent against breast cancer cells, *Int. J. Nanomed.*, **8**, pp. 4399–4413.

24. Han, J. W., Gurunathan, S., Jeong, J. K., Choi, Y. J., Kwon, D. N., Park, J. K., and Kim, J. H. (2014). Oxidative stress mediated cytotoxicity of biologically synthesized silver nanoparticles in human lung epithelial adenocarcinoma cell line, *Nanoscale Res. Lett.*, **9**, p. 459.

25. Laragonzález, J. H., Gomezflores, R., Tamezguerra, P., Monrealcuevas, E., Tamezguerra, R., and Rodríguezpadilla, C. (2013). In vivo antitumor activity of metal silver and silver nanoparticles in the L5178Y-R murine lymphoma model., *Br. J. Med. Med. Res.*, **3**, pp. 1308–1316.

26. Sreekanth, T. V. M., Pandurangan, M., Kim, D. H., and Yong, R. L. (2016). Green synthesis: In-vitro anticancer activity of silver nanoparticles on human cervical cancer cells, *J. Cluster Sci.*, **27**, pp. 671–681.

27. Huo, S., Ma, H., Huang, K., Liu, J., Wei, T., Jin, S., Zhang, J., He, S., and Liang, X. J. (2013). Superior penetration and retention behavior of 50 nm gold nanoparticles in tumors, *Cancer Res.*, **73**, pp. 319–330.

28. Butterworth, K. T., Nicol, J. R., Ghita, M., Rosa, S., Chaudhary, P., McGarry, C. K., McCarthy, H. O., Jimenez-Sanchez, G., Bazzi, R., Roux, S., Tillement, O., Coulter, J. A., and Prise, K. M. (2016). Preclinical evaluation of gold-DTDTPA nanoparticles as theranostic agents in prostate cancer radiotherapy, *Nanomedicine (Lond)*, **11**, pp. 2035–2047.

29. Zhang, Q., Ma, Y., Yang, S., Xu, B., and Fei, X. (2015). Smallsized gold nanoparticles inhibit the proliferation and invasion of SW579 cells, *Mol. Med. Rep.*, **12**, pp. 8313–8319.

30. Gonzalez-Ballesteros, N., Prado-Lopez, S., Rodriguez-Gonzalez, J. B., Lastra, M., and Rodriguez-Arguelles, M. C. (2017). Green synthesis of gold nanoparticles using brown algae Cystoseira baccata: Its activity in colon cancer cells, *Colloids Surf. B Biointerfaces*, **153**, pp. 190–198.

31. Hao, Y., Zhang, B., Zheng, C., Ji, R., Ren, X., Guo, F., Sun, S., Shi, J., Zhang, H., Zhang, Z., Wang, L., and Zhang, Y. (2015). The tumor-targeting core-shell structured DTX-loaded PLGA@Au nanoparticles for chemo-photothermal therapy and X-ray imaging, *J. Control. Release*, **220**, pp. 545–555.

32. Yu, J., Hsu, C. H., Huang, C. C., and Chang, P. Y. (2015). Development of therapeutic Au-methylene blue nanoparticles for targeted photodynamic therapy of cervical cancer cells, *ACS Appl. Mater. Interfaces*, **7**, pp. 432–441.

33. Angela, C., Sandra, P., Fiorenza, M., Isabella, D. A., Vito, L., and Vincenzo, L. (2012). Verbascoside, isoverbascoside, and their derivatives recovered from olive mill wastewater as possible food antioxidants, *J. Agric. Food Chem.*, **60**, pp. 1822–1829.

34. Birkofer, L., Kaiser, C., and Thomas, U. (1968). [Acteoside and neoacteoside: Sugar esters from Syringa vulgaris (L.)], *Zeitschrift Für Naturforschung Teil B Chemie Biochemie Biophysik Biologie.*, **23**, p. 1051.

35. Andary, C., Wylde, R., Laffite, C., Privat, G., and Winternitz, F. (1982). Structures of verbascoside and orobanchoside, caffeic acid sugar esters from Orobanche rapum Gensitae, *Phytochemistry*, **21**, pp. 1123–1127.

36. Alipieva, K., Korkina, L., Orhan, I. E., and Georgiev, M. I. (2014). Verbascoside—a review of its occurrence, (bio)synthesis and pharmacological significance, *Biotechnol. Adv.*, **32**, pp. 1065–1076.

37. Speranza, L., Franceschelli, S., Pesce, M., Reale, M., Menghini, L., Vinciguerra, I., De Lutiis, M. A., Felaco, M., and Grilli, A. (2010). Antiinflammatory effects in THP-1 cells treated with verbascoside, *Phytother. Res.*, **24**, pp. 1398–1404.

38. Peerzada, K. J., Faridi, A. H., Sharma, L., Bhardwaj, S. C., Satti, N. K., Shashi, B., and Tasduq, S. A. (2016). Acteoside-mediates chemoprevention of experimental liver carcinogenesis through STAT-3 regulated oxidative stress and apoptosis, *Environ. Toxicol.*, **31**, pp. 782–798.

39. Debrauwer, L., Maillard, C., Babadjamian, A., Vidal-Ollivier, E., Laget, M., Salmona, G., and Afzal-Raffi, Z. (1989). Study in the chemical constituents of *Plantago cynops* L. and antibacterial evaluation of verbascoside, *Pharm. Acta Helv.*, **64**, pp. 183–187.

40. Zhu, M., Zhu, H., Tan, N., Wang, H., Chu, H., and Zhang, C. (2016). Central anti-fatigue activity of verbascoside, *Neurosci. Lett.*, **616**, pp. 75–79.

41. Razavi, B. M., Zargarani, N., and Hosseinzadeh, H. (2017). Anti-anxiety and hypnotic effects of ethanolic and aqueous extracts of Lippia citriodora leaves and verbascoside in mice, *Avicenna J. Phytomed.*, **7**, pp. 353–365.

42. Zhang, F., Jia, Z., Deng, Z., Wei, Y., Zheng, R., and Yu, L. (2002). In vitro modulation of telomerase activity, telomere length and cell cycle in MKN45 cells by verbascoside, *Planta Med.*, **68**, pp. 115–118.

43. Zhang, Y., Yuan, Y., Wu, H., Xie, Z., Wu, Y., Song, X., Wang, J., Shu, W., Xu, J., Liu, B., Wan, L., Yan, Y., Ding, X., Shi, X., Pan, Y., Li, X., Yang, J., Zhao, X., and Wang, L. (2018). Effect of verbascoside on apoptosis and metastasis in human oral squamous cell carcinoma, *Int. J. Cancer*, **143**, pp. 980–991.

44. Cheimonidi, C., Samara, P., Polychronopoulos, P., Tsakiri, E. N., Nikou, T., Myrianthopoulos, V., Sakellaropoulos, T., Zoumpourlis, V., Mikros, E., Papassideri, I., Argyropoulou, A., Halabalaki, M., Alexopoulos, L. G., Skaltsounis, A. L., Tsitsilonis, O. E., Aligiannis, N. N., and Trougakos, I. P. (2018). Selective cytotoxicity of the herbal substance acteoside against tumor cells and its mechanistic insights, *Redox. Biol.*, **16**, pp. 169–178.

45. Nabiuni, M., Seyfi, D., Behzad, S. B., Parivar, K., Tahmaseb, M., and Amini, E. (2018). Verbascoside attenuates Rac-1 and HIF-1alpha signaling cascade in colorectal cancer cells, *Anticancer Agents Med. Chem.*, **18**, pp. 2149–2155.

46. Attia, Y. M., El-Kersh, D. M., Wagdy, H. A., and Elmazar, M. M. (2018). Verbascoside: Identification, quantification, and potential sensitization of colorectal cancer cells to 5-FU by targeting PI3K/AKT pathway, *Sci. Rep.*, **8**, p. 16939.

47. Zhou, L., Feng, Y., Jin, Y., Liu, X., Sui, H., Chai, N., Chen, X., Liu, N., Ji, Q., Wang, Y., and Li, Q. (2014). Verbascoside promotes apoptosis by regulating HIPK2-p53 signaling in human colorectal cancer, *BMC Cancer*, **14**, p. 747.

48. Jia, W. Q., Wang, Z. T., Zou, M. M., Lin, J. H., Li, Y. H., Zhang, L., and Xu, R. X. (2018). Verbascoside inhibits glioblastoma cell proliferation, migration and invasion while promoting apoptosis through upregulation of protein tyrosine phosphatase SHP-1 and inhibition of STAT3 phosphorylation, *Cell Physiol. Biochem.*, **47**, pp. 1871–1882.

49. El-Halawany, A. M., Abdallah, H. M., Hamed, A. R., Khalil, H. E., and Almohammadi, A. M. (2018). Phenolics from Barleria cristata var. Alba as carcinogenesis blockers against menadione cytotoxicity through induction and protection of quinone reductase, *BMC Complement. Altern. Med.*, **18**, p. 163.

50. Ye, Y. L., Chang, H. S., Tseng, Y. F., and Shi, L. S. (2017). Suppression of IL-8 release by sweet olive ethanolic extract and compounds in WiDr colon adenocarcinoma cells, *J. Food Sci.*, **82**, pp. 1792–1798.

51. Obied, H. K., Prenzler, P. D., Konczak, I., Rehman, A. U., and Robards, K. (2009). Chemistry and bioactivity of olive biophenols in some antioxidant and antiproliferative in vitro bioassays, *Chem. Res. Toxicol.*, **22**, pp. 227–234.

52. He, X. R., Han, S. Y., Li, X. H., Zheng, W. X., Pang, L. N., Jiang, S. T., and Li, P. P. (2017). Chinese medicine Bu-Fei decoction attenuates epithelial-mesenchymal transition of non-small cell lung cancer via inhibition of transforming growth factor beta1 signaling pathway in vitro and in vivo, *J. Ethnopharmacol.*, **204**, pp. 45–57.

53. Sitarek, P., Skala, E., Toma, M., Wielanek, M., Szemraj, J., Nieborowska-Skorska, M., Kolasa, M., Skorski, T., Wysokinska, H., and Sliwinski, T. (2016). A preliminary study of apoptosis induction in glioma cells via alteration of the Bax/Bcl-2-p53 axis by transformed and non-transformed root extracts of Leonurus sibiricus L, *Tumour. Biol.*, **37**, pp. 8753–8764.

54. Liao, Y. F., Rao, Y. K., and Tzeng, Y. M. (2012). Aqueous extract of Anisomeles indica and its purified compound exerts anti-metastatic activity through inhibition of NF-kappaB/AP-1-dependent MMP-9 activation in human breast cancer MCF-7 cells, *Food Chem. Toxicol.*, **50**, pp. 2930–2936.

55. Esposito, E., Dal Toso, R., Pressi, G., Bramanti, P., Meli, R., and Cuzzocrea, S. (2010). Protective effect of verbascoside in activated C6 glioma cells: Possible molecular mechanisms, *Naunyn Schmiedebergs Arch. Pharmacol.*, **381**, pp. 93–105.

56. Rai, M., Ingle, A. P., Birla, S., Yadav, A., and Santos, C. A. (2016). Strategic role of selected noble metal nanoparticles in medicine, *Crit. Rev. Microbiol.*, **42**, pp. 696–719.

57. Cheng, Y., Yin, L., Lin, S., Wiesner, M., Bernhardt, E., and Liu, J. (2011). Toxicity reduction of polymer-stabilized silver nanoparticles by sunlight, *J. Phys. Chem. C*, **115**, pp. 4425–4432.

58. Niidome, T., Yamagata, M., Okamoto, Y., Akiyama, Y., Takahashi, H., Kawano, T., Katayama, Y., and Niidome, Y. (2006). PEG-modified gold

nanorods with a stealth character for in vivo applications, *J. Control. Release*, **114**, pp. 343–347.

59. Newton, J. E., Preece, J. A., and Pollet, B. G. (2012). Control of nanoparticle aggregation in PEMFCs using surfactants, *Int. J. Low-Carbon Technol.*, **7**, pp. 38–43.
60. Sun, X., Wang, Z., Zhai, S., Cheng, Y., Liu, J., and Liu, B. (2013). In vitro cytotoxicity of silver nanoparticles in primary rat hepatic stellate cells, *Mol. Med. Rep.*, **8**, pp. 1365–1372.
61. Vinh Quang, N., Masayuki, I., Yasutaka, M., Shingo, N., Satoko, K., Masanori, F., Hidemi, H., Yasuhiro, K., Takeshi, O., and Yasushi, M. (2013). Preparation of size-controlled silver nanoparticles and chitosan-based composites and their anti-microbial activities, *Biomed. Mater. Eng.*, **23**, pp. 473–483.

Chapter 11

Structural Investigation of Bio-Synthesized Copper Nanoparticles Using Honey

Prerana B. Kane, Priyanka Jagtap, and Ravindra D. Kale

Department of Fibres & Textile Processing Technology
Institute of Chemical Technology, Mumbai 400019, Maharashtra India

rd.kale@ictmumbai.edu.in

This chapter deals with the synthesis of copper nanoparticles (CuNPs) by the solution reduction process of nickel chloride using honey. Particle size analysis revealed the presence of CuNPs in the range of 30 nm. The nanoparticles were further characterized by UV-Vis spectrophotometry, and transmission electron microscopy for morphology studies. In addition, X-ray diffraction analysis was used for the crystallographic determination, and Fourier transform infrared spectroscopy analysis and energy

Nanoparticle Drug Delivery Systems for Cancer Treatment
Edited by Hala Gali-Muhtasib and Racha Chouaib

ISBN 978-981-4800-54-9 (Hardcover), 978-0-429-34125-0 (eBook)
www.jennystanford.com

dispersive X-ray analysis for element determination. The particle size results were in good agreement with the size calculated from XRD, W-H plot and size-strain method.

11.1 Introduction

Nanotechnology is the manipulation of matter on an atomic, molecular, and supramolecular scale. Due to specific characteristics, metal nanoparticles have been thoroughly explored for their different properties such as optical [1], electronic [2], electrochemical [3], antimicrobial [4], catalytic activity [5] and corrosion inhibition using metals such as iron and copper [6].

Among the myriad metal nanoparticles, one of the most interesting materials in the world is copper. It has a phenomenal utility in every possible industry by virtue of its cost effectiveness and a good stability [7–14]. Various methods were used for the synthesis of CuNPs such as metal salt reduction [15], thermal decomposition [16–19], microwave heating [20], radiation methods [21], micro emulsion techniques [22], laser ablation [23], polyol method [24], solvothermal method [25], thermal and sonochemical reduction [26]. Chemical synthesis methods may pose some serious concerns in terms of residual toxic chemicals retained on the surface of the nanoparticles which may create problems in the subsequent medical applications. Thus, it is imperative to synthesize the nanoparticles in the purest form possible. One approach could be their synthesis using microorganisms or plants making it more bio-compatible [27]. Plant extract-based synthesis can be more meritorious than other biological methods by eliminating the need of maintaining cultured cells [28–31]. Valodkar et al. reported biosynthesis of CuNPs [32] by using plant *Euphorbia nivulia* L. with their biological effects on tumor cells. Thus, CuNPs have huge potential in physiological areas such as nanomedicine, biomedical, and pharmaceutical applications.

The aim of this study is to synthesize CuNPs by an eco-friendly method using honey as a reducing agent, with the help of mathematical methodology to find out the optimized conditions. Furthermore, synthesized CuNPs were physically and chemically characterized using UV-Vis spectrophotometer, nanoparticle size analyzer, TEM, XRD and EDX analyzers.

11.2 Experiments

11.2.1 Materials

Phondaghat honey was obtained from a local medical shop in Mumbai, India. Cupric chloride dihydrate $CuCl_2$ (Mol. Wt. 170.48) salt, sodium hydroxide and sulfuric acid were purchased from S. D. Fine-Chem Ltd SDFCL, Mumbai, India. All other reagents used were of analytical grade.

11.2.2 Synthesis of CuNPs

100 ml of 0.01 M Cupric chloride solution was prepared using distilled water in a conical flask. Honey was added to it with continuous stirring in the shaker bath (RossariLabtech, Mumbai, India) at 70 rpm. The effect of temperature, pH and amount of honey was evaluated with the help of experimental software. pH was maintained using 0.1N sulfuric acid and 0.1N sodium hydroxide. Color change of the solution confirmed the nanoparticles formation. After the reduction, the cupric chloride solution was filtered and stored in colloidal form for later characterization.

11.2.3 Experimental Design for Optimization of Process Parameters

Effect of process parameters on the synthesis of CuNPs was analyzed and optimized using central composite design (CCD) for the 3 factors at low, high and the center. The central points were replicated. A total of 20 experiments were performed having 8 cubes, 6 central, and 6 axial points.

At the end of each experiment, absorbance of the solution was measured spectrophotometrically at the maximum wavelength. Absorbance was individually taken as the dependent variable.

11.2.4 Characterization of CuNPs

11.2.4.1 UV-visible analysis

The preliminary characterization of the bio-synthesized CuNPs was done by UV-Vis spectrophotometer (UV-1800 ENG 240 V, Shimadzu, Japan) from 200–800 nm, with a resolution of 1 nm. The bio-reduction of Cu^{+2} in aqueous solution was periodically monitored by taking out the suspension and after 100 times dilution. UV–Visible spectroscopy analysis of CuNPs was carried out with respect to time needed for bio-reduction.

11.2.4.2 Laser diffraction technique

The particle size and particle size distribution were determined through Laser diffraction technique using particle size analyzer (SALD 7500 nano, Shimadzu, Japan).

11.2.4.3 Transmission electron microscopy

The nanoparticles solution was subjected to centrifugation at 4,800 rpm for 5 min and the resulting suspension was redispersed in distilled water. This process was repeated three times. TEM was taken by putting a drop of the suspension on carbon-coated copper grids, after which the water was allowed to evaporate. TEM and selected area electron diffraction data were collected on TEM Model 200 Supertwin STEM (Phillips make) with an accelerating voltage of 200kV and resolution of 0.23 nm.

11.2.4.4 X-Ray diffraction analysis

The purified, centrifuged and redispersed suspension was completely dried at 80°C. The crystallinity and other structural characteristics of the CuNPs was tested using X-ray diffractometer (Shimadzu XRD-6100) with CuKα radiation using the 2θ range of 10–80°.

11.2.4.5 FTIR analysis

The FTIR spectra of honey and the CuNPs were recorded on FTIR instrument (FTIR 8400S Shimadzu, Japan) between 4000–400 cm^{-1}.

11.2.4.6 Energy dispersive X-Ray analysis

The chemical composition of the biosynthesized nanoparticles was done using EDAX (EDX-720, Shimadzu, Japan).

11.3 Results and Discussion

11.3.1 ANOVA Analysis

The relationship between the response and independent variables and the conditions thereof for the maximum CuNPs biosynthesis corresponding to the optimal levels of Temperature (A), pH (B) and amount of honey (C), was established using a second-order polynomial model (Eq. 11.1). The results are tabulated in Table 11.1. The results showed considerable variation in the CuNPs biosynthesis. The maximum CuNPs biosynthesis (absorbance 0.2375) was observed in run no.1 at 100°C, pH 7 using 7.5 g of honey.

Table 11.1 CCD experimental runs for synthesis of CuNPs

Run	Temperature (°C)	pH	Amount of reducing agent (g/100 ml)	Absorbance at 590 nm	
				Experimental	Predicted
1	100	7	7.5	0.2375	0.23377
2	50	7	2.5	0.0874	0.09312
3	100	3	7.5	0.2293	0.22422
4	75	3	5	0.1554	0.14785
5	75	5	2.5	0.1357	0.12965
6	100	3	2.5	0.1657	0.16747
7	75	5	5	0.1643	0.17363
8	75	5	7.5	0.2245	0.22795
9	50	7	7.5	0.2341	0.23297
10	100	7	2.5	0.1867	0.18087
11	75	7	5	0.1458	0.15075

(Continued)

Table 11.1 (*Continued*)

Run	Temperature (°C)	pH	Amount of reducing agent (g/100 ml)	Absorbance at 590 nm	
				Experimental	Predicted
12	100	5	5	0.2079	0.22075
13	50	3	7.5	0.2341	0.24057
14	75	5	5	0.1585	0.17363
15	75	5	5	0.1631	0.17363
16	75	5	5	0.1705	0.17363
17	50	3	2.5	0.0925	0.09687
18	50	5	5	0.2005	0.18505
19	75	5	5	0.1913	0.17363
20	75	5	5	0.1889	0.17363

In the case of absorbance response, the value of the determination coefficient (R^2 = 0.9534), a measure of variability in the response, indicated that 95.34% of the variability in the response was due to the independent variables, and only 4.66% was not explained by it. Also, the value of the adjusted determination coefficient (Adj. R^2 abs = 0.9115) was quite high, signifying high significance of the model. Thus, the analysis of the response trend using the model was reasonable.

The significance of each coefficient was determined by *F*-values and *P*-values using ANOVA and is listed in Table 11.2. *P*-values give information about the coefficients and help in deciphering the nature of the mutual interactions between the variables. The degree of significance proved the linear and quadratic effects of all variables and any variation in their value would alter the rate of the reaction. Also, the probability values of the coefficient suggested that among the three variables studied A (temperature) and C (amount of reducing agent) showed profound effects (0.01–0.05), affecting the model up to 95–99%.

The response surface curves shown in Fig. 11.1 depict the interaction effects of the process parameters keeping one variable at optimum value while changing the other two values.

Figure 11.1a shows that the interactive effect of the factors A and B had no effect on the synthesis of CuNPs. At lower and higher levels of pH, the nanoparticle synthesis was not very significant. However, at the medium levels of pH, it showed a slight increase. Figure 11.1b shows that at higher temperature levels, the nanoparticle synthesis was high. Also, as the amount of reducing agent is increased, the particle synthesis also increased. Figure 11.1c shows that the medium levels of pH and the higher amount of reducing agent led to higher yields of the nanoparticles.

Table 11.2 Analysis of Variance (ANOVA) for the synthesis of CuNPs

Source	Sum of squares	df	Mean square	*F* value	*p*-value Prob > *F*
Model	0.03456	9	0.00384	22.7501	< 0.0001
A-Temperature	0.00319	1	0.00319	18.8789	0.0015
B-pH	2.1×10^{-5}	1	2.1×10^{-5}	0.12458	0.7314
C-Amount of reducing agent	0.02416	1	0.02416	143.136	< 0.0001
AB	0.00015	1	0.00015	0.87136	0.3726
AC	0.00378	1	0.00378	22.3981	0.0008
BC	7.4×10^{-6}	1	7.4×10^{-6}	0.04391	0.8382
A^2	0.00236	1	0.00236	13.9581	0.0039
B^2	0.00163	1	0.00163	9.6468	0.0111
C^2	7.3×10^{-5}	1	7.3×10^{-5}	0.43522	0.5243
Residual	0.00169	10	0.00017	11.4	11.5
Lack of Fit	0.00071	5	0.00014	0.72643	0.6328
Pure Error	0.00098	5	0.0002	11.6	11.7
Total	0.03624	19	11.8	11.9	11.10

$$\text{Absorbance} = 0.047089 - 0.005A + 0.056086B + 0.038438C + 8.58 \times 10^{-5} AB - 0.00035AC - 0.00019BC + 4.68 \times 10^{-5} A^2 - 0.00608B^2 + 0.000827C^2 \quad (11.1)$$

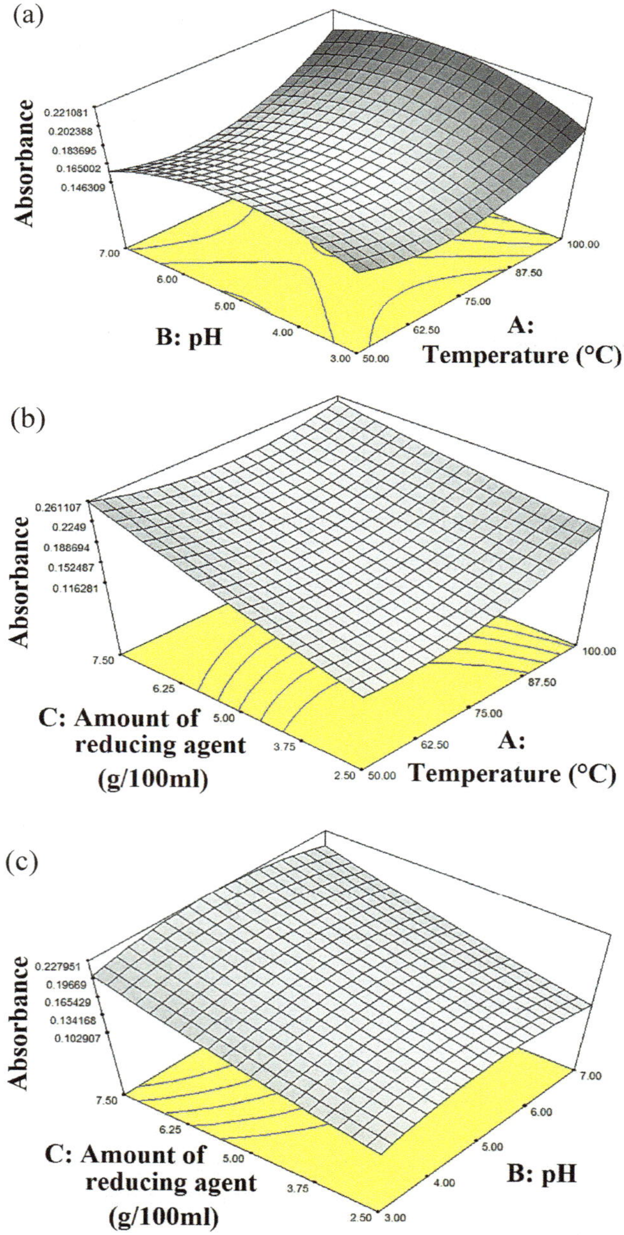

Figure 11.1 Three-dimensional response surface (a–c) showing the interactive effects of the independent variables.

Verification of the model

The experiment was performed using the optimum values as given by the model and compared with predicted data. The measured CuNPs biosynthesis was 0.2375 absorbance at 590 nm, whereas the predicted value was 0.2333. Thus, more than 98.257% was obtained under the practical conditions. The optimized values of the variables for CuNPs biosynthesis using honey were 100°C temperature, pH 4.5 and 7.1 g of reducing agent [33].

11.3.2 Visual Observation

In optimal conditions, the reduction of copper chloride is visually seen by color change from blue to yellowish green and finally to colloidal green (Fig. 11.2). The color of colloidal copper was due to Plasmon absorbance [34]. However, such color changes might also occur due to the variation in the nature, dimension, and morphology of the metal nanoparticles [35].

Figure 11.2 Visual observation of CuNPs at different time intervals.

11.3.3 UV-Vis Spectral Analysis

Figure 11.3 shows the UV-Vis absorption spectrum of CuNPs suspension synthesized using honey with respect to time. As reported previously [36–38], the absorbance peak in the region 580–590 nm validates the formation of CuNPs. The λ_{max} got shifted from 597 to 590 nm with time from 30 to 45 min, and it centered around 590 nm. The nanoparticles synthesis was maximum at 120 min, after which there was gradual decrease in the absorbance.

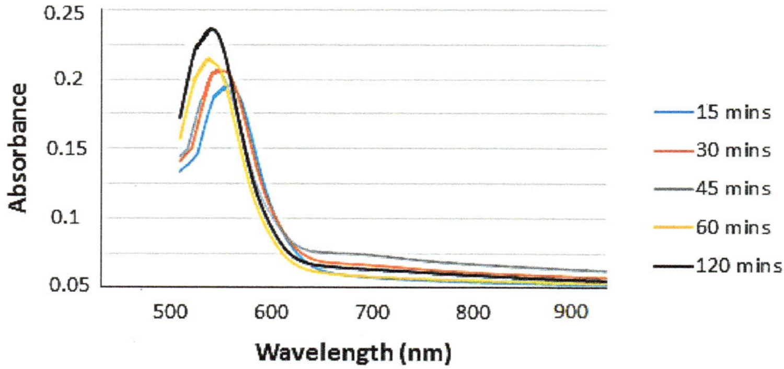

Figure 11.3 UV-Vis absorption spectrum of CuNPs.

11.3.4 Measurement of Particle Size

Figure 11.4 shows the particle size distribution of the CuNPs with a mean particle size of 30 nm. The modal particle size was 29 nm, representing the maximum value of the frequency distribution of the particles. Ninety percent of the particles were below 37 nm size range, 50% below 30 nm, and 10% below 21 nm range (Table 11.3). The particles exhibited normal distribution with standard deviation of 0.081 nm which is very low. This shows that CuNPs of homogenous particle size can be synthesized using honey.

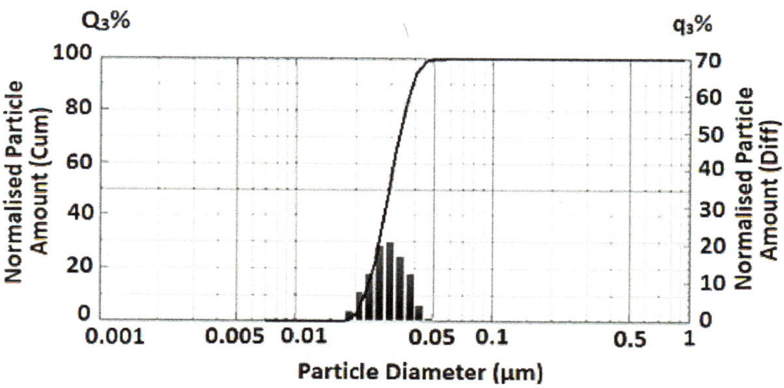

Figure 11.4 Particle size of CuNPs.

Table 11.3 Cumulative particle size distribution of CuNPs

Cumulative %	Diameter (nm)
90	37
80	35
70	34
60	32
50	30
40	28
30	25
20	23
10	21

11.3.5 Transmission Electron Microscopy

Spherical morphology of CuNPs is shown in Fig. 11.5a, with a small percentage being elongated to give ellipsoidal shape. The particle size ranged from 15 to 40 nm with an average diameter of 27 nm. The larger particle sizes indicated the presence of agglomeration, which could be due to possible sedimentation after certain period of time. It can also be observed from the images that the particles were somewhat coated with a thin layer, which could be the capping components from honey [39].

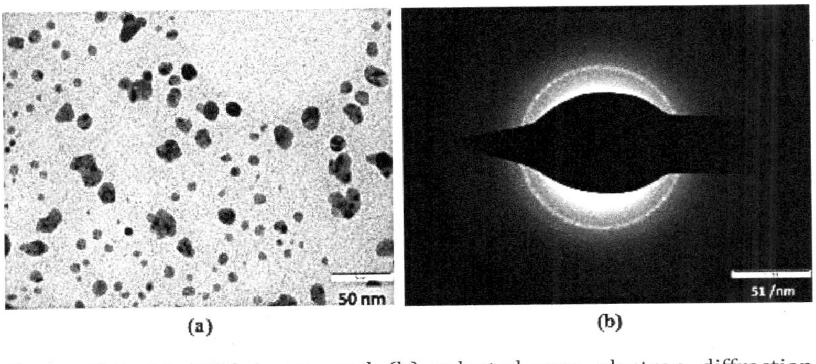

Figure 11.5 (a) TEM image and (b) selected area electron diffraction (SAED) pattern of CuNPs.

The CuNPs were crystalline, as can be seen from the selected area electron diffraction pattern (SAED), Fig. 11.5b which was recorded from one of the nanoparticles. The bright spots were due to different crystallographic planes of elemental copper [40].

11.3.6 X-Ray Diffraction Analysis

Figure 11.6 shows the XRD pattern of CuNPs. Three prominent peaks were observed at 43.4123°, 57.3609°, and 77.2126° for (111), (200), and (220) Miller indices, respectively [41, 42]. These peaks matched with the Joint Committee on powder Diffraction Standards (JCPDS), copper file No. 04-0836.

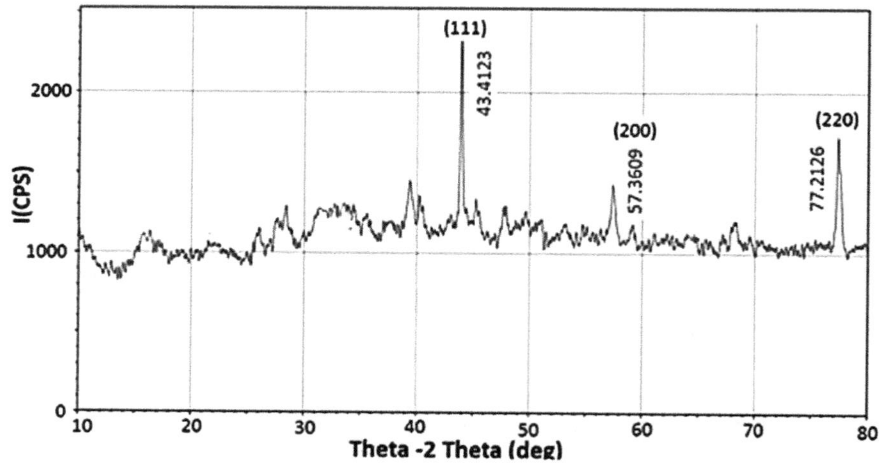

Figure 11.6 X-ray diffraction pattern of biosynthesized CuNPs.

The particles had a crystallinity of 17.24%, which can be gauged from the sharpness of the (111) peak. The plane (111) displaced slightly towards higher diffraction angle (2θ = 0.1173) compared to bulk (2θ = 43.295; JCPDS 04-0836) showing some degree of compressive stress.

Nelson–Riley (N–R) curve was plotted for the calculated a versus $f(\theta)$ (Fig. 11.7). The corrected value of lattice constant a was obtained from the intercept of the y axis by extrapolating the function $f(\theta)$ at zero a_{cal}. The difference in the lattice constant proved that the grains were strained. The broadening of the peak was due to the effect of crystallite size and lattice strain.

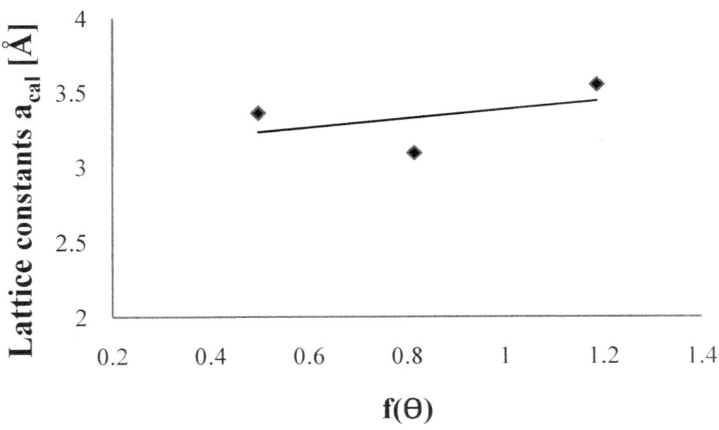

Figure 11.7 Nelson Riley plot of CuNPs.

Particle size calculation

The average particle size of the crystallite was calculated using Debye–Scherrer's formula (Eq. 11.2):

$$D = \frac{0.9\lambda}{\beta \cos\theta}, \tag{11.2}$$

where λ is X-ray wavelength (0.15406 nm), β is full-width at half maximum (FWHM) of the diffraction peak in radians, σ is Bragg's diffraction angle, respectively. The particle size obtained was 25.95 nm [37].

Calculation of average stress in copper grains

The lattice constant of the bulk value (a_0) of copper is 3.615Å (JCPDS: 04-0836) and the corrected lattice constant (a) in this study was 3.0841Å (Table 11.4). Thus, there was some amount of strain in the sample. The average stress grown in the sample is given by Eq. 11.3,

$$S = \left[\frac{a_0 - a}{a_0}\right]\frac{Y}{2\sigma}, \tag{11.3}$$

where a_0 and a are lattice parameters of the bulk and nanoparticles, respectively, Y Young's modulus of the bulk material (for copper = 140 Gpa), σ Poisson's ratio of the bulk material (for copper = 0.34), and the average stress (S) was 30.236 × 10^9 N/m^2 [37].

Table 11.4 Structural parameters of biosynthesized CuNPs

Plane spacing d [Å]	Crystallo-graphic planes *hkl*	Bragg's diffraction angle [2θ]	$f(\theta)=\frac{1}{2}\left(\frac{\cos^2\theta}{\sin\theta}+\frac{\cos^2\theta}{\theta}\right)$	Lattice constant $a_{cal}[Å]=d(h^2+k^2+l^2)^{1/2}$	Lattice constant $a_{corrected}$ [Å]
2.0518	111	43.4123	1.1868	3.5538	3.0841
1.548	200	57.3609	0.8153	3.0960	
1.1888	220	77.2126	0.4973	3.3624	

Dislocation density calculation

The dislocation density (δ) of sample CuNPs was found to be 0.148 × 10^{16} m^{-2}, using Eq. 11.4 [37].

$$\delta=\frac{1}{D^2} \quad (11.4)$$

The calculated crystallite size by the methods cited above was more or less the same (Table 11.5), thus proving the ability of this process.

Table 11.5 Particle size of CuNPs as calculated by different methods

Crystallite size in nm		
Debye–Scherrer's formula	**Laser diffraction method**	**TEM**
25.95	• 30	• 27

11.3.7 FTIR Analysis

The FTIR spectrum of the colloidal CuNPs and honey to identify potential reducing and capping agents is shown in Fig. 11.8. The broad band around 3300 cm^{-1} corresponds to the O–H stretching of the honey components. The small band at 2932 cm^{-1} can be attributed to the C–H amine stretching band of free amino acids and carboxylic acids stretching. These compounds are present in honey at low concentrations. The peaks at 1043 cm^{-1} and 1254 cm^{-1} correspond to the C–O stretch in the C–OH group as well as the C–C stretch in the carbohydrate structure. The absorption band observed at 1027–1045 cm^{-1} can be assigned to the C–O stretching of the glucose/sucrose molecules which might be acting as a major reducing agent. The overall observation confirms the presence of glucose, fructose, and other honey components in the samples of CuNPs. It is reported that proteins are capable of binding to nanoparticles through either free amine groups or carboxylate ions of the amino acid residues. Therefore, surface-bound proteins might be responsible for the stabilization of the nanoparticles. It is interesting to note that not only proteins, but a number of other bioorganic compounds can exist in solution and contribute to the reduction of copper ions and their stabilization by surface capping [39].

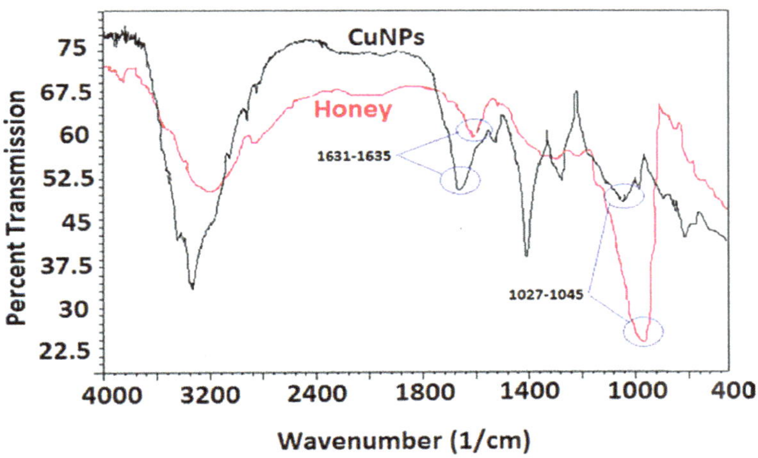

Figure 11.8 FTIR spectrum of CuNP colloid and honey.

11.3.8 Energy Dispersive X-Ray Analysis

Energy-dispersive X-ray spectroscopy (EDX) was used for the confirmation of copper element giving its characteristic signal in the range of 8 keV (Fig. 11.9), which is typical for the metallic copper nano-crystallites [36].

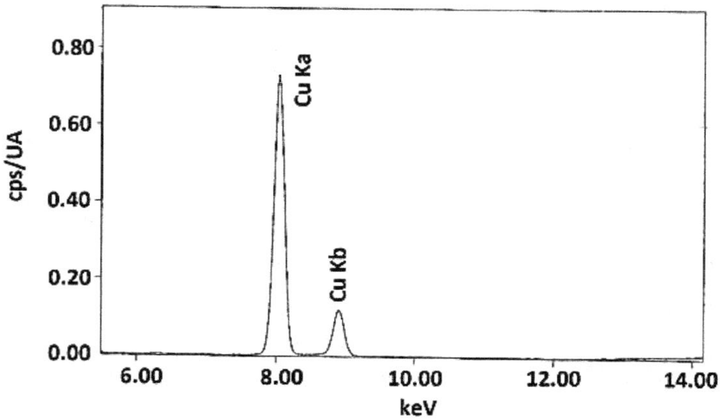

Figure 11.9 EDX pattern of CuNPs synthesized using *Honey.*

11.4 Conclusion

The reduction of copper ions by honey led to the synthesis of CuNPs. TEM results revealed well dispersed CuNPs with an average diameter of 27 nm. The predicted values obtained from the response surface design were in good agreement with the experimental values for CuNPs synthesized. The particle size calculated using different methods was almost the same. This new source seems to be one more promising green technique for the synthesis of CuNPs.

Acknowledgment

The authors would like to acknowledge FIST–DST project, Government of India and World Bank funded TEQIP-II for providing the equipment for this research project.

References

1. Cherepy, N. J., Liston, D. B., Lovejoy, J. A., Deng, H., and Zhang, J. Z. (1998). Ultrafast studies of photoexcited electron dynamics in γ- and α-Fe_2O_3 semiconductor nanoparticles. *The Journal of Physical Chemistry B*, **102**(5), pp. 770–776.

2. Chen, J., Xu, L., Li, W., and Gou, X. (2005). α-Fe_2O_3 nanotubes in gas sensor and lithium—ion battery applications. *Advanced Materials,* **17**(5), pp. 582–586.

3. Wang, X., Chen, X., Gao, L., Zheng, H., Ji, M., Tang, C., Shen, T., and Zhang, Z. (2004). Synthesis of β-FeOOH and α-Fe_2O_3 nanorods and electrochemical properties of β-FeOOH. *Journal of Materials Chemistry,* **14**(5), pp. 905–907.

4. Gopinath, V., Priyadarshini, S., Priyadharsshini, N. M., Pandian, K., and Velusamy, P. (2013). Biogenic synthesis of antibacterial silver chloride nanoparticles using leaf extracts of Cissus quadrangularis Linn. *Materials Letters,* **91**, pp. 224–227.

5. Polsky, R., Gill, R., Kaganovsky, L., and Willner, I. (2006). Nucleic acid-functionalized Pt nanoparticles: Catalytic labels for the amplified electrochemical detection of biomolecules. *Analytical Chemistry,* **78**(7), pp. 2268–2271.

6. Li, D., Chen, S., Zhao, S., and Ma, H. (2006). The corrosion inhibition of the self-assembled Au, and Ag nanoparticles films on the surface of copper. *Colloids and Surfaces A: Physicochemical and Engineering Aspects,* **273**(1–3), pp. 16–23.

7. Ponce, A. A., and Klabunde, K. J. (2005). Chemical and catalytic activity of copper nanoparticles prepared via metal vapor synthesis. *Journal of Molecular Catalysis A: Chemical,* **225**(1), pp. 1–6.

8. Huang, Z., Cui, F., Kang, H., Chen, J., Zhang, X., and Xia, C. (2008). Highly dispersed silica-supported copper nanoparticles prepared by precipitation–gel method: A simple but efficient and stable catalyst for glycerol hydrogenolysis. *Chemistry of Materials,* **20**(15), pp. 5090–5099.

9. Laha, D., Pramanik, A., Maity, J., Mukherjee, A., Pramanik, P., Laskar, A., and Karmakar, P. (2014). Interplay between autophagy and apoptosis mediated by copper oxide nanoparticles in human breast cancer cells MCF7. *Biochimica et Biophysica Acta (BBA)-General Subjects,* **1840**(1), pp. 1–9.

10. Laha, D., Bhattacharya, D., Pramanik, A., Santra, C. R., Pramanik, P., and Karmakar, P. (2012). Evaluation of copper iodide and copper phosphate nanoparticles for their potential cytotoxic effect. *Toxicology Research,* **1**(2), pp. 131–136.

11. Pramanik, A., Laha, D., Chattopadhyay, S., Dash, S. K., Roy, S., Pramanik, P., and Karmakar, P. (2015). Targeted delivery of "copper carbonate" nanoparticles to cancer cells in vivo. *Toxicology Research,* **4**(6), pp. 1604–1612.

12. Nagajyothi, P. C., Muthuraman, P., Sreekanth, T. V. M., Kim, D. H., and Shim, J. (2017). Green synthesis: In-vitro anticancer activity of copper oxide nanoparticles against human cervical carcinoma cells. *Arabian Journal of Chemistry*, **10**(2), pp. 215–225.

13. Wang, Y., Zi, X. Y., Su, J., Zhang, H. X., Zhang, X. R., Zhu, H. Y., J. X. Li, M. Yin, Yang, F., and Hu, Y. P. (2012). Cuprous oxide nanoparticles selectively induce apoptosis of tumor cells. *International Journal of Nanomedicine*, **7**, p. 2641.

14. Guo, Y., Zhang, J., Yang, L., Wang, H., Wang, F., and Zheng, Z. (2010). Syntheses of amorphous and crystalline cupric sulfide nanoparticles and study on the specific activities on different cells. *Chemical Communications*, **46**(20), pp. 3493–3495.

15. Chen, S., and Sommers, J. M. (2001). Alkanethiolate-protected copper nanoparticles: Spectroscopy, electrochemistry, and solid-state morphological evolution. *The Journal of Physical Chemistry B*, **105**(37), pp. 8816–8820.

16. Liu, X., Geng, B., Du, Q., Ma, J., and Liu, X. (2007). Temperature-controlled self-assembled synthesis of CuO, Cu_2O and Cu nanoparticles through a single-precursor route. *Materials Science and Engineering: A*, **448**(1–2), pp. 7–14.

17. Kim, Y. H., Lee, D. K., Jo, B. G., Jeong, J. H., and Kang, Y. S. (2006). Synthesis of oleate capped Cu nanoparticles by thermal decomposition. *Colloids and Surfaces A: Physicochemical and Engineering Aspects*, **284**, pp. 364–368.

18. Nasibulin, A. G., Ahonen, P. P., Richard, O., Kauppinen, E. I., and Altman, I. S. (2001). Copper and copper oxide nanoparticle formation by chemical vapor nucleation from copper (II) acetylacetonate. *Journal of Nanoparticle Research*, **3**(5–6), pp. 383–398.

19. Daroczi, L., Beck, M. T., Beke, D. L., Kis-Varga, M., Harasztosi, L., Takacs, N. (1998). All of the peaks match well with Bragg reflections of the standard cubic cuprite; *MatSci Forum*, **319,** pp. 269–273.

20. Zhu, H., Zhang, C., and Yin, Y. (2005). Novel synthesis of copper nanoparticles: Influence of the synthesis conditions on the particle size. *Nanotechnology*, **16**(12), p. 3079.

21. Joshi, S. S., Patil, S. F., Iyer, V., and Mahumuni, S. (1998). Radiation induced synthesis and characterization of copper nanoparticles. *Nanostructured Materials*, **10**(7), pp. 1135–1144.

22. Pileni, M. P., Ninham, B. W., Gulik-Krzywicki, T., Tanori, J., Lisiecki, I., and Filankembo, A. (1999). Direct relationship between shape and

size of template and synthesis of copper metal particles. *Advanced Materials*, **11**(16), pp. 1358–1362.

23. Song, R. G., Yamaguchi, M., Nishimura, O., and Suzuki, M. (2007). Investigation of metal nanoparticles produced by laser ablation and their catalytic activity. *Applied surface science*, **253**(6), pp. 3093–3097.

24. Park, B. K., Jeong, S., Kim, D., Moon, J., Lim, S., and Kim, J. S. (2007). Synthesis and size control of monodisperse copper nanoparticles by polyol method. *Journal of Colloid and Interface Science*, **311**(2), pp. 417–424.

25. Tang, X. L., Ling, R. E. N., Sun, L. N., Tian, W. G., Cao, M. H., and Hu, C. W. (2006). A solvothermal route to Cu_2O nanocubes and Cu nanoparticles. *Chemical Research in Chinese Universities*, **22**(5), pp. 547–551.

26. Dhas, N. A., Raj, C. P., and Gedanken, A. (1998). Synthesis, characterization, and properties of metallic copper nanoparticles. *Chemistry of Materials*, **10**(5), pp. 1446–1452.

27. Valodkar, M., Nagar, P. S., Jadeja, R. N., Thounaojam, M. C., Devkar, R. V., and Thakore, S. (2011). Euphorbiaceae latex induced green synthesis of non-cytotoxic metallic nanoparticle solutions: A rational approach to antimicrobial applications. *Colloids and Surfaces A: Physicochemical and Engineering Aspects*, **384**(1–3), pp. 337–344.

28. Shankar, S. S., Rai, A., Ahmad, A., and Sastry, M. (2004). Rapid synthesis of Au, Ag, and bimetallic Au core–Ag shell nanoparticles using Neem (Azadirachta indica) leaf broth. *Journal of Colloid and Interface Science*, **275**(2), pp. 496–502.

29. Huang, X., Wu, H., Liao, X., and Shi, B. (2010). One-step, size-controlled synthesis of gold nanoparticles at room temperature using plant tannin. *Green Chemistry*, **12**(3), pp. 395–399.

30. Huang, J., Li, Q., Sun, D., Lu, Y., Su, Y., Yang, X., and Hong, J. (2007). Biosynthesis of silver and gold nanoparticles by novel sundried Cinnamomum camphora leaf. *Nanotechnology*, **18**(10), p. 105104.

31. Bar, H., Bhui, D. K., Sahoo, G. P., Sarkar, P., De, S. P., and Misra, A. (2009). Green synthesis of silver nanoparticles using latex of Jatropha curcas. *Colloids and Surfaces A: Physicochemical and Engineering Aspects*, **339**(1–3), pp. 134–139.

32. Valodkar, M., Jadeja, R. N., Thounaojam, M. C., Devkar, R. V., and Thakore, S. (2011). Biocompatible synthesis of peptide capped copper nanoparticles and their biological effect on tumor cells. *Materials Chemistry and Physics*, **128**(1–2), pp. 83–89.

33. Barabadi, H., Honary, S., Ebrahimi, P., Mohammadi, M. A., Alizadeh, A., and Naghibi, F. (2014). Microbial mediated preparation, characterization and optimization of gold nanoparticles. *Brazilian Journal of Microbiology*, **45**(4), pp. 1493–1501.

34. Krishnaraj, C., Jagan, E. G., Rajasekar, S., Selvakumar, P., Kalaichelvan, P. T., and Mohan, N. (2010). Synthesis of silver nanoparticles using Acalypha indica leaf extracts and its antibacterial activity against water borne pathogens. *Colloids and Surfaces B: Biointerfaces*, **76**(1), pp. 50–56.

35. Shukla, V. K., Pandey, S., and Pandey, A. C. (2010, October). Green synthesis of silver nanoparticles using neem leaf (Azadirachta indica) extract. *In AIP Conference Proceedings*, **1276**(1), pp. 43–49.

36. Caroling, G., Priyadharshini, M. N., Vinodhini, E., Ranjitham, A. M., and Shanthi, P. (2015). biosynthesis of copper nanoparticles using aqueous guava extract-characterization and study of antibacterial effects. *International Journal of Pharmacy and Biological Science*, **5**(2), pp. 25–43.

37. Prabhu, Y. T., Rao, K. V., Kumari, B. S., Kumar, V. S. S., and Pavani, T. (2015). Synthesis of Fe_3O_4 nanoparticles and its antibacterial application. *International Nano Letters*, **5**(2), pp. 85–92.

38. Vasudev, D., Kulkarni, Pramod, S., Kulkarni (2013). Green synthesis of copper nanoparticles using ocimum sanctum leaf extract, *International Journal of Chemical Studies*, **1**(3), p. 1.

39. El-Desouky, T. A., Ammar, H. A. M. (2016). Honey mediated silver nanoparticles and their inhibitory effect on aflatoxins and ochratoxin A, *Journal of Applied Pharmaceutical Science* **6**(06), pp. 083–090.

40. Song, J. Y., and Kim, B. S. (2009). Rapid biological synthesis of silver nanoparticles using plant leaf extracts. *Bioprocess and Biosystems Engineering*, **32**(1), p. 79.

41. Cheirmadurai, K., Biswas, S., Murali, R., and Thanikaivelan, P. (2014). Green synthesis of copper nanoparticles and conducting nanobiocomposites using plant and animal sources. *RSC Advances*, **4**(37), pp. 19507–19511.

42. Aryanejad, S., Bagherzade, G., and Farrokhi, A. (2017). A nanoscale Cu-metal organic framework with Schiff base ligand: Synthesis, characterization and investigation catalytic activity in the oxidation of alcohols. *Inorganic Chemistry Communications*, **81**, pp. 37–42.

Index